CAMBRIDGE LIBRARY COLLECTION

Books of enduring scholarly value

Technology

The focus of this series is engineering, broadly construed. It covers technological innovation from a range of periods and cultures, but centres on the technological achievements of the industrial era in the West, particularly in the nineteenth century, as understood by their contemporaries. Infrastructure is one major focus, covering the building of railways and canals, bridges and tunnels, land drainage, the laying of submarine cables, and the construction of docks and lighthouses. Other key topics include developments in industrial and manufacturing fields such as mining technology, the production of iron and steel, the use of steam power, and chemical processes such as photography and textile dyes.

History of the Manchester Ship Canal from its Inception to its Completion

By the late nineteenth century, charges imposed on Manchester companies for the use of Liverpool's docks and the connecting railway had created an atmosphere of resentment within the business community. The Manchester Ship Canal was to play a major part in the city's regeneration following the depression of the 1870s, but it took a lengthy battle for the scheme to gain the backing of Parliament and for construction to begin in 1887. In this two-volume work of 1907, Sir Bosdin Leech (1836–1912) traces the canal's conception, planning and construction. Volume 2 begins with the project's backers having finally raised the capital necessary to begin construction. The difficult process of building the canal is then detailed. The work includes a large amount of illustrative content, enhancing the light shed on the landscape and notable personalities of Manchester at that time.

Cambridge University Press has long been a pioneer in the reissuing of out-of-print titles from its own backlist, producing digital reprints of books that are still sought after by scholars and students but could not be reprinted economically using traditional technology. The Cambridge Library Collection extends this activity to a wider range of books which are still of importance to researchers and professionals, either for the source material they contain, or as landmarks in the history of their academic discipline.

Drawing from the world-renowned collections in the Cambridge University Library and other partner libraries, and guided by the advice of experts in each subject area, Cambridge University Press is using state-of-the-art scanning machines in its own Printing House to capture the content of each book selected for inclusion. The files are processed to give a consistently clear, crisp image, and the books finished to the high quality standard for which the Press is recognised around the world. The latest print-on-demand technology ensures that the books will remain available indefinitely, and that orders for single or multiple copies can quickly be supplied.

The Cambridge Library Collection brings back to life books of enduring scholarly value (including out-of-copyright works originally issued by other publishers) across a wide range of disciplines in the humanities and social sciences and in science and technology.

History of the
Manchester Ship Canal

from its Inception to its Completion

With Personal Reminiscences

VOLUME 2

BOSDIN LEECH

CAMBRIDGE
UNIVERSITY PRESS

CAMBRIDGE
UNIVERSITY PRESS

University Printing House, Cambridge, CB2 8BS, United Kingdom

Cambridge University Press is part of the University of Cambridge.
It furthers the University's mission by disseminating knowledge in the pursuit of
education, learning and research at the highest international levels of excellence.

www.cambridge.org
Information on this title: www.cambridge.org/9781108071208

© in this compilation Cambridge University Press 2014

This edition first published 1907
This digitally printed version 2014

ISBN 978-1-108-07120-8 Paperback

HISTORY OF
THE MANCHESTER SHIP CANAL

"Floreat Semper Mancunium"

THE RIGHT HON. EARL EGERTON OF TATTON, CHAIRMAN OF THE
MANCHESTER SHIP CANAL, 1887-94.

Franz Baum.   *Frontispiece.*

HISTORY

OF THE

MANCHESTER SHIP CANAL

FROM ITS INCEPTION TO ITS COMPLETION

WITH PERSONAL REMINISCENCES

BY

SIR BOSDIN LEECH

NUMEROUS PLANS, PORTRAITS AND ILLUSTRATIONS

IN TWO VOLUMES

VOL II.

MANCHESTER AND LONDON:

SHERRATT & HUGHES

1907

THE ABERDEEN UNIVERSITY PRESS LIMITED

CONTENTS.

CONTENTS

APPENDICES.

LIST OF ILLUSTRATIONS.

AUTOGRAPH LETTERS, ETC.

PLANS.

PLANS IN POCKET.

PORTRAITS.

PHOTOGRAPHS.

LIST OF ILLUSTRATIONS

CHAPTER XVII.

1887.

CRUSADE FOR CAPITAL—DANIEL ADAMSON RESIGNS—
BOARD OF DIRECTORS STRENGTHENED—LORD EGER-
TON TAKES THE CHAIR—SHIP CANAL MODEL—NEW
BILL TO DIVIDE CAPITAL INTO ORDINARY AND PRE-
FERENCE SHARES — ROTHSCHILDS AND BARINGS
UNDERWRITE PART OF THE CAPITAL ON CERTAIN
TERMS—A RACE TO GET THE MONEY IN TIME—THE
ISSUE A SUCCESS—CUTTING THE FIRST SOD.

I am told by railway men that there is no traffic in the country worked at a lower cost than the traffic between Manchester and Liverpool, and yet we pay the highest rate of carriage in the kingdom.—Sir WILLIAM B. FORWOOD.

THE year 1887 was a momentous one in the history of the Ship Canal. When its most serious difficulties had been overcome, when its prospects seemed bright and fair, then—in July, 1886—there fell upon the promoters a stagger-ing and unexpected blow. They had had the fullest faith that through Messrs. Roths-child & Sons' influence and help the capital would have been easily obtained. When the failure came and the issue of stock had to be withdrawn, when jeers and taunts were being cast, not only on the promoters but on the whole scheme, then the Ship Canal was once more at a very low ebb. This was the position at the dawn of 1887. True the report of the Consultative Committee published on the previous 26th of November had had a beneficial effect, but it was felt that the credit of the Ship Canal must be resuscitated and the confidence of the public restored if the capital was to be raised. It meant preaching a new crusade in Manchester and the chief towns of Lancashire and Yorkshire, and rousing fresh enthusiasm, with but a short time in which to do it. For if £3,000,000 of capital was not subscribed and £600,000 of it actually paid up in cash before the 6th of August, 1887, then the time limit fixed by Parliament would be passed and the powers would lapse.   The labour, trouble and anxiety of years would have been wasted, nearly £200,000 spent in the movement would have been absolutely lost, the promoters made the laughing-stock of the world, and the credit of Manchester seriously damaged.

To impress on the public the great benefits Manchester would derive from the Ship Canal, the *City News* published about this time a series of valuable articles, entitled "Prospects of Manchester as a Port".   In them the possibilities to Manchester were very forcibly and clearly shown.

Sir Edward Watkin, who had at times been a fair critic of water carriage, took occasion at the January meeting of the Manchester, Sheffield and Lincolnshire Railway Company to say :—

He thought that the projected Dore and Chinley Railway and the Manchester Ship Canal were on the same shelf.   Manchester would not make the one, and he was sure the people of Sheffield would not make the other.

The result shows how far wrong a prophet may go.   Within a few years both were made.

We now come to a painful episode in Ship Canal history which led to the retirement of one who had hitherto been the life and soul of the scheme, and who had done magnificent work.   No man better deserved the gratitude of Manchester than Mr. Daniel Adamson.   He had devoted himself with indomitable pluck and energy to make Manchester a port and free Lancashire from the burden of dear freightage, which pressed so heavily on her industries.   But, like most strong men, Mr. Adamson could not brook interference, and was ready to cross swords with any one who attempted it.   He had a great partiality for officials who always moulded themselves to his wishes and were subservient to his desires, whilst those who differed from him in opinion had a rough time of it.   From the period when the question of capital came to the front there had been dissensions in the cabinet; Mr. Adamson's view, and that of a leading official, differing from those of the rest of his colleagues. Many of the latter were of opinion that the goodwill of capitalists must be propitiated, whilst Mr. Adamson pinned his faith to the shillings and pounds of the multitude.

In the report of the Consultative Committee it was made a condition of success that the Board should be reconstituted and strengthened.   To this Mr. Adamson took serious exception; he put his back to the wall and would not move, notwithstanding that Alderman Husband, Mr. William Fletcher, Mr. Richard James and Mr. John Rogerson magnanimously offered to vacate their seats, "solely with the object of facilitating a reconstruction of the Board, and of advancing the interests of the undertaking". At the Directors' meeting prior to the Third Ordinary Meeting of the shareholders, in consequence of a variance between the Chairman and his colleagues, the former declined to sign the report, and it was issued by Sir Joseph Lee (Deputy Chairman) and Messrs. Bailey, Boddington, Jacob Bright, Crossley, Husband, S. R. Platt and J. E. Platt. It recounted the steps taken to raise capital, and the opinion of the Consultative Committee as to reconstituting the Board, and stated that the directors unanimously pledged themselves to do all in their power to carry it into effect. The report went on to say that the Chairman had endeavoured to carry out the promised reconstruction, but finding himself unable to do so had urged the Board to take steps to raise further capital. To this course his colleagues, Sir Joseph Lee and Messrs. Bright and Houldsworth, demurred, as they considered that a pledge had been given to reconstruct before raising more capital. In order to facilitate reconstruction nearly all the directors had expressed a willingness to retire from the Board if necessary. Eventually, at the instance of the Consultative Committee, Lord Egerton of Tatton, Mr. J. K. Bythell, Mr. C. J. Galloway and Mr. Charles Moseley, were invited to join the Board in place of Messrs. Husband, Fletcher, James and Rogerson. If the Committee so reconstructed met with the approval of the shareholders, then a strong effort was to be made to get the capital, and the directors offered to give up their remuneration then due, and work for nothing till the capital was raised.

This report was circulated among the shareholders prior to the meeting which was held on the 1st February, 1887, Mr. Adamson presiding. He said it was almost the universal custom for the Chairman to move the adoption of the report, but he could not do so, not believing it to be a true representation of the facts. Sir Joseph Lee then moved that the report be adopted, and said that it disclosed the entire policy of the Board. He read a joint letter from Messrs. Bright, Houldsworth and himself, addressed to the Chairman and dated the 18th of December, 1886, in which they expressed their intention to resign "unless the Board is reconstituted, and they deprecate the Chairman's policy of appealing to the public for funds before this is done. Failure would be so serious that they are determined not to take part in any further attempt by the present Board to raise capital. The check by the failure of Messrs. Rothschild & Sons' issue had convinced them that a large Manchester backing was essential to success, and if the Board were free to act before reconstruction, it would not be prudent for them to do so."

Sir Joseph Lee went on to say that the signatories to the letter did resign, but in consequence of certain other resignations he, along with Mr. Bright, rejoined the Board on the 31st December. Mr. Houldsworth could not see his way to do so.

Alderman Bailey seconded the report, and endorsed the sentiments in the letter, which he maintained was a wise one.

Mr. Houldsworth declined to go back on the Board till the pledges to the Consultative Committee had been carried out. He paid a high compliment to Mr. Adamson for his past energy and ability; he hoped all differences would be forgotten in the future, and pleaded for a spirit of union.

The *Chairman* would not allow it to go forth that there had been any neglect on his part to accomplish the strengthening of the Board. His view was that a canvass should have been instituted among the capitalists, shopkeepers, artisans, skilled workmen and thrifty people of all degrees, not only throughout Lancashire but throughout England; and if that had been done it was a moral certainty that the necessary capital would have been obtained. If the shareholders passed the report, they should know that it was not a correct statement of what had occurred. Sir Joseph Lee's policy was that of quiescence. If his own policy of action had been pursued a large portion of the money could have been at once secured. He did not desire to introduce personal matter. If Lord Egerton would come on the Board his name would afford great strength, and he would be happy to retire in his Lordship's favour. If the report, as it stood, were adopted, he should retire from the Board, and take no further action in the matter, but would honestly wish them success. If the capital were not found by the 6th of next August, the whole thing would have to be wound up. Therefore he could not join in a policy of quiescence, and he asked the meeting to justify his conduct. If the report were passed, he should admit he had been in the wrong, and would leave the work in future to others.

Mr. Jacob Bright honoured Mr. Adamson for his remarkable courage and extraordinary perseverance in piloting this scheme; he had given it the advantage of his invaluable engineering ability. But when it came to a question of asking the English people for £8,000,000 of money, there were men in this city proposed to be put on the Board who were of greater weight than Mr. Adamson (financially speaking) could possibly be. There was no name in Manchester which would do more for the financial success of this company than the name of Sir Joseph C. Lee.

Other gentlemen supported the report, among them that stalwart champion of the canal, Mr. Mitchell, of the Co-operative Wholesale Society. He trusted the directors would use their best talents till the canal was made, and then they might fall out as much as they liked. He earnestly hoped that Mr. Adamson and Sir Joseph Lee might be induced, at his request, to shake hands.

The report was carried (with one dissentient) amidst much cheering.

Subsequently it was resolved that Mr. Houldsworth be invited to rejoin, and Lord Egerton with Messrs. Bythell, Galloway, and Moseley to join the Board. In a few graceful and touching words Mr. Adamson tendered his resignation:—

I now retire most respectfully from the Board. I wish the canal every success, and I hope the capital may soon be got. I know it will be if the matter is properly pursued, and there will be nobody who will rejoice more than I shall when at last you get to work.

Alderman Bailey, in moving a vote of thanks, said :—

Every shareholder must feel that whenever, in the future history of Lancashire, the Manchester Ship Canal was mentioned, the name of Daniel Adamson would always be associated with it.

On the 10th February Sir Joseph Lee presided at an adjourned meeting, when the new directors were formally elected and Lord Egerton nominated as Chairman. In thanking the shareholders, Lord Egerton said he was not discouraged with the position in which the enterprise now stood, considering that many other great canal undertakings had, in the first instance, encountered difficulties and had not received public confidence. His (Lord Egerton's) kinsman, the Duke of Bridge-water, had to beg every shilling for the Bridgewater Canal, and it was not until by his perseverance and energy he was able to secure the success of what he (Lord Egerton) deemed to be one of the foundations of Manchester's prosperity, that its merits were duly recognised. Again the—at first—discredited Suez Canal had in the end become both successful and remunerative.

In the beginning of March a private circular was issued by the Chairman and Deputy Chairman. It reminded subscribers and supporters :—

1. That they must quickly decide if a Ship Canal was to be made.
2. That London was favourable, but Manchester must lead.
3. That the canal would give an impulse to trade.
4. That the Consultative Committee believed the canal would be remunerative.
5. That the Board had been reconstituted and strengthened.
6. That the directors recommended the undertaking as an investment.

7. That £5,000,000 must be raised before 6th August next, in addition to the Bridgewater purchase.

8. If £3,000,000 was not raised in Lancashire the Ship Canal could not be made.

9. That £750,000 in shares had been taken up and allotted. Many landowners were taking part payment for their land in shares.

10. That a form of application for shares was annexed.

11. That the subscribers' shares incurred no further liability.

12. That as the works would take four years to complete the calling up of capital would be gradual.

13. Interest at 4 per cent. would be paid during construction.

14. The shareholders' liability was absolutely limited to the amount remaining unpaid.

15. A determined effort must be made if the Ship Canal was to become an accomplished fact.

On the 16th March a communication was received from the Prince of Wales that as he was coming to Manchester to open the exhibition and would be staying with Lord Egerton, he was quite willing to cut the first sod of the canal, if it were desired. This honour, however, had to be declined, as arrangements for raising the capital could not be completed in time.

The enthusiasm of the district was now thoroughly roused. The papers teemed with letters not only from working men, but from women, all anxious to contribute their mite to the great work. Much regret was expressed that the Act would not allow of £1 shares. A quotation from a letter will show the spirit of the times :—

Let us show that we in these northern latitudes are not the men and women to be baffled in an honest uphill fight. To have got so far has cost us too much precious energy, time and money to admit of retreat, and for the sake of those who have laboured for us, for the sake of our city, and for the sake of those who will come after us, let us make an energetic effort to carry this scheme to a successful issue.

The concern of Messrs. Rylands & Sons (employing 12,000 hands) through the instrumentality of Mr. Reuben Spencer, instituted a system whereby their employees might have the opportunity of securing shares by small subscriptions, and it was hoped to raise £50,000 by this means to supplement the £50,000 invested by the head of the firm. With the view of convincing the public of the engineering possibilities of the scheme, the directors had a first class model of the canal made, showing a part of the estuary of the Mersey, the entrance at Eastham and the various locks and docks. This was intended for the Jubilee Exhibition, but before being placed there, it was shown in many of the public buildings in Manchester, the first exhibition being in the Art Gallery, where it was explained by Mr. Leader Williams. Subsequently it was exhibited at a crowded meeting of shareholders in St. James's Hall, held on 5th April, when Sir Joseph Lee, Mr. Bythell and others delivered inspiriting addresses, urging additions to the subscription list.   The Chairman said it had reached £1,650,000, but he feared that they must give up the idea of being ready for the Prince of Wales when he opened the Jubilee Exhibition.

By this time the canvass for funds had been thoroughly organised.   There was a repetition of the work done when raising funds for the Bill.   The directors and active friends were told off in detachments to visit the towns and districts of Lancashire and Cheshire to rouse the enthusiasm of the inhabitants and to induce them to take up shares.   Sometimes half a dozen meetings at various places were held on the same night, and the executive had much difficulty on such occasions in finding a sufficient number of competent speakers, especially as they were constantly sending deputations to explain the benefits of the canal to distant parts, such as the Potteries, Birmingham, Glasgow, Dundee, etc.   It frequently happened that I spent four nights a week advocating the canal, and I have known the time when I attended two meetings on the same night, jumping into a cab and driving from one to the other. The meetings as a rule were very encouraging, and they generally ended in an accession of strength.   The working classes were specially anxious to help the movement, but a great obstacle to them was that the shares were £10 each.   To get over the difficulty, in some cases employers acted as treasurers, and gathered weekly shillings ; in others the men formed share societies, and selected their own officials to receive the weekly payments.   Nearly all the limited liability, trading and co-operative societies of the district took up shares, the Co-operative Wholesale Society heading the list with shares to the amount of £20,000.   It was very cheering that this important Society, after an interview with the Ship Canal directors, showed their confidence by taking ordinary rather than preference shares.   This was in marked contrast to the tardy support given by many leading merchants and capitalists of the district, who either held aloof entirely, or contributed the smallest sum that decency would allow them to give.   Referring to the failure of the Rothschild issue of shares in the previous July, Sir Joseph Lee said :—

The fault was not with Messrs. Rothschild but with the people of Manchester and the district.   Very few applications for shares were sent up from Manchester, and the consequence was that Lord Rothschild said, " It is very evident that the people who are to be benefited by the canal do not believe in it ".

On the 27th April a meeting of merchants and traders was convened by the Mayor at the Town Hall to devise means for completing the required capital, and an influential Committee was formed "To take such action as they may consider expedient to raise the capital required, and generally to organise and exert their best endeavours for the purpose of securing the remainder of the capital without delay". This meeting was so crowded that an overflow meeting had to be held. An interesting feature of the meeting was that Mr. Adamson, who since the rift had held aloof, was seen located at the far end of the room. The audience insisted on his coming forward to the platform, they would have no denial. This he did amidst much cheering. He assured the audience that though not now a director, his faith in the undertaking was unshaken; he appealed to them to find the money required, and ended by moving a vote of thanks to the Mayor. This pleasing episode was highly appreciated, and showed the patriotism of the late Chairman who was magnanimous enough to sink private differences for the public good.

It was most exciting to watch the subscription list jump up day by day. On the 29th April it was reported in the Press that within the last twenty-four hours £139,000 worth of share capital had been applied for. Each day brought something considerable. Having broken the ice, Mr. Adamson worked as hard as anybody in attending ward and district meetings. At the ward meeting in Ardwick, Alderman Bennett presided. As a railway director he had got into discredit with the railway world for his support of the canal. He told the meeting that he could not serve two masters, but he had come to the conclusion that if the canal were made, instead of injuring the railways he believed it would be a considerable benefit to them. At a subsequent meeting of the Manchester, Sheffield and Lincolnshire Railway, Mr. Joshua Fielden made a bitter attack on Alderman Bennett for his support of the canal. Sir Edward Watkin's apology, that inasmuch as the Alderman had not taken up any shares, he could not be a danger to the railway interests, did not mend matters.

On the 14th May it was announced that the subscription list exceeded £3,000,000. This was the minimum sum imperatively necessary, though £4,000,000 was wanted. For months there had been rumours that help from London was contingent on Manchester finding at least £3,000,000 of ordinary shares, and now Sir Joseph Lee was free to expose his hand. Through the financial agents of the Ship Canal Company in London (Messrs. Greenwood & Co.), he had approached the two leading London financial houses (Rothschilds and Barings), and had come to an understanding with them that if Lancashire took up at least £3,000,000 of the ordinary shares and Parliament would sanction half the capital, or £4,000,000, to be in ordinary, and the other half in preference shares, then the above-named firms would underwrite the £4,000,000 preference for the sum of 1½ per cent., or £60,000; a further condition being that the contractor should be a man of whom they could approve and with whom they could work.   There can be no doubt that Sir Joseph Lee displayed remarkable diplomatic qualities in securing the agreement, and that he was most ably assisted by Mr. Alexander Henderson of Messrs. Greenwood & Co., himself a master of finance, and of whom it has been said that he had the ability to make money freely without soiling his fingers.

Amongst the convincing agencies of the Ship Canal was the excellent model deposited in the 1887 Jubilee Exhibition.   Located in a prominent position, it was a centre of attraction from the beginning to the end of the Exhibition.   When the Prince of Wales and other high dignitaries visited it, the engineer, Mr. Leader Williams, and his deputy, were generally in attendance to give explanations.   At other times it fell to my lot to take their place, and I believe many shares were secured by explanations given on the spot.   Opponents often came to sneer at the canal, but on being convicted of utter ignorance, they had to admit that many of their difficulties were removed, and that they had been misled by the Press or by the opinion of others.   I made it my duty to collect these objections and deal with them by letters in the Press for the benefit of other unbelievers.   These letters may possibly be some day printed.   My very frequent appearances at the model led to a ludicrous incident.   My wife was one day at a party and met some visitors from a distance when a gentleman asked if she knew who that old gentleman was who was so earnest in explaining the model (whenever he passed), and he was rather embarrassed at the reply, "That is my husband".   After the Exhibition closed the model was taken to Ashton and other towns, and explanations given.

The next step was to ask the holders of original shares to consent to a division of the shares into ordinary and preference, and that the latter should have priority and bear interest at 5 per cent.

The circular stated that the directors had had an offer from responsible capitalists to take the £4,000,000 preference shares, it being a stipulation that when they were offered for subscription, application from ordinary shareholders should have the preference, and that to carry this out Parliamentary powers would be necessary. The circular asked the shareholders to sign a form of assent at once.

A special meeting of shareholders, held on the 20th June, unanimously agreed to the promotion of a Bill entitled, "A Bill to enable the Manchester Ship Canal Company to raise a portion of their capital by means of preference shares, subject to such alterations therein as the directors may approve and Parliament may prescribe". This gave a collective assent and confirmed the assent previously given by means of circulars.   As a security to the ordinary shareholders who would be called upon to pay a deposit of £2 per share before the Bill was passed, all deposits were to be paid to the account of Lord Egerton and Sir Joseph Lee, as trustees, who were to hand them over when the Bill was passed.

The unanimity with which the ordinary shareholders agreed to allow £4,000,000 preference shares to be placed in front of their own holding was remarkable.   Nearly 99 per cent. at once consented, thus smoothing the way for the directors.

Liverpool was not prepared for so much self-abnegation.   The Incorporated Chamber of Commerce at their meeting held on the 10th June referred to the subject, when Sir William Forwood said : "The step which was taken by the promoters of the Ship Canal scheme was not only extraordinary but entirely unprecedented in Parliamentary annals.   It indicated the enterprise was in financial *extremis*.   Lancashire promised Parliament to find the capital, and he did not believe the working men of the district would have invested a penny if they had known it was to go as a guarantee to London capitalists.   The canal would raise a Chinese wall in the heart of Lancashire and Cheshire, and cause a complete severance of the traffic between the counties."   He moved, "That the attention of the House of Commons be drawn to the extreme unfairness of the Manchester Ship Canal Company in introducing this Bill without notice, to take the House by surprise, and thus avoid proper deliberation and inquiry on a matter affecting the interests of a large body of shareholders," etc.   The resolution was not passed unanimously, for Mr. Armour said, "Instead of being an injury, the canal, if it ever were made—and I have great doubts about it—would, I believe, be a source of benefit to Liverpool".

Mr. Houldsworth then moved in the House of Commons to have the Standing Orders suspended that he might bring in a Bill to divide the Ship Canal capital into ordinary and preference shares, and to empower an issue of 4,000,000 of the latter. This was opposed by Liverpool members, but eventually the motion was carried and the Bill passed its first reading.

To the second reading of the Finance Bill it was known there was to be a most strenuous opposition.   The *Liverpool Daily Post* wrote :—

SIR JOSEPH LEE, DEPUTY-CHAIRMAN OF THE MANCHESTER SHIP CANAL,
1886-93.

Litho.—De Verques.

To face page 10.

This proposal, as you know, is going to be fiercely opposed.   Not merely are the Liverpool members against it, but such men as Mr. Sclater-Booth and consistent and punctilious Parliamentarians are of opinion that the proprieties have been improperly violated by the partisans of the Ship Canal.   What is called " Mr. Houldsworth's plant" is regarded with serious disapproval by these eminent authorities.

On *Mr. Houldsworth* moving the suspension of Standing Orders, and that the Bill be read a second time, *Mr. Sclater-Booth* asked that it should be referred back to the examiners of petitions for Private Bills, on the ground that not half the capital had been subscribed, and that the previous consent of the existing shareholders had not been obtained.

Mr. Houldsworth pleaded pressure of time, and said that 70 per cent. of the ordinary shareholders had already given their consent, and the rest would follow shortly.

Mr. Courtney suggested the Bill be referred to a Select Committee of four members.   Eventually this proposition was accepted, and the Bill read a second time.   Lest the Select Committee might dally with their work and endanger the Bill, Sir Henry James moved that they should report in thirteen days, *viz.*, on 27th June, and this, though stoutly opposed by Liverpool and the railways, was carried by 243 to 82.   No time was lost, and on 23rd June a Select Committee, consisting of the Right Hon. H. C. E. Childers (Chairman), Sir John Lubbock, Mr. J. J. Colman and Mr. Gilliatt, met to consider the Ship Canal Preference Share Bill. The opponents were the Liverpool Corporation, the Dock Board Company, and the London and North-Western Railway.

Mr. Pember, *Q.C.*, stated the position—that ordinary shares to the amount of £3,250,000 had been applied for, and that Messrs. Rothschild and Baring jointly were prepared to guarantee the £4,000,000 preference shares.   He showed that the London and North-Western and most other railways had ordinary and preference shares such as he sought for the Canal Company.

Mr. Pope, for the opponents, contended that as nothing had been paid up on 225,000 £10 shares, they might prove to be bogus shares, and quoted from a speech made at a ward meeting : " A boy at the Manchester Grammar School has canvassed his schoolfellows, and sent in applications for £2,000 worth of shares".   Also from another speech : " The Eccles Provident Industrial Society have formed a Ship Canal club to acquire shares by means of small weekly contributions of 1s., and upwards of £5,000 worth of shares have been taken".   In reply it was stated in the first case that Alderman Bailey practically guaranteed the application, and in the second that the Eccles Co-operative Society, a responsible and well-known body, made the application and would see that the money was raised.

Mr. Pope maintained it would be possible to issue £4,000,000 preference shares, because £3,000,000 of ordinary had been issued or allotted without a sixpence having been paid up on the latter. This was the position of the Co-operative Wholesale Society with their £20,000 in shares—there was nothing to bind them.

Sir Joseph Lee, in cross-examination, declined to produce his agreement with the London capitalists, but the Committee decided on its production: it was dated 6th June. By it, if the shareholders agreed to the creation of preference stock, and took up at least £3,000,000 in ordinary shares, the syndicate guaranteed the whole issue of the preference stock. Further, the contractor agreed, if called upon, to accept £500,000 ordinary stock in lieu of cash.

Sir Joseph Lee also stated that a contract had been made with Mr. T. A. Walker on the same lines as the previous one with Messrs. Lucas & Aird. *Mr. Pope* said there was no guarantee that the money would ever be raised, and that the Bill was vicious in principle. He asked the Committee not to encourage it, but to leave Lancashire to find the money, as it promised to do, and not to let it be a preferential speculation of the London Stock Exchange, in the hope of realising hereafter a profit on the shares.

Mr. Bidder, Q.C., asked how it was that Messrs. Lucas & Aird were not to be the contractors? Was it because they would not take the half million of paper money in payment of their work? He brought figures to prove that the capital would be a million too little to complete the canal.

The *Chairman* said they were not going to review the financial estimates of 1885 except as far as they would affect the present Bill.

Mr. Pember claimed all he had to do was to deal with the last million ordinary shares. Of this £500,000 was to be found by the contractor and £100,000 was already subscribed. Was Manchester likely to let the scheme fall through for the sake of £400,000? It would undoubtedly be subscribed.

The Committee found the preamble proved, but required that the 300,000 ordinary shares should not only be allotted but that £2 per share should be paid up. Also that not more than £6 should be called up on preference shares till a like amount had been called up on ordinary shares.

On the 1st July the Bill came before the Standing Orders Committee of the House of Lords, the Duke of Buckingham presiding.   Mr. Coates made the necessary explanations and it passed without opposition.

On the 8th July the Bill was remitted to a Select Committee of Lords, consisting of the Earl of Kimberley (Chairman), Earl Beauchamp, Lord Balfour of Burleigh, Lord Acton and Lord Henley.

Mr. Pember, Q.C., repeated his history of the financial transactions that had taken place as regarded the Bill.   When speaking of the £60,000 to be paid to Messrs. Rothschild and Baring for guaranteeing the issue of preference shares, he said there was one fact in connection with this issue which had only recently come to their notice.   Mr. Pope in the Commons had been astonished at the smallness of the amount, nevertheless it was correct; but there was a process of underwriting shares, and the £60,000 commission to the financiers did not include underwriting, in fact the promoters refused to pay it.   At this hitch or juncture the contractor stepped in, and as he was a very wealthy man, rather than lose the contract, he offered to pay the underwriters' fees of 3½ per cent., or £140,000, out of his own profits on the contract; but this did not affect the liabilities of the company.

Sir Joseph Lee said £3,134,484 had now been taken up in ordinary shares besides the £500,000 by the contractor.

Mr. Bidder, Q.C., in cross-examination, sought to prove that the contractor could not pay £140,000 underwriting, take up £500,000 in shares and have the risk of extra work and cost out of £5,750,000, the amount of his contract.

Throughout the inquiries Mr. Pope had been a model of courtesy, but at this point he referred to the fact that Lord Rothschild was sitting amongst the members of the Committee, and very bluntly asked that he might be told to take his seat amongst the public.   This was refused by the Chairman, who said as a member of the House of Lords his position was correct.

Mr. Henderson, for Greenwood & Co., said they had accepted Mr. Walker's (the contractor) guarantee for £140,000 as good, and his firm would hold themselves responsible if he failed to pay it.

After hearing Mr. Pope and Mr. Bidder, the *Chairman* announced that the preamble of the Bill was proved.

But the last shot was not fired.   The opposing counsel tried to maim the Bill in clauses.

Mr. Bidder asked for the right to appear by counsel before the Board of Trade or Stipendiary when the promoters came to prove that the necessary £5,000,000 of capital had been subscribed.

Mr. Pope made requests that would have perilously delayed the Bill, and on their being declined said testily, "You might as well ask the Committee to pass the Bill without any evidence at all". Eventually the Committee passed the Bill without alterations.

On the 11th July, on the motion of the Duke of Buckingham, the Bill was read a third time, and the next day it received the Royal assent by Commission.

Foreseeing the passing of the Bill, and with a view to save time, the directors at the end of June made arrangements for the conditional allotment of ordinary shares. They fixed the first week in July for the payment of £2 on allotment. The National Provincial Bank of England fitted up St. James's Hall, Oxford Street, as a temporary branch, and had as many as twenty-seven clerks busily engaged in receiving money. The rush to pay commenced by people knocking at the doors long before business hours in the morning, and continued all day with little intermission. This was very wonderful considering there was practically no binding obligation to take up the shares; any one could have pleaded that the creation of preference shares voided his promise. By Saturday, the 9th July, £540,000 out of a possible £629,632 had been deposited, and on Monday a further £32,000 was paid in. Many people who were warned to divide their investment into preference and ordinary could hardly be persuaded to do so. A little incident was related to me which showed the prevailing feeling. An old man of eighty summers being the holder of 400 shares went to pay his £2 per share, or £800, and had then with him in his pockets £4,000, and so desirous was he of promoting the interests of the Canal, that he wanted to pay over the whole of it, and he was much disappointed when it was declined.

On the 28th June the secretary notified the auditors of the condition on which application for shares had been made, *viz.*, "that before the 6th August, 1887, the auditors of the company shall certify in writing that the company has issued or received and accepted applications or guarantees, conditional or otherwise, for a sufficient number of shares to enable the company to satisfy Section 38 of its Act of 1885 which provides that the company shall not execute any works until the capital therein mentioned has been issued and accepted". He informed them that Messrs. Thomas Wade, Guthrie & Co., had undertaken to issue the letters of allotment.

On 13th July the auditors met the directors and gave a certificate that £3,000,000 of ordinary shares had been allotted and £2 paid on each share, and on 19th July

Mr. Meysey Thompson, counsel to the Board of Trade, came down to see that Section 38 had been complied with.

On the 15th July the prospectus of the Manchester Ship Canal Company was issued from London conjointly by Messrs. Baring Brothers & Co. and Messrs. N. M. Rothschild & Sons, for £4,000,000 in preference shares of £10 each.[1] It was announced that the subscription list would open on Tuesday, the 19th, and close on Thursday, the 21st July. As the issue was underwritten the whole of the capital was applied for. Well within the time (6th August) the directors had not only secured the necessary £5,000,000, but £2 per share on over 3,000,000 ordinary shares had been paid up, and on the 4th August, Mr. Calcraft, secretary of the Board of Trade, gave his certificate that two-thirds of the share capital authorised by the Act of 1885 had been issued and accepted, thus complying with the Act of Parliament.

On the 4th July the directors paid for the purchase of the Bridgewater Navigation Company, in one cheque, the sum of £1,710,000, and this valuable property came into the possession of the Ship Canal Company.

At the August meeting of the shareholders, Lord Egerton in the chair, Mr. Daniel Adamson at the commencement asked that the Press be admitted, so that thousands of shareholders, who could not be present might be made aware of the proceedings. The company might now show their hand without fear of their opponents.

The report presented by the Directors reviewed the previous six months' work. An Act to create preference shares had been obtained and had received the Royal assent on the 13th July, being thirty-three days after the introduction of the Bill. the shortest time on record in which an opposed Bill had been passed through Parliament. The report went on to say, "The directors are fully conscious of the great work they now have before them; they are prepared to devote unremitting attention to the interests of the company, and they suggest that the Board could be reduced from fifteen to twelve members without impairing its efficiency". The Chairman (Lord Egerton) in moving the report, congratulated the shareholders on the raising of the capital, and remarked on the great interest Parliament had taken in the scheme. He thought them fortunate in their contractor, Mr. Walker, who had recently carried out in an excellent manner similar works at Preston.

[1] See Appendix No. VI.

Sir Joseph Lee rejoiced they had been able to convince the London financiers that Lancashire was really in earnest and that the canal was a necessity. To show the class of people who had helped, £160,000 was subscribed by people who held fewer than ten shares each; these were chiefly tradesmen, shopkeepers and the general community. On the other hand there were 2,700 capitalists who held between them £5,500,000 in shares. The Bridgewater was a most valuable property; he felt sure Manchester had not only confidence in the scheme, but in the men who would carry it out.

Mr. Adamson believed the work could be done for 10 per cent. less than they were going to pay Mr. Walker, and wanted to know why the contract had not been let by tender. He did not like paying a big price and paying the contractor partly in shares. The directors ought to have put the contract up to competition. This attack on the directors did not find favour with the meeting, and the Chairman said it was neither wise nor profitable to enter into a discussion on rival contractors.

Sir Joseph Lee deprecated a wrangle about contractors. The contract was let, the money got, and now they meant to build the canal. Messrs. Lucas & Aird declined to help the directors financially, whilst Mr. Walker gave the directors valuable assistance. He offered to take £500,000 in paid up shares at a time when the directors, after stumping the country for three months could only raise £3,000,000. He thought the directors had made a good arrangement, and he did not believe the contractor would have any large profit.

Mr. Adamson was understood to say that he got his information from Lucas & Aird, but the audience became impatient and cried "Sit down" so vigorously that he had to resume his seat. The report was passed unanimously.

The provision of the capital and the letting of the contract mark a distinct epoch in the history of the Ship Canal. Since 1882 the promoters had been carrying on a vigorous fight against immense odds, and what was hardest to bear, many of their own leading townsmen were either hostile or lukewarm in the cause. Notwithstanding the enthusiasm for the Ship Canal of the majority of the members of the Manchester Chamber of Commerce, some of the leading directors of that body were suspected of veiled hostility to the project. With one exception the Press declined to commit themselves; their attitude generally was one of masterly inactivity; they would not give a hearty support to the scheme, and certainly their chastening was not joyous but grievous; possibly they thought good fruit would come later on to those who were exercised thereby. The Canal Bill was carried through by a comparatively small band of earnest workers who went into the struggle body, soul and spirit, and undaunted by defeats and disappointments struggled on, notwithstanding that at times they were brought to the brink of despair. As one who never took his hand from the plough during the whole time, I can bear testimony to the wonderful devotion of my colleagues, who faced vicissitudes of no ordinary character. When in 1883 and afterwards in 1884 the Bill was rejected, and when in 1887 Rothschilds' appeal was a fiasco, all appeared to be lost, and it seemed as if wearied men with shattered finances had nothing before them but to collapse and own themselves beaten. But with a little rest, and some outside encouragement, the promoters revived and determined to try once more. Their sustaining power came chiefly from the people who were faithful to the end, however fickle capitalists might be. Having once made up their minds that the canal would benefit trade, cheapen food and bring work into the district, the middle classes were willing to make any sacrifice to achieve success. In comparison to their means they were large contributors to the canal, and in its darkest days they bore their burden without a murmur. The Free Trade movement which made Manchester famous was on a par with the Ship Canal scheme; in both cases the dogged determination of earnest men was the main factor in securing success.

Even before the Bill was passed the engineer and contractor were busily at work arranging their plans, and the land agent, Mr. Dunlop, was in the field purchasing the land.

It was hoped that some member of the Royal family would be induced to cut the first sod, but when this was impossible and winter was drawing on, it was determined that the Chairman, Lord Egerton, should perform that function without ceremony of any kind. So on the 11th November, 1887, the directors, auditors and a few of the officials departed for Liverpool by an early train. It was arranged that neither friends nor reporters should accompany the party. But this order was difficult to carry out. Somehow the reporters had scented their movements, and they came to the side of the steamer at the Liverpool landing stage in force, hoping to go. When they found the way barred they took all possible means to evade the restrictions. They crossed over to the Cheshire side, and in carriages or on bicycles came in hot haste to the rendezvous. One party turned up just as the ceremony was ended, and they at once started to pump information from the officials who had been present.

The spot fixed for cutting the sod was a field about three-quarters of a mile from Eastham Ferry Hotel, which was to become the site of the great lock at the entrance to the canal from the Mersey.   On arriving at the trysting place a new spade and barrow were produced, and Lord Egerton essayed to cut the first sod amid the cheers of those present.   A burly navvy, seeing the noble Earl was not used to delving and was in difficulties, came forward to the rescue, for which interference Mr. Walker, the contractor, chided him.   However at length, with some assistance, the Earl managed to raise the historic sod into the wheelbarrow, and Mr. Leader Williams wheeled and tipped it a few yards away; it was then cut into pieces and carried away by the directors and others.   I have a veritable piece of it yet growing on my lawn as a memento of the event.   Lord Egerton had the spade.   I believe the engineer carried off the barrow in triumph as a trophy to be regarded as an heirloom in his family.   The group present at this interesting ceremony, in addition to the Chairman and Deputy Chairman, were Alderman W. H. Bailey, Mr. Henry Boddington, Mr. J. K. Bythell, Mr. W. J. Crossley, Mr. C. J. Galloway, Mr. Joseph Leigh and Mr. S. R. Platt, directors; Mr. George Hicks and Mr. Bosdin T. Leech, auditors; Mr. Leader Williams, engineer; Mr. Abernethy, consulting engineer; Mr. W. H. Hunter, assistant engineer; Mr. A. M. Dunlop, land agent; Mr. A. H. Whitworth, secretary; Mr. W. J. Saxon, solicitor to the company; and Mr. Marshall Stevens, the company's manager.

Subsequently the directors inspected the contractor's work and plant.   They were astonished at the immense quantity of the latter he had already got on the job, and were told the Ashbury Railway Carriage Company had contracted to deliver 100 waggons weekly for the next six months.   The stream of labour finding its way to the new works was so great that to save the nuisance of inquiry by tramps, the authorities put up finger-posts at the various cross-roads to guide people to the works.   It was with affected equanimity that the Liverpool Press spoke of the work of the day.   One paper said the first sod of the big ditch had been cut.   But would the great steamers struggle up the 37 miles of narrow waterway in preference to discharging cargo at Liverpool, whence railways could distribute it everywhere?   Would the railway magnates allow business to be taken out of their hands without any effort?   The railways could reduce their charges below those of the Ship Canal without feeling the pinch, and the knowledge of this had deterred even sanguine men from putting money into the Ship Canal.

The *Liverpool Mercury* alone took a generous view of the position :—

If the canal is a success it will doubtless bring business to all the districts where it touches—and we would not except Liverpool from the list—more work, more commerce, more people and more wealth.   No change was ever affected without some one losing by it; but in the end the result should be a very great good to a very great number.

CHARLES J. GALLOWAY, DIRECTOR OF THE SHIP CANAL; CHAIRMAN
OF BRIDGEWATER COMMITTEE, 1893-1904.

To face page 18.

On the 2nd December the first steam navvy was set to work at Eastham, and several more were on the ground ready to start: each of these removed soil at the rate of 1,200 cubic yards per day. Over 700 men were on the job. Later on operations were commenced near Mode Wheel on land purchased from Lord Egerton. The old river was used for the conveyance of timber, plant, etc. Mr. L. P. Nott, the contractor's agent at Manchester, had no difficulty in getting any quantity of labour.

Reviewing the position at the end of the year we find that the construction of the canal had been started in earnest. To facilitate matters the route had been divided into nine sections: Eastham, Runcorn, Norton, Warrington, Latchford, Irlam, Barton and Manchester, with a railway deviation section at Warrington and Irlam. Each section had an engineer and assistant engineer, all under the direct orders of Mr. Leader Williams. Mr. Walker, the contractor, had made a similar division of his staff, so that each section could work independently of, and vie with the other. Mr. Walker had as a residence taken Dunham Hall in order to be near the line of work. At four points the work had been started, *viz.*, Eastham, Warrington, Warburton and Salford, and huts for the men had been erected. Many landholders had been settled with. Where exorbitant sums had been demanded for land, possession had been taken under the powers of the Lands Clauses Act, leaving the ultimate price to be paid to be fixed by arbitration, as provided by that Act. This was the case with the site of the future Manchester Docks, to be constructed on land belonging to Lord Egerton. Thus the year ended with the construction of the canal in full swing.

CHAPTER XVIII.

1888.

THE CONTRACTOR STARTS IN EARNEST—HIS SYSTEM OF WORKING — PARLIAMENTARY BILL — DESCRIPTION OF WORK ON EACH SECTION—GENERAL PROGRESS—T. A. WALKER—ARBITRATIONS.

If they wanted trade to flourish, and manufactures of every description to rise round them, they must make the Queen's Highway such as every ship on the sea could find its way to Middlesbro'.—Sir WILLIAM CUBITT.

A BRIGHT horizon greeted A.D. 1888. The capital had as nearly as possible been obtained, and the work was well in hand at several different points. The only cloud was that the shares, instead of going to a premium, were at a discount of a few shillings. Of course this was a disappointment to many, yet it was not to be wondered at considering the enormous number of shareholders. The list published on 1st December, 1887, showed close on 39,000 shareholders, a number far exceeding any original proprietary list ever before published. It is a most interesting document, as it demonstrates how widespread the enthusiasm had been. No doubt some would find it inconvenient to meet the calls, and when their shares were put on the market they were compelled to take a lower price than that at which unissued shares could be bought.

It is often a matter of curiosity as to which are the most popular British surnames and Christian names. Dissecting the list of shareholders, I find there were 346 Jones, 443 Taylors and 540 Smiths. Subdividing the Smiths there were 62 John Smiths, 34 Thomas Smiths, 46 William Smiths. This goes to prove that John Smith is the most common English name. Almost all the subscribers to the preliminary fund took up ordinary shares.

The third week in January was devoted to the payment of the second £2 call, and again the St. James's Hall was crowded with depositors from early morn till late

Killon.

RUSTON AND PROCTOR'S STEAM NAVVY.

Killon.
2
2

To face page 20.

at night. Many paid up the shares in full. At the close of the week it was found that out of 340,000 allotted shares the call had been paid on 332,000. On the remaining 8,000 the first call of £1 only had been paid.

Strange to say, two leading men at Mr. Adamson's 1882 meeting (both of whom spoke as representing Warrington) withdrew from the canal, when funds were most wanted, and did not even take shares for the money advanced towards preliminary expenses. When the Warrington meeting of congratulation to Mr. Adamson was held, Mr. Henry Boddington, Junior, did not forget to give names and proclaim the fact.

On the 24th January Lord Egerton summoned a conference of all the authorities in the watersheds of the Mersey and Irwell, with a view of trying to stop river pollution. He said it was absolutely necessary, if the Ship Canal was to be carried out successfully, that steps should be taken at once to purify the streams which would supply the Manchester and Salford Docks, and he hoped the different local authorities would carry out some distinct sewage system, and make the water so pure as not to damage the health of people working on the canal or residing near it. The sewage of towns ought not to be put into the streams. Sir R. Kane, an authority on sanitation, had described the Irwell as the hardest worked and foulest stream in the world. He moved that the authorities on the Mersey and Irwell take the needful steps to prevent the pollution of the rivers and their tributaries.

Alderman Thompson hoped the Corporation of Manchester, with the help of the Local Government Board, would' be able to carry out the construction of their sewage scheme long before the canal was opened. The present position of the river was a scandal. They intended to precipitate the sewage by chemicals and then pass the effluent over the soil.

Alderman Walmsley, of Salford, did not want to cross a lance with Alderman Thompson, but he regarded conferences as worthless unless there was compulsion. In Salford they had spent £250,000 on an intercepting sewer to keep sewage out of the river. Manchester must not consider sewage non-deleterious because it was in a running stream. He had already threatened the directors of the canal, that if they impounded water in their docks in the filthy, disgusting and insanitary condition in which it now came down, he should ask that the 1876 Pollution Act should be put in force against them as ruthlessly as against any one else. He moved that the several authorities in the watershed should, before 25th March next, state what practical steps they were taking to keep their sewage from polluting the river.

Sir Joseph Lee said it would be unfortunate if the canal, after costing £10,000,000 should become unpopular with traders on account of the fouling of the water by sewage.

Lord Egerton closed by saying that he would try and see the subject satisfactorily dealt with before the canal was opened.

Preparation for the advent of many thousand workmen along a course of 35 miles was a work of considerable magnitude. It was necessary to provide huts or housing accommodation for most of the men, also their bodily and spiritual necessities had to be considered, and schools must be found for the children. Large rooms had to be provided for newsrooms and services on the Sunday. Sick people and those who meet with accidents would need looking after, and dispensaries or infirmaries would have to be provided.

Mr. Walker, the contractor, in his Severn Tunnel and other works had had great experience, and he believed in making his workmen healthy and happy. He had a large heart and strong religious convictions, so one of his first acts was to provide adequately for his men. His huts were well built and sanitary, he put up meeting rooms, hospitals and schools of a good type, and arranged for a staff of missionaries, nurses, and medical men all along the line of route. These must have been a heavy item of expenditure to him.

As before mentioned, the route was divided at first into nine sections which were subsequently reduced to eight. Certainly the Eastham end was by far the most attractive. It almost seemed desecration to invade such a pretty rural spot with its beautiful woods and foliage stretching down to the estuary. Well might Eastham be termed the " Richmond of the Mersey ". Here the Liverpool people came in thousands at holiday times to ramble in the fields and woods. An enterprising hotel proprietor, Mr. Thompson, ran his own boats from Liverpool to the ferry close by, and had laid out an adjacent piece of land for dancing and other amusements. To add to the attractions, he had an aviary and a collection of wild animals. My second visit was in the spring of 1888, when with a party of Corporation friends, I went to inspect the works, the Ferry Hotel being our headquarters. The cutting was about half a mile distant, and the scene at night was one not to be forgotten. Aided by a good supply of Wells' pneumatic oil lights, hundreds of men seemed as if they were working against time to remove a mountain of soil that stood in their way. Steam navvies of the Ruston and Proctor type (commonly but incorrectly called American devils) were driving their sharp teeth into the hillside and removing tons of soil into long strings

WARBURTON CUTTING.   GEOLOGICAL STRATA.

Birtles, Warrington.

To face page 22.

of waggons, which as soon as possible were drawn away to a tip by small engines. At another place Priestman's grabs were engaged in a similar way, and there were numbers of men filling waggons by hand.   One could hardly find a safe place out of the way of engines and trucks, and what with the screeching of the former, the shouting of the men, and the strangeness of the surrounding, lit up as it was by a lurid light, it was one of the weirdest sights I ever witnessed.   We afterwards walked along the course of the canal to Daresbury, near Warrington, taking three days to make the inspection.   No wonder that at holiday times crowds of people from Liverpool crossed to see the works, occasionally paying a visit to Eastham Church— a very interesting structure, built in A.D. 1152.   There is a yew-tree in the church- yard, and records show that in A.D. 1300 a sum was paid to the monks to protect it.   All kinds of people patronised the Ferry Hotel, where tea could be provided for 1,500 people at one time.   On Whit Monday 40,000 people visited Eastham.

Later on, 23rd June, I acted as guide to 170 members and friends of the Manchester Geographical Society who visited the Eastham section.   I then found it an exceedingly difficult matter to keep many of the ladies from harm ; several of them nearly got run over by the engines or trucks.

On Bank Holiday in August I walked to the Warburton cutting, in which a flood had occurred, submerging the machinery and stopping the work.   It was a re- markable sight, and I was much struck with the variegated strata on the side of the cutting, similar to that which is to be seen on the Norfolk coast.

At the Bath meeting of the British Association, Mr. Lionel Wells read a paper on the Ship Canal and its appliances.   There was an amusing scene, when, in the discussion that followed, an old gentleman (who said he was a railway shareholder) denounced the Ship Canal as destructive to his interests, also a fraud on the rate- payers.   In his fury he sat down and smashed his chair, cutting a ludicrous figure with his legs in the air.

Mr. Walker lost no time in putting down contractor's lines for the distribution of material.   This reminds me that on one of my earliest trips along the line from Eastham *via* Pool Hall with some friends, our seats in the waggon consisted of boxes, which we were afterwards horrified to learn contained dynamite, and as my friends had been smoking on the top of this explosive we were all thankful no accident had occurred.

During the excavations an old Roman footpath was laid bare at Pool Hall, composed of boulders and edged with red sandstone : it was found 15 feet below the surface, and on the top of it were very large trees. Antiquaries are of opinion that the old road was part of one that crossed the estuary from Stanlow in Cheshire to Hale in Lancashire. At one time the river Mersey was fordable in carts, and it would appear that from time immemorial there had been a direct communication from Warrington to Hale then across the river and over the Wirral peninsula to Parkgate, continuing across the Dee to Holywell. Strange to say the ancient family of Stanlow had a pew in Hale Church on the other side of the river, and Stanlow itself is now in the parish of Hale. Pool Hall Brook was found to be too low for the canal, so a culvert 60 feet deep on the rock had to be made to take the water under the canal to the estuary.

From Eastham to Pool Hall the canal skirts the shore, then it runs inland, leaving a promontory between the canal and the estuary. On this land the contractor dumped the soil out of the cuttings, with the result that a small mountain was formed and called "Mount Manisty," after the contractor's agent on the section. Now that the canal is completed, this forms an important feature in the landscape.

Farther on the shores of the estuary are again reached, and the canal runs in front of Ellesmere Port, which it cuts off from the river by a huge embankment about $1\frac{1}{2}$ miles long.

By midsummer the whole section was opened out, the last portion being the Booston Wood length, and as in many places the red sandstone had been reached, the character of the machinery had to be changed.

Accidents of all kinds were numerous. An engineer had both his feet taken off, and an old hawker, Mary Murphy, had a marvellous escape. She walked right in front of an engine coming up from behind her and was instantly knocked down, the engine pushing her along the sleepers. When it was stopped she was half way under it and every one thought that she was killed, but it was found that apart from being bruised and shaken only her toes were injured. She was sent to hospital and was soon fit to return to her home in Liverpool.

The making of the canal opened out many curious old places, with histories that had almost been forgotten. Among the rest Pool Hall, a fine example of the houses of the gentry in Henry VIII.'s time; it is within a stone's throw of the canal. The "Pool" who built it was the son of a certain Sir William who was High Sheriff in Henry VIII.'s time. A deed relating to the manor is enrolled in the Cheshire Domesday Book dated "the Thursday before the feast next after the return of Earl Randle from Jerusalem," which event took place in 1220. Some years ago a number of warlike implements were found in a pit near the Hall, supposed to have been buried during the Civil Wars.

Leaving the other sections to be described later on, I pass to the February shareholders' meeting. Lord Egerton, after describing the magnitude and variety of the machinery at work on the canal, announced that Sir Edward Jenkinson had been elected director in place of the late Mr. Charles Moseley, whose death he deplored.

In the middle of the month the London and North-Western Bill to make two new cuts or waterways to their Garston Docks, came before the Court of Referees, and the Ship Canal Company applied for a *locus*. Mr. Pember was able to pay them off in their own coin as he said, "What is sauce for the goose is sauce for the gander". He told them they were now asking for what they themselves pleaded meant destruction to the estuary when the Ship Canal wanted dredging powers in the previous year. People get strange bedfellows at times, and here the Dock Board and the Ship Canal Company were co-petitioners, but that there was no love lost was evident from the speech of the Dock Board's Parliamentary Agent, Mr. Rees :—

I protest in the strongest way possible, on behalf of the Dock Board, at its being supposed, directly or indirectly, that we represent the interests of the Ship Canal Company. On the contrary, we hate them. (Laughter.) We shall look after ourselves.

The *locus* was given to both applicants.

When the Bill came before the Commons Committee the London and North-Western Railway withdrew the clause as regarded the new channels to Garston.

At the meeting of the Iron and Steel Institute, of which Mr. Daniel Adamson was this year's president, that gentleman was presented by Sir Henry Bessemer with the Bessemer Gold Medal, given annually as a mark of distinction for inventive services as regards iron and steel. When other boilermakers were deriding the use of steel plates for boilers, Mr. Adamson, as a result of his researches, formed an opposite opinion, and in 1858 constructed the first steam boiler made of Bessemer steel for Messrs. Platt Brothers, of Oldham. In consequence of certain alarming statements, this boiler was inspected twenty-one years afterwards, and was found still at work and showing no signs of corrosion. The same courage and indomitable pluck that had carried the Ship Canal to success had enabled him to establish the superiority of steel over iron. A little later in the year, another staunch supporter of the Ship Canal, Alderman Harwood, had conferred on him the well-merited honour of knighthood for services rendered to the city.

Sir Humphrey de Trafford, of Trafford Park, had throughout displayed a very strong antipathy to the Ship Canal.   He did not care to have his domestic peace invaded, or the home of his ancestors (who had lived there from the time of the Conqueror) disturbed.   He had done all in his power to keep the canal at arm's length, and though his estate consisted of pasture fields, he had driven the engineer to plan docks on the race-course and on land belonging to Lord Egerton and partially covered with buildings.   Sir Humphrey had practically succeeded in keeping the canal out of Stretford.   After his death in 1886, however, negotiations were reopened with his successor, resulting in the whole plan of the docks being recast and a new Bill being brought into Parliament.   Application ought to have been lodged in November, 1887.   It was 10th May, 1888, however, before the new Bill was ready, and if it was to go through, the Standing Orders must be suspended and the sanction of Parliament obtained.   The agreement come to with Manchester, Salford and Stretford enabled the Ship Canal, amongst other changes,[1]

1. To alter the size and shape of the triangular dock south of the Manchester race-course.

2. To make the Mode Wheel locks south of the Salford Cemetery instead of close to the race-course.

3. To do away with two large docks close to Messrs. Goodwin's soap works.

4. To widen the channel up to Trafford Bridge.   To construct a swing bridge at Trafford Road.

5. To build an alternative opening bridge at Fairbrother Street, and by means of a semaphore to acquaint the public which bridge was open.[2]

It had been a very difficult matter to secure agreement between the parties interested.   Eventually Salford gave up about 16 acres of dock area, and Stretford, in return for about 1⅛ miles of additional dock frontage, consented that the line of canal should be so drawn that about 40 acres of Stretford would be thrown on the Salford side of the river, and be transferred to that borough.   Further, the Ship Canal Company arranged that a small dock of 3½ acres should be made next to Messrs. Goodwin Brothers' soap works.   Sir Joseph Lee estimated the changes detailed in the Bill would effect a saving of £83,000 to the Ship Canal.

After the Bill had been approved by the Examiner of Standing Orders, it was passed on to a Select Committee, of which Sir John Mowbray was Chairman, and they consented to suspend Standing Orders.

As it was necessary that the proposed Bill should be approved of by a special meeting of shareholders, one was called, Sir Joseph Lee, who was supported by Mr. Walker, the contractor, being in the chair, and here an unexpected difficulty arose.

[1] See Plan 10 in Pocket.   [2] This bridge was afterwards abandoned.

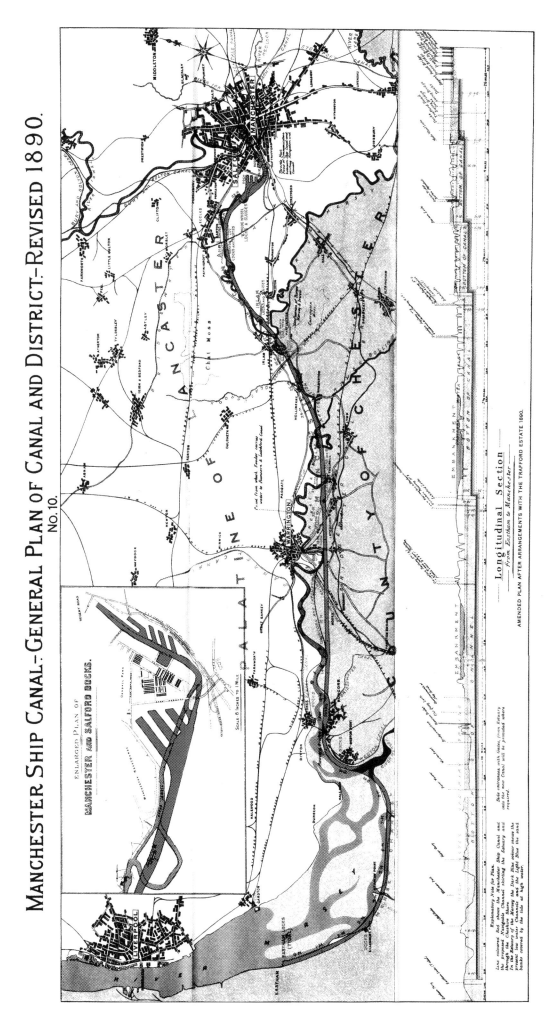

MANCHESTER SHIP CANAL.—GENERAL PLAN OF CANAL AND DISTRICT.—REVISED 1890.
NO. 10.

ENLARGED PLAN OF
MANCHESTER AND SALFORD DOCKS.
SCALE 6 INCHES TO 1 MILE

Longitudinal Section
from Eastham to Manchester
AMENDED PLAN AFTER ARRANGEMENTS WITH THE TRAFFORD ESTATE 1890.

The material originally positioned here is too large for reproduction in this
reissue. A PDF can be downloaded from the web address given on page iv
of this book, by clicking on 'Resources Available'.

Mr. Samuel Kelsall, a shareholder, and an overseer of Stretford, protested against the transfer of any land to Salford. He urged that the company should leave the boundary question to be fought out between Stretford and Salford, and ended by hoping that Manchester, Salford and Stretford would all be united in one grand municipality, which would solve all difficulties.

The *Chairman*, Sir Joseph Lee, alluded to the rumour of Mr. Walker's ill health, and to another that he was a myth. Well, Mr. Walker was with them now, and he thought they would admit that he was a very good-looking sort of an Englishman, and in sound health. He (Sir Joseph) said that 6,000 men were now at work on the canal, and no extraordinary difficulties had been encountered. The contractor was to remove 1,000,000 yards of soil per month and in May that quantity had been exceeded.

Mr. Walker then spoke a few words. He was amused at the rumours about his health that had induced the White Star Line to offer him a passage to America. He hoped in a few months to double the work done, and to complete the canal in $3\frac{1}{2}$ years.

At the subsequent Stretford Vestry Meeting, which was very thinly attended, Mr. Kelsall carried a resolution to oppose the Bill, and thus the overseers came in direct conflict with the Stretford Local Board, who had, as they thought, protected the interests of Stretford in the arrangement made. This conflict of authority caused some angry correspondence, and a letter headed "The Porkhampton Burlesque," poked fun at "Swineopolis being disturbed at the loss of £10 15s. 3d. per annum in rates".

On the 29th June, the Bill came before a Select Committee of the Commons consisting of Mr. Marum (Chairman), Sir T. Lawrence, Mr. J. Stuart and Mr. Joseph Howard. Unfriendly petitions had been lodged against it by the overseers of Stretford, the Manchester Race-course Company and Mr. Thomas Chadwick. The latter claimed that the smoke from passing ships would prevent him carrying on his paper business, and that he would be in constant danger of fire from sparks, etc. The Race-course Company pleaded they would not have as good access to the canal or river as they had before.

Mr. Littler, for the Race-course Company, argued that the Bill of 1885 gave them access to the canal as frontagers, which, now the course of the canal was being altered, would be taken away. In reply, Colonel Clowes' agent, who had sold the land to the Race-course Company, gave evidence that there was a tow-path between the race-course and the river over which the owners of the former had no rights.

Mr. George Shaw, assistant overseer of Stretford, was called, and asked if when he got a petition signed against the Bill on the ground of 40 acres being taken away from Stretford, he had ever let it be known that 50 acres were being given in exchange. Mr. Shaw did not deny that he had been silent on the point, and confessed it would be a great advantage to Stretford if the proposed alterations were carried out.

After two days' hearing the Committee found the preamble proved. The Bill then went before the Select Committee on Standing Orders in the Lords, and the Duke of Buckingham allowed the Bill to proceed.

On the 16th July the Bill went before a Select Committee of the House of Lords, Earl Beauchamp being Chairman. All the opponents, except the Race-course Company had dropped away. The Stretford Overseers, at a ratepayers' meeting, had been told very plainly that they must not spend any more money in opposing a Bill, of which their own Local Board approved. Mr. Littler, Q.C., for the Race-course Company, struggled hard to show they had had access to the river Irwell, whilst the owner of the land, Mr. Clowes, declared there was a width of 48 feet intervening, which he had actually sold to the Ship Canal Company.

The Committee passed the preamble, adding a rider : " But the said Race-course Company shall be entitled only to the same rights, if any, over the substituted works as they are now entitled to under Section 62 of the Manchester Ship Canal Act of 1885 ".

As before stated the contractor commenced at the Eastham end, or *No.* 1. *Section* previously described ; afterwards he opened out at various points in the route. Mr. Walker lost no time in connecting one point with another by a contractor's line, and established a direct communication between Eastham and Manchester. Over this a supply train ran each day. The course lay on the top of the bank, and became known as "the overland route".

No. 2 *Section* extended from Ellesmere Port to the pretty village of Ince with its lighthouse among the trees. The canal runs alongside the estuary, divided from it by the Ellesmere and Ince embankments, and between them is the land through which the river Gowy runs into the Mersey. Here the canal dips inland, cutting off the Stanlow promontory, on which stand the remains of a Cistercian monastery endowed with "the land of Stanlow and the Villa of Manricaceston and Staneye". To this Abbey belongs a long history. It was damaged by an inundation in 1279; the great tower of the Church was overthrown by a violent storm in 1286, and in 1289 the Abbey was partly burnt down. What remains is now occupied as a farmhouse and barn. There are subterranean passages leading to the Gowy, and there is much besides to interest the antiquary, who should consult Ormerod's *Cheshire*. A large property owner on this length was the late Captain Park Yates, and directly the contractor's works were commenced, that gentleman got a temporary injunction to prevent him taking engines and waggons over his land. It was of no avail that the contractor had arranged with the tenant to go over the fields, and was quite willing to give security for any damage he might do. In consequence of this vexatious obstruction the material used in the works often had to be brought by water at an extra cost instead of being taken along the roads.

Ince is a queer, old-fashioned village, where there used to be a number of manors which in the days of King Athelstan paid tribute to the monastery of secular canons at Chester. The Manor House of Ince is in a wonderful state of preservation and the old church is worthy of a visit. The work in this section did not start till April, and by June Mr. Walker had built a large mission room, to be used as a reading-room on week-days, and huts for his men. On Stanlow peninsula he created a village where previously had been one solitary house. Before the navvies reached the rock they had much trouble with a depth of overlying peat, which kept sinking, and carrying the railway with it into the cutting below. Another difficulty in this section was the River Gowy, which had be to taken by a syphon under the canal and into the estuary. To save two syphons, the Thornton Brook was first turned into the Gowy inland by a special diversion which was constructed for the purpose.

The next section, *No.* 3, extending from Ince to Weston, runs entirely inland, passing through Frodsham Score, a kind of open common which the sea invades at high tides. On the portion between the canal and the estuary are powder magazines belonging to the Chilworth Company, on which they professed to place great value, because of future difficulty in getting permission to store explosives near large towns. At Weston Point are important docks, and here the Bridgewater and Weaver Navigations exchange traffic. This was the last section to get under weigh, as the Act of 1885 contained a provision that the canal should be completed to a defined point near Weston Point, and a subsidiary channel with a lock therein (now known as Weston Marsh Lock) formed to the Weston Canal, before any works should be commenced abreast of the Weston Point Docks.

This section, with its soft peaty soil, was always difficult to work in wet weather, as the place then became a complete puddle. There was a length of 2 miles of this soft soil, and, when they could get a good foothold, the German excavator and other navvies did excellent work. At the Weaver estuary, when the tide was low, it was possible to cross on planks supported by posts. Here immense stocks of timber had been located in order to start pile driving for the sluices in the following spring.

I now proceed to give the state of the works a year after their commencement, when I walked the whole length from Eastham to Manchester. Also a short description of some of the districts through which the canal passed.

Sections 1, 2, 3, *Eastham to Weston Point*, suffice it to say that the July rains gave much trouble, flooding the deep cuttings and impeding the work. In other respects good progress had been made. Curiosity, and the charm of a summer ramble, brought numbers of visitors to the scene, who often not only impeded the work but put themselves in great peril. They rambled all over the place regardless of the engines that were flitting about, and numberless were the hair-breadth escapes. One poor fellow ventured too near the edge of an embankment, which gave way. He was precipitated to the bottom, dislocated his neck and died instantly. Mr. Manisty, the engineer, generously paid for his funeral. About the same time a rather exciting incident took place. Some timber having got adrift at Eastham, three men jumped into a boat with the hope of saving it, but the wind and current being too strong to allow them to guide the boat with the one oar they possessed, they were carried 6 miles down the river. Eventually their cries attracted the attention of the *Mersey King* and they were rescued.

Section 4, *Weston Point to Norton*, commonly called the Runcorn section. This begins at the pumping station, just below the Bridgewater Docks, and extends for 5 miles in the direction of Warrington, as far as Moore. For the first 2 miles the canal runs alongside the Mersey, from which it is separated by an embankment; then, passing under the London and North-Western bridge at Runcorn, it runs inland through the Astmoor Marshes, till it comes close to the river at Randles sluices, but soon leaves it and again goes inland to Moore. It passes by historic ground. "It was here, at Runcorn," observes Leycester in his *Antiquities*, "that magnanimous woman, Elflede, built a town and castle in 916. Lady Elflede, or Ethelfled, was a sister of King Edward the Elder, and of her it is said that she governed the Mercians in most dangerous and troublous times for a period of eight years." The canal runs close by the castle rock which was once surrounded by a deep ditch 12 feet wide, and commanded the river. Farther on it passes below Halton Castle, said to have been a favourite place of recreation with John of Gaunt. In the reign of Elizabeth it was converted into a prison for Popish recusants. It was possessed in turns by the Royalists and Parliamentarians, and afterwards became a debtors' prison. It is now in ruins. Norton is half-way along the section, and abreast of it, on the other side of the river, are the tall chimneys of Widnes, called in derision "Little Hell," because the fumes from the chemical works are destructive to both vegetable and animal life. All materials for the works came by the Runcorn and Latchford Canal; this waterway was in time partially absorbed by the greater undertaking. The change was imperative, inasmuch as the Ship Canal crosses the Runcorn Canal no fewer than three times within a mile.

The terminus of this section, the quaint village of Moore, is supposed to have belonged to the priory of Norton, and after the dissolution of the Abbeys in the reign of Henry VIII. the manor was bought by Richard Brooke from the Crown. An old chronicler says it got its name because it signifies, "A more barren ground than marshes be, a miry and moorish soil yet serveth not to get turf thereon".

Work on this section was started in November, 1887, and soon got into full swing. A depôt at Norton was formed, with offices, store-rooms, engine sheds, blacksmith shops, etc., and nearly 1,000 men were employed: and here also were deposited immense quantities of timber and machinery. A singular accident occurred on the railway soon after it was opened. A switch lad was carelessly looking about him, and failed to observe in time a train coming from the opposite direction. The plough of the engine struck him a fearful blow and broke his arm at the elbow. By an unhappy coincidence the lad's father was the driver of the engine, "The Ince." Mr. Walker, missioner, took charge of the lad and conveyed him to the Warrington Infirmary.

When the contractor came to deal with the canal under the railway bridge at Runcorn, the railway company opposed any deepening on the ground that it might affect the foundations of the bridge, and their action delayed the work. Near Randles sluices a large bed of peat was encountered, greatly adding to the difficulty of securing good foundations for the sluices to be erected there. The peat was of an unusual character, consisting entirely of little twigs, and had not the varied strata usually found in bogs. This quaking bog, with quicksand underlying, seems to have been continued more or less under the adjacent Mersey; for when the Liverpool engineers came to take the tunnel for the Vyrnwy water-pipe under it they were completely nonplussed. The difficulty ended in a celebrated lawsuit, when Mr. Hawksley, the well-known engineer, obtained £35,000 and all costs from the Liverpool Corporation, because without just cause they had taken the work out of his hands. I well remember seeing the compressed air (which ought to have worked the machinery for driving) bubbling up all the way across the river.

In the case of an adjacent spoil-heap, the weight caused the bog to sink, and the land rose up some distance away.

In this stuff, and in heavy clay, it was found necessary to use Priestman's grabs. These are very handy tools, and can be employed for lifting, dredging, and a variety of purposes.   They are capable of raising a great quantity of soil at each lift, and are not unlike a steam navvy, but instead of being fitted with a bucket, or scoop, they have a chain, at the end of which is a grab.   They act as if a man's hands were bound together at the wrist inside to inside.   With the fingers extended the grab is driven into the soil, and then the fingers are drawn together, point to point.   When enclosed, the jib crane winds the grab over a waggon or cart and then its clutch is let loose, and the soil falls into the receptacle beneath.

During July there were many landslips on this section which seriously retarded the work.

In excavating in soft material near Runcorn, an immense boulder of grey granite, weighing 6 tons, was unearthed, and has been an object of much interest.

Some delay arose at the commencement of the operations, as before the contractor could begin work in the estuary, it was necessary that plans of the whole works therein should be approved· by the Acting Conservator of the Mersey.   At last consent was given, and soon the district assumed a busy appearance.   Another important event occurred in August, *viz.*, the closing and drainage of part of the Latchford Canal.   This might have led to the stoppage of Messrs. Wigg Brothers & Steele's works, at Runcorn, and was the cause of considerable alarm.   However, the Ship Canal Company temporarily arranged to keep these works supplied with water by means of pipes laid from the upper portions of the Latchford Canal, and carried on to Messrs. Wigg's premises.

Whilst busy with the Latchford section, traces were discovered of an old Roman road, and some fragments of Roman pottery and a broken "Quern" or hand-mill were found.

Section 5, Norton to Latchford.—This section is one of the most important on the line of the canal.   It has a length of about 4 miles and runs parallel to the Runcorn and Latchford Canal, and between the Ship Canal and the estuary of the river are Norton and Richmond Marshes.   Here again were found treacherous mosses that needed a very solid embankment, to make which heavy stuff had to be tipped.   There was always danger of the ground sinking and causing a rise elsewhere. On one occasion a subsidence of 9 feet took place in one night.   On this length, too,

LATCHFORD CUTTING AND VIADUCT.

Birtles, Warrington.

were several railway crossings of great importance.   The contractor had between 600 and 700 men at work at the commencement of the year.   This section passes through Sir Richard Brooke's estate, and the interest of the navvies was considerably aroused by cutting through a large number of rabbit burrows : indeed in one case the excavator scooped up along with the earth a nest of young rabbits, which fell an easy prey to their captors.

Section 6, Latchford to Warburton.—This length is about 5 miles.   It passes close to Thelwall.   In 923 A.D. King Edward the Elder founded a city here, and called it Thelwall, but all that now remains of it are "The Pickering Arms" and three or four farm buildings and cottages.[1]   Close by is the quaint Thelwall Church with the tombstone of a former landlord of the hotel, on which is inscribed :—

> He always was a neighbour's friend,
> Deny it if you can,
> He never paid a lawyer's bill,
> Nor caused another man.

On this length the contractor started the year with over 700 men.   He made it a centre, and built on it the usual equipment, besides three long streets of huts. When the excavators got down about 14 feet, they came across several fossil trees, which were dug out.   By July great changes had taken place, several Priestman's grabs, ordinary steam navvies, and Whittaker's excavators were at work, and a German navvy was in course of construction.   The clay out of the cutting was being converted close by into bricks, by Murray's patent machinery.   Here I saw for the first time horse barrow roads, and very strange they look when at work. In special cuttings a plank is laid from the top to the bottom, at an angle of about 45°.   Up this plank a man and barrow are drawn by a gin-horse attached to a rope and pulley.   Going up, the man pulls backward to steady himself; he could not walk up if he were not holding the barrow.   Going down, with an empty barrow, he crouches with the barrow behind him, and slides down the plank, using his feet as a brake.   In this way 100 barrowfuls of earth can be lifted in an hour.   A man goes up and down nearly twice in a minute.   When strangers saw half a dozen of these

---

[1] The *Anglo-Saxon Chronicle* says : " A.D. 923—In this year, after harvest, King Edward went with his forces to Thelwall and commanded the town to be built and occupied and manned." Edward the Elder, King of the English, A.D. 901-925, built the fortress and town of Thelwall and sojourned there.   Victorious over the Danes, he preserved the blessings of Christianity and established supremacy over all Britain.   He was crowned at Kingston-on-Thames on Whit-Sunday, A.D. 902, by Plegmund, Archbishop of Canterbury, himself a Cheshire man, and lies buried in Winchester Cathedral.

"horse roads" in full work, side by side, they looked in astonishment at the exciting scene, and said it was one of the most curious sights on the canal.   At the depôt here the British Workmen's Public House Company had established tea and coffee rooms, where 200 workmen could sit down at a time.   A church was also built to accommodate 600 persons, to which day and Sunday schools were attached.

During the last week in August one of Ruston & Proctor's steam navvies did a record week's work: it filled 557 waggons in 12 hours, equal to 2,500 tons in weight.   There was a healthy rivalry between the men who worked the navvies, and the following week a driver from Latchford wrote to the papers and stated that in a 12 hours' day he had filled 640 waggons with 2,900 tons of soil.

Section 7, Warburton to Barton.—This section had a length of some 5 miles. On it were some important locks and sluices.   Strange to say, before the Ship Canal was made, the only bridge and the only works between Barton and Warrington were on this length.   The Warburton Bridge was a private one, for crossing which a toll was paid.   Close by was a good straight reach on the river, and here in my youth the Warrington Regattas took place, and afforded a pleasant afternoon's picnic for Manchester people.   Nearer Irlam were Ockleston's Paper Mills and Mill Bank Hall. When the owners claimed compensation from the Ship Canal, one item was that a rookery, which was one of the features of the Hall, would be disturbed.   Not less singular was the claim as regarded the works.   It was contended that paper made with the filthy and chemically polluted waters of the Irwell had the remarkable property of preventing steel from rusting; therefore it found favour in the Sheffield market: and to give pure water in its place would be a serious loss.   Needless to say such contentions were not taken seriously.   Work on this section was not commenced till early in 1888.   Great progress was being made when a tremendous flood occurred, which submerged the whole of the excavations at Stickens Island, where a cutting of 40 feet had been made, and stopped blasting operations for some time.

No. 8 Section, Barton to Manchester.—This section extends from Stickens Cut, Barton, to Woden Street, Hulme.   Without doubt here is the most important part of the canal, for on it, besides the Barton and Mode Wheel locks, are two important swing bridges, the Barton aqueduct and the whole of the terminal docks.   On this section alone there were 2,800 men at work, 26 horses, 30 locomotives, 7 Ruston & Proctor steam navvies, 3 Whittaker's crane extractors and 7 French excavators.   Whilst excavating at Salt Eye, near the Barton aqueduct, and at a depth of 10 or 12 feet under the clay, a huge well was discovered, 260 feet deep and with a mouth about 8 feet in diameter.   When an iron tube was inserted it yielded about 5,000 gallons of water

Killon.

BARTON LOCKS, SIXTY-FIVE FEET GATE.

To face page 34.

per hour. The surmise is that an attempt had been made in time past to bore for coal; no other reason for the shaft can be suggested.

A novelty in the shape of a mechanical concrete mixer had been introduced at the Pomona Docks, which did three times as much work as could be done by hand, and at the same cost—another proof of the enterprise of the contractor.

So heavy was the work on this section that it was found necessary to make a division, and henceforth No. 8 extended from Stickens Island to Mode Wheel, with Mr. Bourke as engineer, while No. 9 took dock work only, Mr. Kyle being the resident engineer.

HUTS.

In the vicinity of towns such as Runcorn and Warrington navvies are able to find lodgings for themselves, but in out-of-the-way districts a contractor is bound to provide accommodation for his men, usually in the shape of rows of wooden houses. Within these rough structures hundreds of men with their families are housed. It generally happens that a ganger with a capable wife rents a large hut and takes in a number of single men as lodgers. The genuine navvies are stalwart men who follow jobs from place to place, and seem to enjoy change of scenery and hard work. Keep them from drink and they are good-hearted, generous men, who are merciful though they are strong, and do not give much trouble to the police. They have their meals in a large common room, and enjoy their pipe and social chat after a hard day's work. Much depends on the supervision exercised, but in some places the huts are a trouble to the excise, inasmuch as the keepers make profit by selling beer to the inmates; when found out heavy fines follow. The average rent of a hut is 6s. to 8s. per week, which includes a sufficient allowance of coal; the lodger pays 13s. per week for food and lodging. At Eastham Mr. Morris, the head ganger, who lived among his men, was known as "The Bricklayer Poet," and had published a volume of poems. Here, too, the mission room was the scene of many entertainments got up to brighten the lives of the navvies and their families. The contractor's agent of the Eastham section, Mr. Manisty (son of Mr. Justice Manisty), and his good wife, with the help of the vicar, Mr. Torr, and other friends often got up very creditable entertainments, and the gratitude of the men was poured out in the following lines :—

THE EASTHAM CHILDREN TO MRS. MANISTY.

Thanks, lady, thanks, for thy very generous gift
  We all received with gratitude and love,
And when at eventide our hands uplift,
  For blessings on thy future we will move.

So little sunshine on our way hath smiled,
    That we are wont to value much the "treat".
Few trouble much about a Navvy's child,
    Ill-mannered, rough, and oft with naked feet.

But we can grateful feel for favours shown,
    And such kind acts you never will repent ;
For we shall love you, claim you as our own,
    Your kindness we accept as only lent.

And Mr. Manisty, we thank him much,
    His orange scrambles gave us much delight,
The race for sweets, and all the sports were such
    That gave us joy and made our faces bright.

We told our mothers when we got us home,
    As on the table we poured out our store,
And for the little ones, who couldn't come,
    We raced for bits of cake upon the floor.

We'll mind your words, and when next year comes round,
    We hope all will be spared again to meet,
Then Eastham Woods shall with one noise resound ;
    Till then we'll all be good, and win your treat.

THE NAVVY

I'm a Navvy, I work on the Ship Canal ;
I'm a tipper, and live in a hut with my Sal ;
If ever you come to Eastham, call at Sea Rough Wood,
    There's a hearty cheer, without the beer,
    And "tommy" that's always good.

We have lodgers, a splendid lot of young men,
In the evening around the tables are ten.
Some of the lads are strangers, never been out before,
    But some can tell of a long, long spell,
    Tramping the country o'er.

Tales of daring the older ones would tell,
And tales of want, when with nowhere to dwell.
They'd nestle in a hayrick, or shelter in a barn,
    And tales like these us navvies please,
    When not o'erdone the yarn.

Our wives are rough, but yet no one can say
They're hard-hearted: none are sent empty away;
If they only look like navvies, they're welcome to a share,
And told to lay by the fire till day
If there's no bed to spare.

Our work is hard, and dangers are always near,
And lucky are we if safely through life we steer;
But still the life of a navvy, with its many change of scene,
With a dear old wife, is just the life,
That suits old Nobby Green.

—SHIP CANAL POET.

Without doubt a great help in the progress of the work was the use of a light produced by the combustion of oil gas (the refuse of gas tar after distillation), and called "Wells' Light," from the name of the patentee. A pure white and steady light is produced of great candle power; the cost is not over 3d. per hour, and by its aid men can work by night just as well as by day.

MR. T. A. WALKER.

From the beginning to the end of the sections, wherever you went, the letters "T. A. W." stared you in the face; they were stamped on everything you looked at and handled; every truck had these conspicuous letters on its side. They were the initials of the contractor, Thomas Andrew Walker, a very remarkable man. Born in 1828, we find him sixteen years later busy in the survey of new railway lines. In 1847 he was on the staff of the well-known contractor, Mr. Thomas Brassey, and that gentleman placed such reliance in him that in 1852 he was sent to Canada to take up a responsible position on the Grand Trunk Railway. He obtained the confidence of the people there and started business on his own account, constructing Government railways. He left Canada in order to survey and construct railways in Russia and Egypt. On his return to England in 1865 he took a prominent part (under Mr. Fowler the engineer) in the construction of the London underground railways. Along with his brother, Mr. Charles Walker, he had in hand the very difficult task of applying Brunel's Thames Tunnel to railway purposes. In 1879 he carried out the Severn Tunnel under almost superhuman difficulties, an achievement which has made his name famous all over the world. In this huge work he employed 3,000 men and boys, but it was a small work compared with the Ship Canal, in which 16,000 men were engaged.

ARBITRATIONS.

During the whole of the year Mr. Dunlop, the land agent, was busily engaged in purchasing land. In many cases he was able to come to terms, in others he was forced to arbitration. Amongst the cases adjudicated upon were the following (*see opposite page*).

It had been intended to purchase the works of Messrs. Roberts, Dale & Co., Cornbrook. To the surprise of the Ship Canal Company that firm had, in anticipation, obtained from the old Bridgewater Navigation Company a ten years' lease for a portion of the land. It is stated that the vendors asked £155,000 for their works, and that the Ship Canal Company at once closed the negotiation.

Besides land arbitrations, the company had to encounter rating appeals. The Runcorn Union sought to increase the ratable value of the Bridgewater Canal in Runcorn Township from £7,097 to £16,152, but on an appeal to the Quarter Sessions the rate was fixed at £7,050, or £47 lower than before. In consequence the Union had to pay all costs. It was a case of the biter bitten.

ODDS AND ENDS.

As time went on and the Ship Canal shares declined in price, some few holders tried to rid themselves of their responsibilities, and in a case on the 9th August, the Ship Canal Company *v.* Vaughan, the defendant pleaded that though he had applied for shares and had had them allotted, he was not liable, because he had not paid a single penny hitherto. The Recorder gave a verdict for the company, with costs.

During the year the directors thought it advisable to make the traders and scientists of the district familiar with the canal, and it was visited by members of the Iron and Steel Institute, the Master Builders' Association, the Geographical and Geological Societies, and numberless other public bodies.

An animated discussion arose this year in the pages of *Fairplay*, which contained dolorous prophecies about the future of the canal. These prophecies are only of interest now as showing how easy it is to become unduly prejudiced, and how difficult it is to anticipate results.

Consul Hale, the United States Vice-Consul in Manchester, published a report to his Government which was complimentary to the new project, and he no doubt did much to familiarise his countrymen with the merits of the proposed canal.

In October this year Messrs. Lewis & Co., of Liverpool, sent to Mr. Gladstone a copy of their prize essay on the future of Liverpool, and received from him the post card, dated 20th October, 1888, which appears on page 40.

JOHN K. BYTHELL, CHAIRMAN OF THE MANCHESTER SHIP CANAL,
1894 *Seq.*

Lafayette, Ltd.

To face page 38.

Vendor.	Claim. Description of Property.	Umpire.	Evidence Given.		Statutory Offer.	Award.	Remarks.
			For Claimant.	For Company.			
Shropshire Union Railways and Canal Company.	29 acres land, 17 acres fore-shore, Ellesmere Port	W. J. Beadel, Esq., M.P., London	£25,559	£3,395	£5,613	£11,732	
Captain Park Yates, Ince Hall, Ince	155 acres pasture land, 24 acres arable land. Compulsory sale, injury to mansion and park	W. J. Beadel, Esq., M.P., London	£59,377	£25,816	—	£37,620	Strange to say, Mr. Dunlop, the Ship Canal valuer, was also trustee for the vendor's wife.
R. C. Naylor, Esq., Hooton Hall	121 acres of land and fore-shore. Damage and severance of estate from River Mersey	W. J. Beadel, Esq., M.P., London	£82,812	£21,384	£22,000	£28,924	Mr. Naylor spent over £50,000 on the Hall. Then his wife's health broke down and he had to live at his Northamptonshire seat. The authorities wanted to increase his rates, and this so annoyed him that he removed all his furniture and the Hall lay empty ever after, paying no rates.
Lord Egerton of Tatton	56½ acres of land at Throstle Nest, east and west of Trafford Road. Dock site, 10 per cent. for compulsory purchase	John Clutton, Esq., London	£85,923, being 3¾d. to 4d. per yd. according to situation	£42,447	£55,000	£63,240	
Peter P. Brownrigg, Esq.	Moiety of 4 acres, 2 roods, 33 poles land near Ellesmere Port, including bed of clay 1½ yards thick. Damage and severance	His Honour Judge Horatio Lloyd, Assessor and Jury	£4,678 claim £29,203, or £6,500 per acre	£363	£400	£500	Some few years previously the Duke of Westminster had sold 19 acres of land (of which this formed part) to the vendors, including mines and minerals, for £810, or a little over £42 per acre. The clay had not then been opened out.
James Reilly	Pomona Gardens, Manchester. 21½ acres of land with buildings, including agricultural hall and chair works. Goodwill and compulsory sale	Daniel Watney, Esq., London	£141,774	£49,681	£57,500	£70,352	A most costly arbitration, and the case was adjourned repeatedly at the instance of counsel.

A FAMOUS POST CARD

On other side is the Right Hon. W. E. Gladstone's opinion of Lewis's
Prize Essays on the Future of Liverpool, in *fac-simile*.

COPY OF POST CARD FROM MR. W. E. GLADSTONE TO MESSRS. LEWIS & CO., LIVERPOOL, *RE*
SHIP CANAL, 20TH OCTOBER, 1888.

CHAPTER XIX

1889

DISPUTE ABOUT TIDAL OPENINGS—INJUNCTION—APPLICA-
TION TO PARLIAMENT NECESSARY—DEATH OF MR.
WALKER—SHAREHOLDERS' MEETINGS—ISSUE OF DE-
BENTURES—VISITORS—SPEECH BY MR. GLADSTONE—
DESCRIPTION OF SECTIONS AND PLANT—INCIDENTS
AND ACCIDENTS—ARBITRATIONS.

*The country is fined many millions a year from want of water carriage. If our mine
owners and manufacturers were relieved to the extent to which water carriage would relieve
them, it would turn the delicately balanced scale of competition in our favour in a majority
of instances. The attempt to force all the goods traffic on the railway is, in fact, as wise as
if a gentleman insisted upon carrying all the produce of his estate to market in his carriage.—*
Sir ARTHUR COTTON.

THE new year opened well. Every section along the Ship Canal was in active
operation, and the whole work had been systematised. Wherever machinery
could save manual labour it was applied, regardless of cost. Experienced
hands were plentiful, and on reckoning up the cubic yards excavated during the
previous year, it was found that if the same rate of progress could be maintained,
the work ought to be finished pretty well up to time. It was admitted that the
bottom of the trench would often be of harder material than was anticipated, and the
height to which the excavated material would have to be raised would be greater,
but to counteract this, the general arrangements were on a sound basis, and various
economies of time and labour were being exercised. But there are cataracts and
breaks that often upset men's best calculations, and the Ship Canal Directors found
this out to their cost.

The Mersey Dock Board were in no pleasant frame of mind after their Parliamentary defeat, and they kept a keen look out to see the clauses in the Act of

Parliament were strictly carried out.   At last they got their opportunity.   On the plea that the canal, as proposed by the promoters, meant a reduction of estuarial space, they had induced Mr. Forster to push the canal inland as far as possible. At first when the canal almost formed part of the estuary, it was thought necessary that the water should run freely in and out at certain states of the tide, and forty-five apertures, each 100 feet wide, were shown upon the original sections which were deposited in connection with the Bill of 1885.   When the Eastham to Runcorn section was pushed inland, the tidal openings were absolutely useless, and it was an omission not to have excluded them from the Act of Parliament.

On the commencement of active operations the Ship Canal Company, with the sanction of the Mersey Conservancy, proposed to substitute three openings of 600 feet each in place of the forty-five openings, considering that the three openings, taken in conjunction with sluices at the Weaver and at Randles would be more than sufficient for all practical purposes.   The alert Dock Board officials, however, demanded that no change of any kind should be made.   It was useless to plead the large number of openings were not only unnecessary but would be an obstruction to navigation.   It was futile to argue that the Conservator of the Mersey was agreeable to the change.   There was the Act of Parliament, and they would have it carried out to the letter.   So to law they went.

In April, 1889, the action of the Attorney-General (on the relation of the Mersey Docks and Harbour Board) v. the Manchester Ship Canal came on in the Chancery Court before Mr. Justice Chitty.   A motion was made asking for an injunction to restrain the defendants from making openings in the embankments otherwise than in accordance with the 1885 plans and sections.

After several adjournments the motion came on for hearing on 16th July.   The *Attorney-General* alleged that the Ship Canal Company were endeavouring to break a statutory bargain arrived at after a most solemn discussion and much careful consideration.   He stated the Dock Board believed that the maintenance of the port of Liverpool depended on the scour on the estuary being maintained, and recited a statement by the promoters :—

We quite recognise the value to the estuary of this volume of water, and accordingly we are going to provide a very large number of openings, which will allow the water as it rises to flow into the canal, and as it falls to flow out of the canal.

The Attorney-General went on to say that members of the Dock Board were there to call attention to a most serious departure from the engagement made for the protection of the estuary, and he submitted that the Mersey Commissioners had no right to allow the defendant company to depart from a statutory bargain, or vary the way the works should be carried out. The Acting Conservator, Sir George Richards, had no right to consent to tidal openings being omitted, even if this was subject to their being put in afterwards if they were found necessary. This remark also applied to Sir John Coode, the engineer, who had been called in by Sir George Richards to advise with him. The Attorney-General asked that the statute might not be deviated from.

Mr. Henn Collins, K.C., counsel for the Canal Company, contended that they were discretionary openings, to be put in if necessary or expedient, and not even shown on the plans. True, the sections showed pictures of an embankment indented in places, but nothing more.

Mr. Justice Chitty said he had come to the conclusion that the consent of the Mersey Commissioners was of no avail to the defendants; that the tidal openings were works defined by an Act of Parliament, and that the company were bound to execute them. He granted the injunction, subject to conditions. As regarded the line of embankments the action was dismissed.

The operation of the injunction was suspended for six months.

When Mr. Squarey, solicitor to the Dock Board, informed the Runcorn Board of Commissioners that Admiral Richards, the Acting Conservator, was withholding his consent to raising the tidal openings 1 foot 10 inches, and favoured reducing the number and length of the tidal openings, that body, instead of supporting the view of the Dock Board, resolved to ask the Conservator to grant the application of Manchester as being of advantage to Runcorn.

Mr. Justice Chitty's decision compelled the Ship Canal Company to go to Parliament for a Bill to deal with the tidal openings. The Bill became law, so that beyond delaying the work the action of the Dock Board did not serve their purpose.

In a previous session the Dock Board had had a Bill in Parliament, which was withdrawn because of threatened opposition. This year they came again with an Omnibus Bill, whereupon the Corporations of Lancashire determined to try and get the share of representation contemplated when the Dock Board was formed. They therefore asked that, inasmuch as it was a national and not a local trust, and that all traders throughout the kingdom passing goods through Liverpool paid for its support the conditions of voting should enable them to vote and have representation.

They asked further :—

1. That the residential qualification of a Member of the Board should be extended to 50 miles.

2. That a wider publicity should be given to the fact that payers of £10 per annum in dues were entitled to a vote.

3. That the power of voting by proxy should be restored.

4. That the day for the revision of the voters' list should be made public.

5. That there should be an interval of at least seven days between the nomination and the poll.

An instruction to the Committee to carry out these changes was moved in the House of Commons by Sir William Houldsworth, and caused a long discussion. Eventually, on the suggestion of Mr. Courtney, the instruction was withdrawn, and the matter referred to a hybrid Committee.

In due time the Select Committee of the Commons met, with Mr. Childers as Chairman. Sir Joseph Heron was a witness for the united Corporations, and this was his last appearance before a Parliamentary Committee, as he died shortly afterwards. He gave his evidence with admirable clearness, explaining the part he took in the formation of the Dock Trust, and said that it was always the intention that taxation and representation should go together, which was not carried out when outsiders, who paid the greatest share of the dues, had only 400 votes, whilst the agents, brokers, etc., in Liverpool had 2,089 votes.[1] Whatever might be the fate of the Ship Canal, Liverpool would be, to a great extent, the port for Lancashire, and the county would continue to be always interested in the prosperity of the Mersey Docks.

In addition to the Corporations of Lancashire, the Ship Canal Company also petitioned against the Bill in respect of the powers sought to be vested in the Dock Board as to clearing the river of wrecks and obstructions. Their opposition, how-ever, was withdrawn on the Dock Board agreeing to free the portion of the estuary which lies between Bromborough Pool and Eastham, and which forms the approach to the Ship Canal, and any portion of the estuary over which the Canal Company had control, from the operation of the proposed clause. They also came to an arrangement as to anchorage, navigation, etc. Evidence was heard on both sides, and eventually the Committee decided to pass the preamble of the Mersey Docks and Harbour Bill. They declined to alter the area of residence, or restore voting by proxy, but decided that, in future, the owner of the goods paying duty should be placed on the voting list, and not the agent only; and that proper notice should be given of the time of the revision of the voters' lists. Also that there should

---

[1] A merchant writing to the *Liverpool Journal of Commerce* in 1906 said: "It is the custom to pay dues for every clerk and employee down to the office boys, to make them into voters, and that hundreds of 'faggot' voters are so made with the view of illegitimately influencing elections".

be an extended time between the nomination and election of members of the Board.

A renewed difficulty the Canal Company had to face was the closing of part of the Runcorn and Latchford Canal, whereby the water supply of Messrs. Wigg Brothers & Steele, of Runcorn, was diminished. For a long time it lay in the balance whether the Company would not be compelled to purchase these large works. The firm early took the alarm, and on the 21st June obtained an interim injunction to prevent their supply of water being interfered with until satisfactory arrangements had been made to supply the works from other sources.

After two postponements by agreement, and an arrangement for a temporary supply of water, the case came before Baron Pollock, sitting in the Chancery Court on the 4th September. Counsel for Messrs. Wigg Brothers & Steele complained that the Canal Company had placed a level crossing over the road to their works. Counsel for the Canal Company in reply stated the freehold of the road was theirs, and the plaintiffs had only a right of way over it; further, that if rails might not be put across the road, the works of a big undertaking would be seriously impeded, and that this was an attempt to put undue pressure on his clients. Eventually a further temporary agreement was come to, and the case was postponed until the next Michaelmas sittings.

In November of this year the Canal Company deposited two Bills in Parliament.

1. *A Various Powers Bill.*—To enable them to make various alterations as regarded land and works; to purchase land; to make amendments to previous Acts, and to raise additional capital, share and loan, with priority of dividend and interest.

2. *Tidal Openings Bill.*—To enable them to alter the embankments and tidal openings, and to do and deviate other works connected with the canal.

At this juncture there occurred the greatest disaster that could possibly have befallen the Ship Canal undertaking. Mr. T. A. Walker, the contractor, and the master-mind in the carrying out of the works, died on the 25th November, at his residence, Mount Baron, near Chepstow. For some time he had been suffering from Bright's disease, and was now suddenly struck down by paralysis, at the age of sixty-two, leaving a widow and four children.

Mr. Walker, as I first knew him, was a robust, good-looking man, with determination of character stamped on every feature. He was brusque, yet kind and thoughtful for all his employees, by whom he was beloved. When the Barry Docks were opened, he preferred to rejoice and dine with his 2,000 navvies rather than take part in a grand banquet given by the directors. In religion he was a member of the Plymouth Brethren.   He was buried at Caerwent Church, Chepstow.   Mr. Walker has been termed the "Prince of Contractors," but as space will not allow me to deal with his career, it must suffice to say that when he took in hand the Ship Canal it was felt on all sides that the right man had got the job, and his appointment as contractor gave confidence to everybody.   His will was sworn at something under £600,000, and it was quickly known that, anticipating a fatal result of his illness, he had made arrangements for the Ship Canal works to be carried on without interruption, and had given ample powers for this purpose to his executors, Messrs. T. J. Reeves, L. P. Nott and C. H. Walker.

Warrington has always been a fickle supporter of the canal, but I am pleased to chronicle that when the General Purposes Committee of the Council proposed to ask for the repayment of £1,000 subscribed towards the preliminary expenses of the canal, Alderman Burgess urged that the motion be rejected, and pointed out that Warrington had already benefited considerably by the Ship Canal, and would benefit still more.   The proposal, he said, arose out of a suggestion made by Salford, but he thought Salford had enough on her hands at present without meddling with a matter of that sort, and he resented the interference.   Other members, including Mr. H. Roberts, said they regarded the amount as a donation, and the only explanation given was that of the Deputy Town Clerk, who said the letter from Salford was an inquiry as to what Warrington intended to do.   Alderman Burgess carried his amendment without opposition.

At the first meeting of the Ship Canal shareholders Lord Egerton said the canal was no longer a matter of faith but of visible works.   He congratulated them on the fact that a fourth of the work had been done in a fourth of the estimated time, and that all the land had now been secured.   He lamented the death of Mr. John Rylands, a director, a staunch supporter, and a shareholder to the extent of £60,000, and announced that Mr. Henderson, of London, an even larger shareholder, had joined the Board. Sir Joseph Lee stated that the contractor had £700,000 worth of plant on the job and at least 11,000 men; these in time would be increased to 20,000.   He foreshadowed the issue of debenture bonds, and said this could be undertaken when £5 had been paid on each ordinary share allotted.   He thanked Mr. James Reilly for the use of St. James's Hall, condoled with him on his being driven away from Pomona Gardens, and hoped he would live to see ships where formerly he had made chairs.

Shortly afterwards Messrs. Baring Brothers and Messrs. Rothschild & Sons issued £1,359,000 (part of £1,812,000 sanctioned by Parliament) 4 per cent. debenture stock, redeemable or convertible into 3½ per cent. stock in January,

Thomas A. Walker, Contractor for the Manchester Ship Canal.
By permission of Messrs. Macmillan & Co., Ltd.,   *To face page* 46.
London.

1896.   In the circular it stated that the £4,000,000 preference stock and all the ordinary stock, except £617,990, had been taken up, and of the latter the contractor was under obligation to take up £500,000 if required to do so.

The result of the debenture issue was very satisfactory.   The list was opened on Tuesday, 19th February, and was advertised to remain open till Thursday the 21st. It was, however, closed early on Wednesday morning, as the amount required had been over-subscribed the previous day.

When the midsummer meeting of the shareholders was held, the Chairman gave an encouraging report.   He said that out of a total of 44,000,000 cubic yards to be removed, 20,000,000 had been dealt with and the contractor was within time.

Sir Joseph Lee referred to the visits that had been paid to the canal by the Chairmen and Directors of most of the principal railways in the country, and believed the result would be a better understanding with the railway companies who now saw that the canal was no chimerical scheme, but one that must become a great factor in the trade of the country.   He quite believed that they would all feel it to be in the interests of their companies to connect their lines and work in unison with the Ship Canal.   There had been other visitors to the canal who were not in sympathy with the undertaking.   The Mersey Docks and Harbour Board had visited it, and the result was that they were still breathing out "threatenings and slaughter".   But the directors were prepared to meet them, and their opponents had gained nothing by the expenditure of money in Chancery.   The directors had great confidence in their legal advisers; he believed the position the company had taken up was a secure one, and they would apply for additional powers in November.

Subsequent to this meeting the state of affairs between the executors of the contractor and the directors caused some uneasiness.   The former brought a heavy claim on the ground of delay in getting possession of portions of the land, and for the loss due to machinery being kept idle, also for extra rock cutting, etc.   The directors, too, had their grievances.   Mr. Walker at the start undertook to bring on the ground £300,000 worth of second-hand plant to be approved by the Ship Canal engineers, and on this the Ship Canal were to advance cash at 4 per cent. interest.   A further £400,000 worth of new machinery was to be provided by the Company.   The whole £700,000 was to be repaid by the contractor in forty-seven instalments which were to be deducted from the monthly payments made to him. Complications soon began to arise through Mr. Walker wishing to use less of the old machinery and to have more new, and also asking for a suspension of the instalments. Eventually it was arranged that the executors should waive their claim for losses sustained by delays, and that the balance still owing for machinery should be paid off in five payments of £100,000 each.   An arrangement was also made as to interest.

VISITORS.

As the fame of the Ship Canal became spread abroad visitors flocked from all sides to see the works, among the rest Lord Colville of Culross, and other Directors of the Great Northern Railway Company; the Manchester Architectural Society; the Manchester Field Naturalists' Society; the Manchester Archæologists' Society; the Liverpool Geological Society; the Liverpool Engineering Society; and a party of Liverpool Cotton Brokers.

On 8th June about thirty London journalists, accompanied by Sir Joseph Lee, Mr. Jacob Bright, Mr. Leader Williams and others, inspected the works.   They went some in open trucks and the rest in Mr. Walker's own saloon.   They were taken down a siding and run into the bottom of the canal to have a good view of the works. At one of the temporary wooden bridges an accident occurred.   In consequence of the saloon being too tall it came into collision with the crown of the bridge and had its top taken off.   Fortunately no one was hurt.   Farther on a still more serious accident happened at another bridge where repairs were in progress.   Here the trucks dashed into the scaffolding and brought it down on to the top of the first truck, but the occupants, having timely warning lay down on the bottom, and again were more frightened than hurt.   The whole of the workings from Pomona Gardens to Eastham were fully explained.   The party eventually returned *via* Liverpool, but before doing so they had dinner at the Eastham Hotel, where Mr. Bythell, referring to the greatness of the scheme, ventured to prophesy that while Manchester would certainly benefit by the canal, Liverpool would not be a loser; there was no reason to fear that the Mersey Docks would become a breeding place for gulls, or that the owl would hoot amid the grandeur of desolation arising from neglected and crumbling warehouses along the river bank.   The veteran engineer, Mr. Thomas Hawksley, was one of the party.

The following day a party of American engineers was taken to Irlam to inspect the Manchester end of the works, and they had the whole explained to them by Mr. Leader Williams, the engineer, including the proposed swinging aqueduct at Barton.   They expressed themselves highly delighted with all they saw.

In the middle of July the Shah of Persia came to Manchester and a visit to the canal was arranged for him. Great preparations were made. A marquee was provided in which the large model of the canal was placed for the Shah's inspection. Flags and streamers floated on all sides, and the directors and a distinguished company awaited his arrival. As usual His Imperial Majesty was very late in putting in an appearance. He seemed to think that to keep his hosts waiting was of no importance.

After introductions and an explanation of the model by the engineer, the Shah, accompanied by Lord Egerton and other directors, proceeded along the works as far as Irlam, where the train was transferred to the Cheshire Lines railway, and so returned to the Central Station. The Shah was amazed at the magnitude of the undertaking.

On 4th December, Manchester was honoured by a visit from Mr. Gladstone. At a luncheon given by Alderman Mark, the Mayor, he thus spoke of the canal—a model of which he had carefully inspected :—

I may be told the canal is not directly connected with municipal life, nor is it ; but it is in my opinion very directly connected, indeed, with the energetic action of public spirit and desire of improvement, of which the development of municipal life has been one of the main causes. The Ship Canal is not only local but it may claim the honour of being national. Now by my own family and by our recollections I have been much associated with the great city of Liverpool, and for my own part I am perfectly convinced that Liverpool will suffer no damage whatever. Great difficulties have been encountered. Those difficulties have tested the spirit of the undertakers. There is something of courage, there is something of persistence, there is something almost of heroism in the energy and determination by which these difficulties have been surmounted, and I heartily hope the results may be satisfactory. I am quite certain they will be so to the nation at large, and I heartily hope they may be so to all concerned.

DESCRIPTION OF THE SECTIONS AND PLANT.

In order to follow the progress of the works, I propose to describe the state of affairs in 1889 of the nine sections into which the canal was divided.

No. 1 Section extended from Eastham to Ellesmere Port. Contractor's agent, Mr. E. Manisty ; Ship Canal resident engineer, Mr. W. Elliott. Length 4 miles 20 chains. The principal works on this section were the dredging of the entrance to the channel, and the construction of the Eastham Locks, and the Pool Hall embankment, including three syphons under the canal. In the beginning of January, 1889, about 1,200,000 cubic yards, chiefly of hard clay, had been removed, mostly from the site of the Eastham Locks. In consequence of difficulties relating to the tidal openings, there had been delay in completing the Pool Hall and Ellesmere

Port embankments, both of which impinged on the estuary; indeed these works after a time were stopped by injunction. To give an idea of their magnitude, it may be said that the Pool Hall embankment was 1 mile long, varying in height from 10 to 40 feet. In this, an aperture in the gantry had to be left. The embankment was built of stone with a hearting of clay, and had a breadth of 30 feet at the top, with a batter of 1 to 1 inside and $1\frac{1}{2}$ to 1 outside. At this time the Ellesmere Port embankment had made small progress, but it afterwards extended across the bay and deprived Ellesmere Port of its old access to the river, except that an aperture was left for small vessels, across which the contractor's railway was carried by a swing bridge.

As it was imperative that the canal should run in front of the port, which would then be inaccessible from the estuary, it was arranged that on completion all ships of the size usually trading there should be entitled to the shorter and safer passage by the Ship Canal, free of charge.

At Eastham a huge shed had been erected in which to build the dock gates. It was a question for some time whether they should be made of steel or of some hard wood. It was eventually decided to build them of greenheart (*Nectandra Rodiæi*), which is a native of British Guiana. The tree grows near the sea to a height of 80 or 90 feet, the first 40 or 50 feet unbroken by a single lateral branch; and it has a circumference of 9 to 12 feet. This wood has the advantage over steel of being very hard and tough, and it will stand a great deal of rough usage without breaking. It is one of the few woods that will sink in water, having a specific gravity of 1·14 and an average weight of about 72 lbs. per cubic foot. It is impervious to the attack of insects. The building of these gates was a most interesting sight, and they needed good joiners and sharp tools. A pair of gates for the 80 feet lock weighed 540 tons in air.

No. 2 Section, Ellesmere Port to Ince.—Contractor's chief agent, Mr. A. C. Brown; Ship Canal resident engineer, Mr. H. W. Abernethy. There were no locks or docks, the chief work being an embankment. The soil in general was reclaimed bog over bluish clay, thoroughly saturated and very subject to landslips. It would appear from the roots and trunks of trees dug up, that at one time this part of the district, now full of peat, had been forest land, and the sudden changes in strata were very remarkable. Up to January, 1889, 1,000,000 cubic yards had been excavated, of which 30,000 were rock. The whole section was full of water and pumps were constantly at work. One of these, brought from the Severn Tunnel

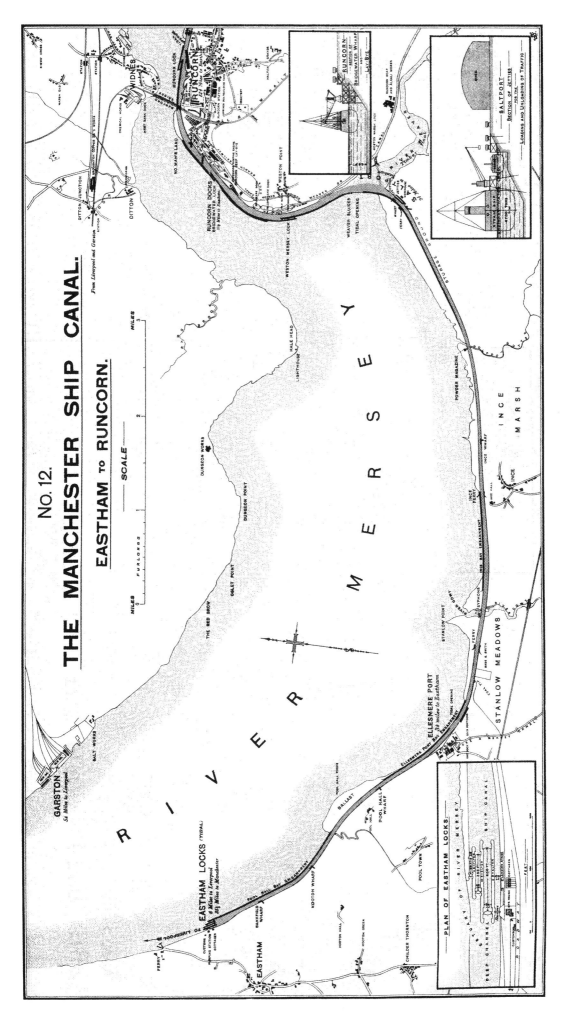

The material originally positioned here is too large for reproduction in this reissue. A PDF can be downloaded from the web address given on page iv of this book, by clicking on 'Resources Available'.

lifted 300 gallons a stroke to a height of 85 feet.   At Ince bay the canal again fringed the estuary, necessitating an embankment.   Near Ince, as on the Eastham section, nature had provided the contractor with an admirable supply of stone for the embankments and for facing purposes; and the ample supplies thus obtained repaid the extra cost of excavating.

No. 2 section produced a poet who penned an amusing description of the staff and plant, and signed himself a "Waggon Fettler".   He ended with :—

> Our staff here are good and work all they can
> For the good and the comfort of each working man ;
> In the mission, the school, the concert room too,
> We want earnest workers, there's plenty to do.
>
> So now kind friends, I'll bid you adieu,
> And wish success to the canal all through.
> May T. A. W. successful be
> In making a canal that leads to the sea.

At the beginning of October the estuary of the Mersey was visited by a furious gale.   In three separate places the canal trenches on it, and large embankments with railways on the top were in the course of construction to separate the canal from the estuary.   Upon these embankments the tide, lashed into fury by the gale, swept with great power, especially along the easterly portion of the Ince section.   For a mile the havoc was continuous, breaches being made in the bank, the railway destroyed, the cuttings flooded, and the waggons on the top washed away.   Two excavators in the bottom had 15 to 20 feet of water over them, and one of the pile drivers was overturned.   Fortunately the Eastham section, where the bank was solidified, escaped with little damage, though the water outside was so high that it began to trickle over.   Here was a busy scene, large gangs of men trying to repel the invasion of the waters.   Fortunately the gale modified just in time, and the position was saved.

The Gowy excavation and syphon sustained no damage.   This is the only portion of the section where rock is not reached, and inasmuch as two immense cast-iron tubes, each 12 feet in diameter, had to be taken under the canal on made foundations, it was one of the most difficult and dangerous works in the entire undertaking.

No. 3 Section, Ince to Weston Point.—Contractor's chief agent, Mr. J. Weston ; Ship Canal resident engineer, Mr. H. W. Abernethy.   Half this section from Frodsham Marshes to Weston Point was river work, extending across the mouth of the Weaver. Through marsh land which was covered at high tide, a part of the cutting was in clean sand, and a part in grey slimy clay saturated with water. It was difficult to find places solid enough to support a heavy dredger weighing 60 tons, and worked from a base of 4 feet $8\frac{1}{2}$ inches. Up to January, 1889, 500,000 cubic yards had been excavated. Not only had ten sluices, each 30 feet long, to be built across the Weaver, but also an embankment to divide the canal from the estuary was required; and in addition to these a lock was constructed at Weston Marsh to enable ships to be lifted into the Weaver Canal. The Holpool Gutter was diverted into the canal, so that the old channel with its black ooze of forbidding aspect, which at high tide formed a waterway to the powder magazine, was cut off from the canal. At this point the Priestman's grab did good work; it drove its fangs into the soft material and nothing could resist it. For a visitor to get through this slough of despond was no easy matter, and a slip on one of the many planks to be crossed would have been fraught with most unpleasant consequences.

Under the charge of Mr. Weston, many hundred navvies had been housed on the Frodsham Marshes. The settlement was named Marshville, and was supplied with water from an artesian well.

In this section some remarkable work was done in June. A German excavator, working ten hours a day, filled 2,903 waggons in four days, each waggon containing four cubic yards of earth. The maximum for one day was 3,060 cubic yards.

The chief work was on the sluices at the mouth of the Weaver. Here ten Stoney's sluices, similar to those at Randles, were being put down to convey the waters of the Weaver into the estuary. Each had a clear opening of 30 feet, or a combined opening of 300 feet. The makers were Messrs. Ransome & Rapier, of Ipswich. To cope with the enormous body of water flowing up and down the tidal portion of the Weaver in flood-time, it was essential that the dam on which the sluices were to be built should be of the most substantial character. A platform, resting upon piles 400 yards long and containing 15,000 cubic feet of timber, had been formed half-way across the stream for conveying plant, material, etc. From this a pilot dam was driven parallel to the platform, then twelve smaller dams or bays 36 feet square, each division or bay being separated by a line of closely driven piles. When the bays were completed, concrete foundations were laid, and on these large cylinders were placed to form the base of the sluice piers. Further description may be dispensed with; suffice it to say the foundations were of marvellous strength. Though the

NO. 13.

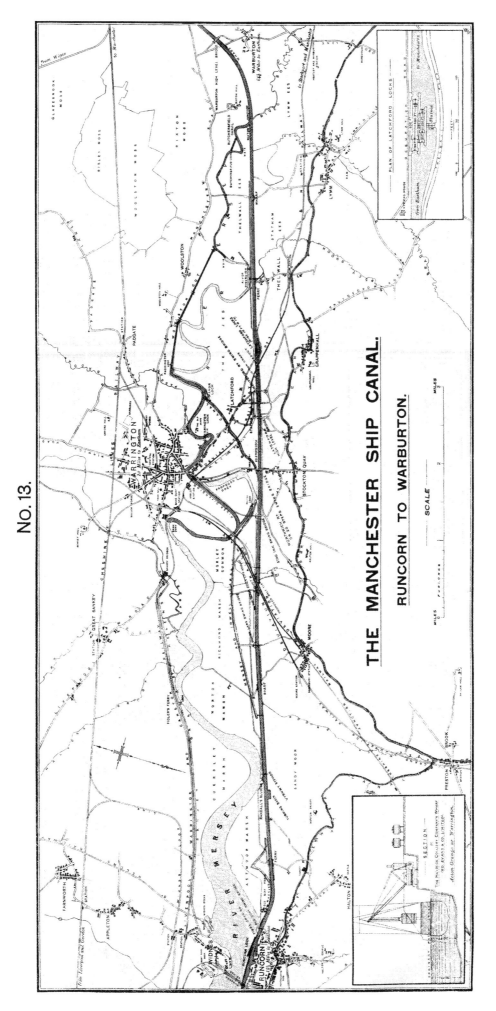

THE MANCHESTER SHIP CANAL.

RUNCORN TO WARBURTON.

SCALE

The material originally positioned here is too large for reproduction in this reissue. A PDF can be downloaded from the web address given on page iv of this book, by clicking on 'Resources Available'.

RANDLES SLUICES.

Birtles, Warrington.

To face page 52.

sluice gates were each to weigh about 34 tons, they were to be in balance and to move up and down with the greatest ease.

No. 4 Section, Weston Point to Norton.—Contractor's chief agent, Mr. C. A. F. Gregson; Ship Canal resident engineer, Mr. H. W. Abernethy. River works. This was a heavy section, both as regards river and land works. The land cuttings were through soft clay, specially deep, varying from 60 to 70 feet. From Weston Point to a small island (No Man's Land) an embankment 45 feet wide had to be built to divide the estuary from the canal; this had a top width of 45 feet, slope 1 to 1, was built of rock and clay, and in some parts was 50 feet high from the foundation. From No Man's Land to the old quay dock, in consequence of the reduced width of the river, a narrower but solid concrete wall was being built by the aid of a coffer dam. In this wall small locks were to be built. Opposite Runcorn and Weston Point two other sets of Locks were to be constructed, and at Randles sluices the Stoney's gates were being made to regulate the discharge of water from the canal. The total amount of excavation up to the end of January was about 1,000,000 cubic yards. This section was full of engineering difficulties. It was necessary to build an embankment between the port of Runcorn and the estuary, and at the same time to keep a large business going. It was at Runcorn that the Duke of Bridgewater and his wonderful engineer, James Brindley, won their spurs, and it was here, too, that the London and North-Western Railway Company built a bridge, in 1866, to connect Lancashire and Cheshire, which was then thought high enough to admit of large ships passing underneath, if ever they should come up the river. A clear headway of 75 feet at high water of spring tides had been stipulated for.

At Randles sluices the canal runs close to the river, so that the cutting in which the sluices were placed, and which now forms the connecting channel between the canal and the estuary, was only about 50 yards in length. It was necessary to put in piles 40 feet long to keep the masonry for the sluices from settling bodily, in consequence of the boggy and slippery nature of the earth. By the middle of May the work was completed and the two sluice gates put in position. Each had a clear span of 30 feet with a depth of 16 feet. They are capable of being raised 10 feet. When each gate is moved, a dead weight of 511 tons has to be dealt with, so they are indeed a substantial piece of mechanism.

No. 5 Section, Norton to Latchford.—Contractor's chief agent, Mr. C. H. Walker; Ship Canal resident engineers for the canal, Mr. J. F. Dixon and Mr. O. G.

Brooke; for the railway deviations, Mr. S. H. Hownan Meek. This section consisted chiefly of loose soil and sand, with red sandstone under. The cuttings were about 60 feet deep. On the section were deviation railways No. 1 and No. 2 close to one another, also the diversion of the Mersey near Arpley Station. This latter involved the cutting of a new river course three-quarters of a mile long.

Up to the end of January 1,350,000 cubic yards of earth and 350,000 cubic yards of rock had been removed. On one day a steam navvy, built by Ruston & Proctor, filled 640 waggons in twelve hours, i.e., removed 2,500 cubic yards of sand in the working day.

No. 6 Section, Latchford to Warburton.—Contractor's chief agent, Mr. John Price; Ship Canal resident engineer for canal works, Mr. G. C. H. Brown. This section was chiefly soil and clay. It included the construction of the Latchford Locks and sluices, and the No. 3 railway embankment, and was crossed five times by the windings of the Mersey. Up to the end of January 2,000,000 cubic yards of soil, and 250,000 yards of rock had been excavated. Work was prosecuted night and day, there being no suspension whatever.

At the Warburton end of this section an extensive landslip took place during April, which displaced many thousand tons of earth, and sunk a steam navvy. Some dozen waggons were smashed and buried in the débris. Fortunately this happened in the night, or it might have entailed serious loss of life. The damage done amounted to about £2,000.

Another accident took place in October. Two trains, one loaded with granite and the other with bricks, came into collision at the Morris Brook Grove curve through gross carelessness; they were both going at a good speed; the drivers of the "Garston" and "Salford," when they saw a smash inevitable, both jumped off into a cutting 35 feet deep. There was considerable wreckage both of engines and waggons, the damage being about £150; but happily, no lives were lost.

The cutting which was the scene of the landslip previously described has a peculiar geological interest. Here were found the well-known keuper marls, and these marls, when the cutting was at the deepest, showed finer ripple marks of a past age than any that had yet been seen. Looking from the bottom of the cutting the sides seemed to have rainbow colours graded in a most promiscuous way, and mixed in streaks.

No. 7 Section, Warburton to Barton.—Chief agent for contractor, Mr. C. J. Wills; contractor's engineer, Mr. W. W. Strover; Ship Canal resident engineers, Mr.

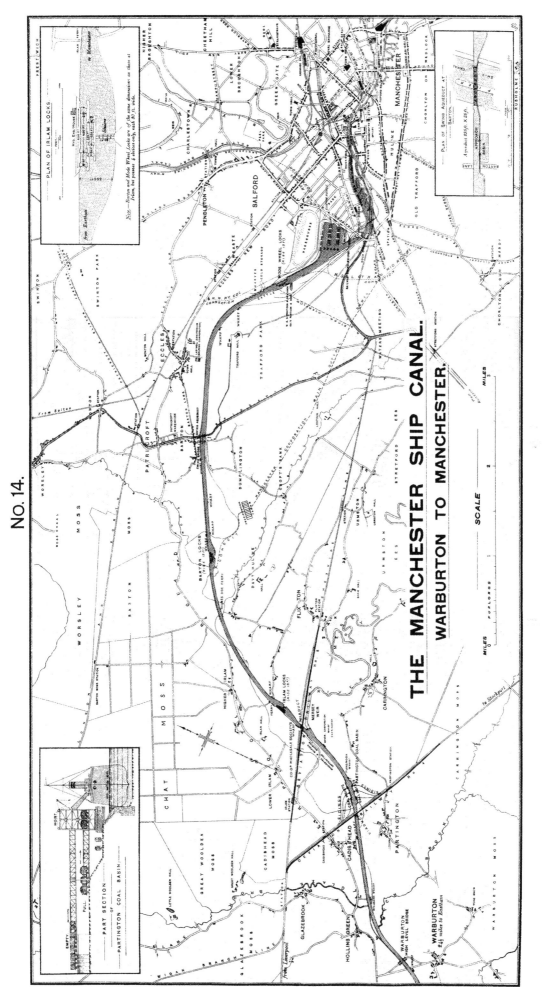

NO. 14.

THE MANCHESTER SHIP CANAL.
WARBURTON TO MANCHESTER.

The material originally positioned here is too large for reproduction in this reissue. A PDF can be downloaded from the web address given on page iv of this book, by clicking on 'Resources Available'.

W. O. Meade-King and Mr. L. H. Moorsom. This section passes through the alluvial valleys of the Mersey and the Irwell. The line of the canal crossed these rivers fifteen times. The chief works were the Irlam Locks, the Partington coal basins, and the No. 4 and 5 railway deviations. These latter were most extensive, including long approaches, new stations at Partington, Cadishead, and Irlam; and near Irlam Locks, a long river diversion (800 yards), by which the Mersey was temporarily turned into the Irwell. Great difficulty was experienced in excavating for the Irlam Locks, the soil being full of water and so like quicksand that it cost almost as much for tipping as for excavating. As a result, about 10,000 tons of cement were required to provide a solid foundation for the locks. Up to the end of January 1,250,000 cubic yards had been excavated in this section.

Sections 8 and 9, Barton to Manchester.—Agent for contractor, Mr. L. P. Nott. Ship Canal engineers, Messrs. Bourke and Kyle. This section is 6 miles in length, and includes the Mode Wheel and Barton Locks, and the Manchester and Salford Docks. In eight weeks after the work was commenced, the progress made was almost incredible. The finest types of excavating machinery were to be found here; English, French and German steam navvies, besides Whittaker's English steam crane navvies, and Priestman's grabs. The Ruston & Proctor navvy is a very useful machine; it is worked from the bottom, and is supposed to remove about 1,000 tons per day. The Whittaker navvy works in a similar manner. The German navvy is a more complex machine. It works from the top, and needs a covered shed, under which the waggon stands to receive the soil that is brought up by a projecting ladder dredger, fitted with a series of buckets. It has been timed to fill a truck in one minute and eleven seconds. This however, is out-distanced by the French navvy, which is supposed, when working under favourable conditions, to remove 3,000 tons per day, and it cuts through the hardest of clay with apparent ease. The only French navvy employed was working near Stickens Island, and though it was operating on tough material full of stones, it scooped out upwards of 2,000 tons per day.

The difficulties above Barton were much aggravated by the fact that the canal many times crosses the old course of the river.

This section possesses considerable historic interest. Here was the famous aqueduct of Brindley. Close by is the Roman Catholic Church, a masterpiece of Pugin, and not far off are the home and works of Nasmyth, the inventor of the steam hammer. Here, too, was to be placed the product of Mr. Leader Williams' fertile brain, which was to carry the Bridgewater Canal over the Ship Canal by a swinging aqueduct.

At Barton the old Corn Mill near the bridge, which had been quite a landmark in the district, was demolished in order to make way for the new aqueduct.

The contractor had constructed an immense shed, similar to the one at Eastham, for making the greenheart dock gates. So far the concrete walling had advanced rapidly, and in some parts the fender course was being put on.

To facilitate excavations it was proposed to use dredgers in the Salford Docks, and an order had been placed with Messrs. Simons & Co., of Renfrew, for four screw hopper dredgers each of 850 tons capacity.

INCIDENTS AND ACCIDENTS.

From my diary it appears I paid many visits to the works during the year. On the 12th April Messrs. Hoy, A. E. Lloyd, Murray and Dr. Woodcock accompanied me on a two days' inspection. We started at Eastham and intended to walk to Runcorn, but when part of the way there, one of the party fell lame, and we had to get a lift on an engine. We saw the large dock gates being built, visited Ellesmere Port, Ince, and its lighthouse, and the work in progress at Holpool Gutter, crossed the Weaver on planks, and then were rowed up the canal to Weston Point. Here we halted at an admirable hostel where we got a beef-steak pie, voted by all of us to be the best we had ever tasted. Then we walked through the docks to Runcorn, where we took train home. The progress of the works since our last visit astonished us all.

A fortnight afterwards I walked the length from Runcorn to Latchford, and on the way visited Randles sluices ; the patentee was there and explained everything to me. The sluice gates weigh 24 tons, and two boxes filled with concrete weigh about the same. These were connected by wire ropes, and formed an equipoise which a child could work by turning a handle. The structure was difficult to erect, because the sluice gates were fixed where the canal cut through a singular bog. Instead of the usual layers of light and dark coloured moss, this bog was formed entirely of small branches, mostly birch. From the number of tree trunks laid bare, it must be assumed that this was the site of a submerged forest.

All great works are subject to a variety of incidents and accidents and the Ship Canal was no exception.

The year started with a wedding in camp which a son of the soil thus described in verse :—

BARTON AQUEDUCT.   IN WORK.

Birtles, Warrington.

To face page 56.

'Twas a navvy's wedding—jovial, hearty,
Ne'er was merrier wedding party,
May their joy for long remain,
Love each other may the twain.
For each other's weal to strive,
Be the object of their lives.

The Humane Society awarded a medal to Thomas Edwards, chainman, for bravely jumping into the Mersey and saving the life of a young woman, who, along with her sister, had fallen in. He swam about for half an hour to try and find the sister, but failed. The medal was presented in the carpenter's shop by the chief agent, later on in the year.

The people of Latchford and the vicinity were alarmed about the end of May by a serious explosion that shook the whole neighbourhood. It seems that for blasting purposes about a hundredweight of dynamite had been brought from the magazine and placed in a chest in the middle of the works, being in the charge of two boys. Whether they had been smoking, or a spark from a passing engine was the cause is uncertain, but a terrific explosion took place, frightening the whole district and causing much destruction, yet strange to say, nobody was killed.

Near Barton an ancient boat or canoe was dug up. It was at first mistaken for a simple tree, and would have been destroyed had not Mr. Bourke, the resident engineer, recognised its true nature, and had it preserved. It was found in a bed of fine river sand, at a depth of 25 feet from the surface, and 12 feet under the present river level. It was 13 feet 8 inches long and 2 feet 6 inches wide, and was hewn out of a single trunk of oak. The stem and stern are round, and the workmanship, for a boat of the kind, is of a very high order. This relic of antiquity, after being fixed in a frame and submitted to a strengthening process, was presented to the Manchester Museum, at Owens College.

At a meeting of the Gloucester engineers after a visit to the Ship Canal, Mr. Cullis, a member of that body, said the earth and material excavated on the canal would make a wall, 6 feet high and 2 feet wide, round the earth at the equator. Speaking of the Wells' light, he said it was produced from cheap petroleum, and brilliant effects were obtained at a cost of only threepence per hour.

There are always people who seek to evade their obligations, and in a few cases it became necessary to compel shareholders to pay arrears of calls. One almost humorous instance occurred. A Rochdale man was summoned for five guineas, and was confronted with his own application, but declared he could not say if the signature was his or not. He confessed it looked much like his own writing, but he never remembered signing it. After prevaricating and trying the patience of the court, defendant had to admit the signature, and the judge told him his conduct had been so bad he would have to pay the arrears and costs at once, and no time would be given him.

On the 28th July, whilst the quarrymen at Barton were dealing with a block of stone weighing 10 tons, they were amazed to find a live frog in a small pool of water which had percolated through the red sandstone, and settled in a cavity. The frog was sightless, had a film over its mouth, and changed colour after exposure to the air. How it got there, and how it existed in its prison-house is a mystery.

On the 8th December a great fire took place at the Ship Canal depôt, Carriers Dock, Liverpool, doing damage to the extent of £40,000, and at one time it was feared it would assume even larger proportions. Unfortunately two firemen were injured, and had to be taken to the hospital. The Canal Company were insured.

I am permitted through the kindness of Sir Leader Williams to give a facsimile of the letter Edwin Waugh wrote to him after a visit to the canal, which shows how deeply interested the poet was in the great work. This letter will be found at the end of this chapter.

At the beginning of the year it was decided that the bridges required at different points should be built of steel, and the order was placed with Mr. Arrol of Glasgow, the well-known contractor, who built the Forth Bridge.

Early in February a deputation visited Bristol to inspect dredgers, and to ascertain the best means of working them.

At a meeting of the Geological Society, held at Owens College, an excellent photograph was handed round of the top stone of an ancient hand corn mill, found 17 feet below the surface of the ground, in the Salt Eye cutting.

The fact that a deal of shebeening was going on in the Ship Canal huts was well known to the police, but the difficulty was to catch the culprits. About the middle of April a posse of police secreted themselves in a furniture van, which, when it came near the huts, managed to get its wheels hopelessly fast in the mud, and assistance was sent for. To the surprise and mortification of the navvies, who came with horses to pull it out, they discovered on arrival at the huts that they had been befriending their mortal enemies, the police.

Birtles, Warrington.

BARTON ROAD SWING BRIDGE.

To face page 58.

The police were able to make a sudden raid on the huts, with the result that they found four of the keepers defrauding the revenue, and these were brought before the magistrates, when heavy fines, with costs, were inflicted.

One of the earliest arbitrations of the year was to assess the compensation to be paid to Sir Gilbert Greenall, Bart., for 41½ acres of land at Lower Walton, near Warrington.   Mr. Beadel, M.P, was arbitrator.

The valuation for the vendor	£16,521
Statutory offer by Ship Canal Company	10,500
The award of the arbitrator	10,920

It is fair to say the vendor's claim included the stoppage of a water course, which eventually the Ship Canal agreed not to interfere with.

The compensation payable to Mr. W. J. Legh of Lyme Hall for land at Warrington was referred to a special jury, and Mr. E. T. Wilson, under Sheriff of Lancashire, presided at the inquiry.

The claim of the vendor	£1,779
Statutory offer of the Ship Canal Company	650
The award of the jury	853

COPY OF LETTER FROM EDWIN WAUGH TO SIR LEADER WILLIAMS, 30TH SEPTEMBER, 1889.

THE HOLLIES,
NEW BRIGHTON,
CHESHIRE.

30th Sep, 1889.

Dear Williams,

I have read our friend Kay's lively and interesting pamphlet with great pleasure — If I remember right you desired me to return it; and, there it is

This gives me a chance of thanking you again for all your kind attentions during that day of marvels when you led us like an engineering Moses through that wilderness of wonderful dislocations, which is rapidly growing into one of the grandest pieces of orderly and constructive skill which the world has ever seen × ×× Thanks to the far-sighted men who originated the startling

scheme; thanks to the men who have
fought for it, with such a noble
persistence, through evil report
and good report, through extraordinary
difficulties and against all odd,
thanks to your own professional
skill and daring ingenuity as
an engineer; and lastly, thanks
to the great contractor, who is carrying
out your bold conceptions in
such a wonderful way. — I
believe the day is coming when
the whole world will thank you,
and that you will be remembered
in connection with this remark
able enterprise long after you
have been gathered to your
fathers

I could not walk with you
through all the great cuttings of
the works on that day; and there
fore I lost the advantage of
your own explanations and com-
ments; but I saw the whole
thing, — and I saw it well;
and I tell you now as I told
you before that it was one of
the most interesting days of

my life   It was a thing to be
remembered vividly through a long
life; and of course, I shall
not have time to forget it. —
I can only say, now, that when
this great work is completed, I
hope I may be there to see
it; and to you duly rewarded
for your eminent services, the
benefits of which will stretch
far into the coming time
   I wish you health and long
life!

   Believe me dear Williams
      yours faithfully,
         Edwin Waugh

E Leader Williams, Esqr. C E

CHAPTER XX.

1890

DEATH OF DANIEL ADAMSON—PROGRESS OF WORKS—DIS-
ASTROUS FLOODS—CANAL INUNDATED—TIDAL OPEN-
INGS BILL—MERSEY BAR—WORKS AGAIN FLOODED—
ACCIDENTS—SHAREHOLDERS' MEETING—SETTLEMENT
WITH MR. WALKER'S TRUSTEES—CURIOUS FINDS—
VISITORS—ARBITRATIONS.

I dare not withhold from this great design whatever influence I possess, because I feel the results of its accomplishment might make all the difference between widening industries, between growing prosperity and, in the course of years, something very much resembling decay. —JACOB BRIGHT, M.P.

A T the opening of the new year the operations required for the carrying out of the contract for the construction of the canal were in a state of suspended animation. The concern was moving on minus its head. Nominally the works were being carried out by the trustees, with Mr. Topham as director, just as in Mr. Walker's lifetime, but the moving spirit had been taken away. An autocrat was almost a necessity for the proper progress of such large works. As is often the case, misfortunes did not come singly, for floods and storms in the first few months of the year did much damage, and considerably retarded the work.

One of the first difficulties the directors had was with the Salford Corporation, who insisted on a bridge being made over the canal at Fairbrother Street. On investi-gation, however, it was found that in the agreement no suitable approaches had been provided for on the Manchester side, and the matter dropped.

In the second week in January the death occurred of Mr. Daniel Adamson, ex-Chairman of the Ship Canal Company, and without whose dogged determination the Bill would not have been obtained, and the canal in all probability never would have been made. Manchester owes him a debt of gratitude which I hope she will some

(63)

day seek to recognise by perpetuating his name in connection with the undertaking. Born in Durham, and of parents who were not too largely blessed with worldly goods, he came in early life into the Manchester district and finally settled down at Hyde, where he built works and soon got into a large way of business, making steel boilers, which old-fashioned people then condemned.   He was not content with making boilers; he founded the Frodingham Iron Works for producing the material of which they were made, and I had, from his own lips, the history of how he came to start the business, and it shows the shrewdness of the man.   Once he travelled with a rough tenant farmer who had a few lumps of heavy material tied in a handkerchief.   He began to show these to the people in the carriage, saying he had found them on his farm, and he wanted to know what they were.   Mr. Adamson looked at them, and at once made up his mind they marked the presence of iron ore.   He afterwards paid a visit to the farm, and assured himself that his surmise was correct.   The result was that a company was formed to establish the Frodingham Iron Works.   Of one part of his career he always spoke with pride and pleasure, and that was his early connection with Timothy Hackworth, and the Stockton and Darlington line.   Mr. Adamson did not originate the idea of a Ship Canal, but when a leader was wanted, no one so quickly grasped its possibilities, and he threw himself into the forefront of the struggle with an earnestness and impetuosity that would take no denial.   He had his failings, of which, however, I should like to speak gently.   He could not brook opposition, and soon lost his temper with those who did not see eye to eye with him, or had not his quick perception.   From the day of the first meeting at the Towers, to his death, I was always working side by side with him in the cause of the Ship Canal, and I have many of his letters, which I prize highly, thanking me for services rendered.

Mr. Adamson was both mentally and physically far above his fellows, and it was wonderful how he could sway an audience by his north-country eloquence.   People appreciated his fine presence, his rugged oratory, and the scathing speeches which he made when under the fervour of excitement.   No speaker could hit harder, and he indulged in invective regardless of consequences.   A long residence in Lancashire had no effect in toning down his Northumbrian burr, and to the last he always spoke of the "Canaul".   Whenever Mr. Adamson was announced to speak, a full meeting was assured.   He died in his seventy-first year, and was interred at the Southern Cemetery.   Every one would have rejoiced had he lived to see his great project carried to a successful issue.

Birtles, Warrington.

THE GREAT FLOODS AT LATCHFORD.

To face page 64.

Birtles, Warrington.

After the Flood at Latchford. Frozen up.

To face page 64.

The last week of January was disastrous to the Ship Canal works. In consequence of prolonged rains both the Irwell and Mersey overflowed their banks and completely flooded the district. Where the canal cut across the course of the old river, the ends of the latter were closed by means of dams, and the rush of water was so great that it washed away the dams and filled the new deviations. The greatest damage was done on the Latchford section, near Thelwall and Lymm; at the former place a great embankment of sand, about 30 feet high, had been raised to prevent the river flowing into the adjacent cutting. On the Sunday night, the 26th January, the watchman heard a great noise and found the embankment had partly given way, and the result was that the plant in the cutting, including a steam navvy and forty waggons, were under water. The soil underneath the railway had been washed away, and only the iron metals were left to bridge the gap. The work of months had been destroyed in one night, which meant both a serious loss to the contractor and delay in the work.

A party of directors visited the works after the flood and found 5 feet of water on the line. This convinced them there need be no fear in the future about a sufficiency of water for the canal.

At the end of January considerable progress had been made in pumping the water out of the cuttings, and immense brick piers to carry the London and North-Western Railway over the canal had been commenced. Quantities of granite had been received for the coping of the lock walls, and the piling for the sea wall to separate Ellesmere Port from the estuary was assuming gigantic proportions.

This huge embankment was started from the centre, nearly opposite the old dock entrance, and it is estimated that 15,000 piles of 12 x 12 inch timber, each 35 feet long were driven; thus 650,000 cubic feet of timber were buried under this one embankment. Whilst the work was proceeding a piling pontoon foundered, precipitating all on board into the water, but they escaped with no damage beyond a wetting.

One of the most forward portions of the canal was the Little Bolton cutting, west of the Mode Wheel Locks. When the before-mentioned flood took place, all the navvies had been removed, and the length was ready, except that a few rails at the bottom wanted clearing out. The barrier on the east side gave way, and in twenty minutes 65,000,000 gallons of water rushed into the cutting and filled it. All this water had to be pumped out again. At the Pomona Garden end, where 6,000 piles (each 12 x 12 inch) had to be driven for a temporary coffer dam, the work was going on night and day.

On visiting Ellesmere Port a month later, I found that the pile-driving at the jetty had made wonderful progress; indeed on the western side preparations were being made to place a swing bridge over the aperture left open to carry on the trade of the port. When completed, shipping would enter through the Eastham Locks. At Norton, when the flood water was pumped out, the cutting had the appearance of a complete wreck, and the excavation for the Liverpool pipeway had all to be done over again. Many people visited the flooded part out of mere curiosity; but some cracks in the cutting at Warburton also possessed much geological interest.

So difficult did it become to find lodgings for the men at Eastham that the contractor was compelled to run a boat and take them to, and from, Liverpool and Birkenhead. The owner of the land would not allow any houses to be built, and the men were put to much cost and inconvenience every night and morning. For 9d. a man was supplied with twelve boat tickets, and at the end of nine weeks, if he had regularly attended his work, the whole amount was returned to him by the contractor.

In the case of the claim of the joint owners of the Birkenhead Railway against the Ship Canal Company, it was officially announced that Lord Balfour of Burleigh had been appointed arbitrator, and that he would begin the inquiry on the 30th May.

On the 24th April a Select Committee of the Lords met to consider the Tidal Openings Bill. It consisted of Earl Camperdown, Chairman, the Earl of Arran, Lord Saltoun, Viscount Hood and the Earl of Rosse. The Dock Board were the chief opponents, and they objected to the proposed reduction of 2,400 feet in the area of the tidal openings.

Mr. Pember for the promoters explained that now the canal was carried further inland, openings as originally designed were not required, that they would cause cross currents in the canal dangerous to shipping, and that the Randles sluices and the ten sluices at the Weaver were in substitution of those first intended, and would render them unnecessary.

Sir George Richards, the Acting Conservator for the Mersey, who came forward as an independent witness, said he agreed with his engineering adviser, Sir John Coode, that less damage would be done to the navigation of the river by the changes proposed in the Bill, than would follow if the Parliamentary obligations imposed on the Ship Canal Company were carried out. He declined, however, to relieve the company of the responsibility of working the sluices. Mr. Pope for the Dock Board remarked, " *We* are not anxious to take any responsibility whatever for the works of the Ship Canal". In reply Mr. Balfour Browne reminded the Committee that the

Birtles, Warrington.

Transporter at Warburton.

To face page 66.

Canal Company incurred a penalty of £500 per day if anything they did caused serious damage to the estuary. Eventually the Committee decided to pass the preamble provided the company's penalties under the 1885 Act was also made applicable to the present Act. In the same session a Select Committee, presided over by Lord Basing, passed the Cheshire Lines Bill, enabling them to make connections with the Ship Canal Docks.

The Lancashire and Yorkshire Railway Company also appeared before a Committee of the Commons with a Bill empowering them to obtain access to the Ship Canal; this was vigorously opposed by Salford on account of interference with streets, but in the end the Bill was passed.

On the 30th April Mr. Ismay, Chairman of the White Star Line, wrote a letter to the Chairman of the Dock Board pregnant with facts about the bar. He said plainly that the insufficiency of the facilities at Liverpool for large steamers was an undeniable fact. The *Majestic* coming from America on her last journey waited five hours outside the bar, and at midnight had to land her passengers in a tender because the steamer could not dock. Though she paid £308 tonnage dues for dockage, yet she had to coal, discharge, and load in the river. He pleaded strongly for the removal of the bar and the improvement of dock accommodation, and warned the Dock Board if they did not bestir themselves, shipping would certainly go elsewhere. This letter created quite a sensation, and no doubt led to the subsequent operations on the bar.

In the Sloyne, near the mouth of the canal, and in the fairway of the channel lay several powder ships under licence from Government. In prospect of Ship Canal traffic this anchorage was thought to be unsuitable and unsafe, and an application was made for their removal. In consequence an inquiry was held by Colonel Majendie, C.B., Government Inspector. It was contended by Liverpool that if the ships were removed the powder trade with Africa would probably go to Germany, and it was urged that the Canal Company ought to pay the cost of removal. Colonel Majendie, after an inquiry, reported that the location of the magazines was neither unsafe nor unsuitable on the grounds urged by the Canal Company, but that it was so for the reason that the water, having shoaled under the magazines, the explosives in each magazine would not at all times be below the level of the surrounding water, which was necessary to comply with the regulations. He therefore ordered them to be removed to a place within the legalised area, where there would always be a sufficiency of water, and where they would be less in the way of Eastham traffic.

On the 20th June the Tidal Openings Bill came before a Committee of the Commons, with Sir Charles Dalrymple as Chairman. The witnesses gave evidence similar to that given in the Lords, but it was a shorter inquiry. Mr. Bidder complained of the promoters wanting to reduce the number of sluices from forty-five in number and 4,500 feet in length to three in number and 1,800 feet in length, and said there was no obligation upon them to use even these. Mr. Pember, in concluding a powerful address, said :—

He understood the hostility of the Dock Board to the Ship Canal. The fox at the end of a long run liked to die with his teeth in the hounds, and the Mersey Dock Board was dying with its teeth in the canal. They had hunted well in the six or eight good runs they had had. He did not blame them, it was the nature of the animal (laughter). Seriously, he did not complain about the Dock Board seeing the matter through. That was creditable, though it did not add to the strength of the contention of the Board; it only made the maintenance of the contention respectable.

After a few minutes' consideration the Committee passed the preamble of the Bill.

The Various Powers Bill, which dealt with alterations of roads, footpaths, swing bridges, etc., also gave the Canal Company power to construct and maintain railways and provide warehouses, and authorised the raising of £600,000 by debentures or mortgages. It passed both Houses with little opposition. Warrington was practically the only opponent, and she got a clause to satisfy her demands.

Meanwhile, the whole of the canal was being pushed on with vigour, but with various mishaps. At Stickens Lock the work was delayed through a ship, which was bringing Cornish granite, going to the bottom. Several landslips occurred opposite the Mill Bank paper works, and on one occasion some men who incautiously went too near were carried down with the slip, but fortunately managed to scramble out. At the Little Bolton bank near Mode Wheel, which was washed away in January, 2,000 loads of sandstone, rubble, and clay were required to fill up the breach.

At Warburton Messrs. Fleming & Fergusson had begun to erect a pontoon dredger in the excavated canal, so that when water came it would be able to work its own way out. The construction at Pomona of a coffer dam, formed of two rows of piling with a heart of clay between them, soon rendered it possible to complete the construction of the Pomona Docks, and the deepening of the channel of the Irwell, alongside the docks. At the railway bridges blue brick piers of great strength

Birtles, Warrington.

PIERS FOR THE LATCHFORD VIADUCT.

To face page 68.

EMBANKMENT, WITH GANTRY, ELLESMERE PORT.

Birtles, Warrington.

To face page 69.

were gradually raising their heads, and at Eastham the lock wall next to the river had been completed.

Up to September all the work done was up the river; now for the first time a dredger, the *Manchester*, commenced to cut a way from the Eastham approach of the canal entrance. This dredger was built by Simons & Co., Renfrew, and could remove 1,000 tons per hour. At Ellesmere Port the double piling of the embankment was well advanced, and the tipping of clay, rubble, etc., to fill the intervening space, was proceeding at the rate of fifty train loads a day, and there were twenty-five pile-driving engines busy at work every tide. To strengthen the embankment, the exterior and interior piles were tied together by hundreds of tie-rods $2\frac{1}{2}$ inches in diameter and 36 feet long. These were buried in the hearting and added to its stability. At Eastham Locks Messrs. Ransome & Rapier, of Ipswich, were well forward with Stoney's lifting sluices, as also with the equalising sluices in the dock walls; the former had both a tidal and river pressure, and to ensure stability were only 20 feet wide against 30 feet at the other lock sluices.

On the Ellesmere and Runcorn length the stone work to the Weaver sluices was being rapidly filled in and gave evidence of a most substantial undertaking.

Between Runcorn and Latchford over 200 waggon loads of boulder clay were being tipped daily into the embankment near No Man's Land, and a pipeway (as per agreement) was being laid from an upper reach of the Runcorn and Latchford Canal in order to supply Messrs. Wigg Brothers & Steele with water.

The railway deviations between Runcorn and Barton were making rapid progress, and Messrs. Arrol & Co. were delivering girders for the railway and canal bridges. The Little Bolton cutting near Mode Wheel had been cleared of water. It was found that the water which Nature had supplied in twenty minutes, took Art (by means of a trough) many days to pump out. The Barton aqueduct was advancing slowly.

The great demand for labour brought with it serious difficulties. Agitators almost caused a strike both among the boatmen and the navvies. Meetings of both were held at Runcorn and Manchester respectively. The men walked in procession to their meeting, headed by brass bands. Mr. Price, agent of the Labourers' Union, was surprised that the navvies were content with earning $4\frac{1}{2}$d. to $4\frac{3}{4}$d. an hour, when the price in London was 6d., and urged they ought to have 7d. an hour for an eight hours' day. Further they ought to have compensation, if, through American navvies and other labour-saving appliances, their means of earning a livelihood were taken

away from them. If publicans were to have compensation, why not a labourer?
Mr. Ward,[1] president of the Union, said 4½d. an hour was not even boys' pay. In
case of accidents, why did they not have a navvy on the jury? They were going to
have an election, and with 700 voters at, and about Eccles, they could carry a man
of their own choice. A strike was only just averted, but the cost of labour went to
a figure in excess of that calculated upon by the contractors.

In November the troubles of the Ship Canal recommenced. To all outward
appearance the death of Mr. Walker had so far made no difference in the progress of
the work, and every one expected the canal would be completed at the specified time
and at the estimated cost; for the contractor was to receive £100 per day in case of
earlier completion, and to sacrifice £100 per day if behind time.

The first discordant note was struck by the *Liverpool Mercury* in an able
article. The editor had sent a competent man to gauge the time it would take
to complete the canal under favourable circumstances; and he showed that it was
a physical impossibility to complete any one section within nine months of the time
named, and that in all probability it would take much longer. The report went on to
say :—

It must be borne in mind, however, with rare exceptions the weather has been con-
spicuously favourable ever since the work of construction commenced. There have been no
severe winters and no excessive weight of rain at any period. A severe winter, followed by
a cold wet spring and summer would make a difference of some months probably in the date
of completion.

Prophetic words, as the sequel will show!

Then the *Mercury* attacked the correctness of Sir Joseph Lee's estimates as
regarded cost of completion, and maintained that there would be an insufficient
water supply.

Strange to say, within ten days of the appearance of the above article, the windows
of heaven opened, the storm came down, and miles of the canal were devastated by
floods. On Friday, the 7th of November, and again on 24th November, occurred
two of the most disastrous storms on record. There were repeated hurricanes
of rain, such as had not been experienced for years. From Latchford to Warrington the canal intersected the Mersey a great many times. As the river had
to be kept open, dams were left to keep the water out of the excavated canal.
In other cases a straight cut was substituted for a bend, and thus the river was

---

[1] Now M.P. for Stoke. At one time he worked as a navvy on the Ship Canal works.

WESTON MARSH LOCKS ENTRANCE, 17TH JULY, 1890.

Birtles, Warrington.

brought close to the canal cutting. Brooks were often carried temporarily by troughs over the canal into the river. Ordinary precautions had been taken to prevent mishaps, but they were utterly inadequate to meet an abnormal strain. The torrential rain and floods alarmed the engineer, and he tried by using cement to strengthen all the weak places. But this was of no avail, for late one night, when the river had been greatly increased by tributary brooks, the pressure on the banks became so severe that they burst, and the water broke into the canal cuttings with terrific force. The head of the wave (8 or 10 feet high) rushed forward and filled the partly excavated canal. Fortunately this happened in the night, or the loss of life would have been appalling. One poor fellow, John Williams, an engine driver, was in the cutting, and before he could escape, the tremendous onrush of water caught him and he was drowned. Other men had narrow escapes. The water (in some parts 40 feet deep) submerged steam navvies, locomotives, rolling stock, workmen's tools and materials of all kinds. In various places the tops of cranes were to be seen projecting a few inches out of the water. The damage done was enormous, and to repair it, and to pump the water out, cost at least £100,000, besides delaying the work for months. Not less than 6 miles of canal were flooded; bridges and temporary erections were washed away, and a tunnel that was being constructed under the London and North-Western Railway at Thelwall collapsed, carrying with it a portion of the line and making a diversion of traffic necessary. Many of the canal slopes which had not had time to solidify were more or less washed away. The country was inundated for miles round the canal, and until the water subsided there was no possibility of estimating the amount of mischief done. There were many marvellous escapes. Four men on the locomotive "Lancaster" saw the man Williams carried away by the flood, but were powerless to render help. When a huge wave was close upon them, with a noise like thunder, they just managed to scramble high enough up the bank to avoid its force, and in five minutes their engine was submerged. William Roberts, a watchman, with a cork leg, was in the cutting and heard Williams shout "the water is upon us". With all the speed in his power he tried to scramble up a 3-inch pipe on the side of the embankment, but he would probably have been carried away if a mate had not got hold of him by the hair of his head and hauled him upwards.

Before steps could be taken to repair the damage, the neighbourhood was visited by a second flood, even more disastrous than the first, though the locality was not exactly the same.

On the 23rd November very heavy rains had again caused the Irwell to overflow its banks at Barton, and the whole of the low-lying meadows were flooded, and fears were entertained that the Barton Locks would be drowned out.   The water had risen within 3 feet of the top of the bank, notwithstanding all efforts to battle with it.   Pumping engines had been going day and night.   Men would risk their lives no longer; the track and works at the railway bridges were submerged, and only the roof of the watchman's hut could be seen on the Ship Canal Railway.   At 6 P.M. the flood was nearly as high as in 1866, and the river was still rising.   During the night the rainfall never ceased.   The waters gradually topped the embankments, and then millions of gallons (at first in a small stream) poured from the new waterway into the old bed from which it had been diverted, just above the Barton aqueduct.   So great was the rush that in fifteen minutes the section on either side of the bridge was filled.   In spite of the efforts of gangs of men, who were trying to strengthen or raise the embankments, the waters flooded everywhere.   The Trafford Park section, which was as nearly as possible completed, was the last to be drowned out.   The dam here was very strong, but at last the rushing waters tore a huge piece out, and in a few minutes the canal had disappeared: there was only a wide, deep, expanse of water.

The whole canal from Latchford to the Trafford Docks being more or less submerged, work was necessarily suspended.   A large quantity of the working tools and machinery lay buried under the seething waters.   The cutting was so nearly completed, that a few weeks' work in clearing up the bottom and finishing the beaching on the sides would have made it ready for filling.

By the floods 3,000 men were thrown out of work, and all calculations as to the completion of the undertaking were upset.   The occupiers of the huts on the Lymm brick-fields were in sore straits, and they would all have been drowned if rafts had not been procured, on which they floated away to places of safety.

Though the engineer in his report took a hopeful view of the situation, the floods of November were a staggering blow to the directors.   They were beginning to have doubts as to completion, and all hope of finishing in the specified time had to be abandoned.   Just when Mr. Walker's master-mind was most wanted, it was no longer available.

No time was lost in getting pumping machinery to work.   The pumps started as soon as the river had gone down, and in three weeks they had pumped several lengths dry, and enabled work to be resumed, notably at the Latchford, Irlam and Barton Locks.

Birtles, Warrington.

FLOOD AT THELWALL, 1890.

To face page 72.

To add to the directors' anxieties, the Guardians of the Runcorn Union, fearing that their sewers would be interfered with, took the company into the Chancery Court, and obtained an injunction to prevent certain work being proceeded with, without their consent.

ACCIDENTS.

The year did not pass without the usual crop of accidents, and the hospitals were often requisitioned.

Early in the year a watchman, a most civil man, and one who had followed Mr. Walker to various works, fell down dead at his post, and in April a tenter on one of the pumping engines at Rixton was found dead at the bottom of the cutting with a shot through his breast. His mysterious death was never cleared up, but it is supposed he was the victim of foul play.

On the 18th July a destructive fire broke out at the large Ship Canal warehouse, Duke's Dock, Liverpool, and in two hours destroyed the building, which was full of cotton, grain, and Manchester goods of every kind. The fire illumined the whole of Liverpool. The damage, estimated at about £100,000, was fortunately covered by insurance.

About the end of July a railway collision occurred near Ellesmere Port, which caused the loss of three lives. Through culpable neglect, two engines, the "Deal" and the "Rhymney," with their trucks, smashed into one another. The "Deal" acted as the mail, *i.e.*, conveyed the trucks in which the men went down the line to their work. The driver stated that there were six people besides himself and the fireman on the engine. Five jumped off when they saw the collision was inevitable, but three remained on and were all killed. The driver of the "Rhymney" was arrested on the ground that the mail had always the right of road. Eventually, however, he was discharged. At the inquest the driver of the "Deal" was asked why he allowed people to ride on his engine instead of being in the trucks? and his reply was that two of the persons with him had lost arms and could not get into the trucks.

This reminds me that Mr. Walker, in the kindness of his heart, would never let a man want who had been injured in his service: therefore many of his watchmen and caretakers were maimed men, indeed "Walker's fragments" were known far and wide, and many of them were most faithful at their posts. A tale is told of a geologist who had a comical experience when he wanted to go to the bottom of the Eastham cutting to examine some boulder clay.   He had scarcely started when a gruff watch-man, with one foot in the grave and a timber substitute, shouted to him and forbade him to descend the practicable path.   The geologist showed his pass, but was told by the one-legged man "his orders is that nobody's to go down that there cuttin', and he don't care about passes".   Expostulation was useless, and seeing that reading was not one of the watchman's accomplishments, the geologist thought, as he had a couple of yards start, he would try old timber-toes' mettle at a "sprint," but that worthy was pre-pared for the emergency, and promptly whistled to a mate below, who was fully equipped in the matter of legs, but minus one arm.   As between them they had a full set of limbs, and two over, the baffled geologist, finding the odds too heavy against him, passed them some compliments on certain cranio-metric peculiarities which he professed to have discovered, and went his way to a distant ladder which afforded him a passage safe from pursuit by fragmentary foes, though, by the way, he saw a wooden-legged man safely descend a ladder at Stanlow with astonishing celerity.

In consequence of recent rains a mass of the embankment near the brickyard on the Lymm section gave way, leaving the contractor's railway line along the canal suspended in the air, but this was soon repaired.

<h3 style="text-align:center">SHAREHOLDERS' MEETINGS.</h3>

The shareholders at their meeting in February were told by Sir Joseph Lee, who presided as Chairman in the absence of Lord Egerton, that the hydraulic machinery at the docks had been ordered from Messrs. Armstrong & Co., of New-castle, and that it was the intention of the contractors to work night and day to make up for past delays caused by frost and flood.   He further said they had a sum of £4,000,000 (including Bridgewater savings) with which to finish the canal, and this should dispose of the rumours that they were getting short of capital.   He congratu-lated the meeting on the way the calls were being met, and said that after inspecting the whole length of the works he saw no reason to doubt for one moment that the canal would be finished within the contractor's specified time.   He deplored the loss of Mr. Adamson and Mr. Walker, but said the latter had left men quite equal to carry the struggle to a successful conclusion.   At the extraordinary meeting which followed, Mr. Bythell explained the Bills before Parliament, and said they were necessary to give them control of their own property, not one of the least important being, that they would be able to charge terminals for the use of their railways at the docks.

FLOODS AT FLIXTON, 1890.   WAGGONS OVERTURNED.

Killon.

To face page 74.

At the August meeting Sir Joseph Lee explained that the capital authorised in 1885 was £10,000,000. In consequence of payment of interest out of capital and a proportional reduction in borrowing powers, this was reduced to £9,812,000. They had yet to receive £2,952,487, and they had just had granted to them borrowing powers for roads and equipment to the extent of £600,000. Of the total amount £704,000 had already been spent. They had also £300,000 accumulated profit of the Bridgewater in hand. Of unissued shares there were £305,430, but these the contractor was bound to take up by 15th October. When these were taken up, the company could issue £43,000 more mortgage debentures.

Later in the month an extraordinary general meeting of the shareholders was held, and power was given the directors to issue £600,000 of debentures, interest not to exceed 4 per cent. These debentures were to rank before the ordinary and preference shares. Another resolution empowered the company to pay off the debentures in 1914 or to convert them into $3\frac{1}{2}$ per cent. perpetual debentures.

The arrangement made in August, 1889, with Mr. Walker's trustees did not work satisfactorily in its financial details, and throughout the year there was constant friction between them and the Ship Canal Board as to the progress of the works and other matters. The contractors demurred to taking up the remaining £250,000 ordinary shares on the 15th October as previously arranged, and as a set-off made a large claim for extras, alleging that the engineer, Mr. Leader Williams, had not given certificates for the full amount of work done, and had classed at the low price of dirt, the soft red sandstone which they expected to be paid for at the rate for stone. They also said they had sustained serious loss in consequence of their machinery being idle, through possession of the land being delayed. Having obtained the assistance of an eminent engineer they sent in a bill of extras amounting to £495,000.

The position being serious, and the work not progressing satisfactorily, the directors decided it was the best policy to effect a settlement with Mr. Walker's trustees, and to take the completion of the contract into their own hands.

The terms were that the trustees, in lieu of their claim for matters in dispute, were to accept £180,000, and that £70,000 was to be considered the balance due for work done, the total, £250,000, being placed against the £250,000 of shares which the contractor was bound to take up. As regarded plant, which had cost from £900,000 to £1,000,000, it was agreed that it should be taken by the company as worth £460,000, and that if it realised more at the completion of the works, the trustees were to have a portion of any surplus. All remaining contracts were to be taken over by the company, and all debts due by Mr. Walker's trustees were to be paid by them.

The company, therefore, from the 24th November, took the entire control of the completion of the works into their own hands, and were free to act, whilst the trustees had been severely hampered by their executorial position. It was arranged to retain the officials employed by the late Mr. Walker, with Mr. Topham and Mr. Reeves, his confidential advisers, at their head. Mr. Gillies, who had gone through Mr. Walker's books for the company, was to commence a new set of books, and to audit them in future.

It was arranged that the settlement arrived at should not be made public. Everything was to go on just as if the executors of Mr. Walker remained in command. It was feared that if the cancellation of the contract leaked out, complications with sub-contractors might arise and the whole staff be disorganised.

Much blame was subsequently thrown on Sir Joseph Lee and the directors for cancelling the contract and taking over so great a responsibility. It must be remembered, however, that they were in a serious dilemma, and chose what they thought to be the lesser of two evils. They could not foresee the disastrous weather which commenced directly they took control of the works. It is a question if the hampered trustees would have faced the difficulties that subsequently arose without appealing to the directors. Anyway litigation was avoided.

CURIOUS FINDS.

In the Trafford cutting at Eccles the shaft of a runic cross was dug up: it seemed to have some connection with the hollowed log previously found at a depth of 15 feet, and considered by some to be a trough, and by others an ancient coffin. The cross is similar to those found in the Isle of Man, and is supposed to belong to the eleventh or twelfth century.

During April a second dug-out canoe was unearthed near Partington, in a bed of gravel containing much driftwood, and with it some hazel-nuts bearing marks of squirrels' teeth. Not far away the workmen came upon a spring, the waters of which were impregnated with salt. This may account for the local name " Salt Eye "

VISITORS.

The policy of the Board being to show shipowners and the various traders of Lancashire the facilities for traffic on the canal, deputations from time to time were personally conducted over the works. Amongst the first visitors, however, were two inquisitive Liverpool boys, each aged 13, who were so anxious to see the canal that they had tramped all the way from Liverpool for that purpose. They were found wandering about by the police, and restored to their parents.

At the end of March the officials were informed that a Belgian nobleman, Comte d'Oultremont, accompanied by his secretary, wished to visit the canal. Mr. Leader Williams, on being introduced, at once recognised his visitor as the King of the Belgians, with whom not many months previously he had had the honour of an audience in Brussels. The King, who was travelling incognito, had not forgotten the engineer of the Ship Canal, and smilingly whispered to him, "Do you know me?" "Yes, your Majesty," replied Mr. Williams in another whisper. His Majesty was then shown the chief works at the Manchester end. He took an unusually intelligent interest in the engineering details, and complimented the directors and engineer on the utility and prospects of their great enterprise, and on the skill with which it was being carried out.

During May the canal was visited by the Liverpool and Manchester Fruit Buyers' Association, the Liverpool General Produce Brokers, the leading timber merchants of Liverpool and Manchester, one of whom was the Mayor of Liverpool, and by a number of coasting steamship owners. Also by Sir Edward Watkin and several friends, including Mr. Herbert Gladstone, M.P.

The Manchester Field Naturalists, after a visit to the canal, were told by Alderman Bailey that the canal would reduce by one-half the cost of the carriage to and from Liverpool, and that he felt sure that the land between the two cities "would become the busiest on the earth; it would be a wharf over 30 miles long, and the sun of commercial prosperity which shone on Manchester would likewise warm Liverpool". Evidently there were happier days in store for the latter city than they expected. He instanced the extraordinary fact that goods could now be carried 3,000 miles on water as cheaply as 30 miles by railway, and pointed out there was no wear and tear on water, whilst rails and rolling stock rapidly depreciated.

Following quickly one after the other, influential representatives of the following trades visited the canal: The Colliery Proprietors of South Lancashire and Cheshire; the Iron Masters from various parts of England; the Liverpool Engineering Society; the United Cotton Spinners' Association; Continental Steam Shipowners; Lancashire Oil Merchants; deputation of the Shipbuilding Firms of Great Britain; the British Silk Manufacturers; Manchester Association of Engineers; Directors of the Manchester Exchange; members of Lloyds' Register of Shipping and of the

Mersey Dock Board; Sugar Refiners and Merchants; representatives of the Paper Trade; Bank Directors, Managers, and their friends; Stock and Sharebrokers of the principal English towns; Manchester and Liverpool Provision Trade Association and wholesale and retail traders; the Directors of the Oldham Cotton Buying Company; the East Coast Shipowners; The Lord Mayor of London and the Sheriffs, the Lord Provost of Glasgow, and Mr. Ritchie, M.P., President of the Local Government Board.

When deputations visited the works they were accompanied by some of the directors and taken in saloon carriages (by the permanent way) along the whole length of the canal, having lunch generally at Bridgewater House, Runcorn. Complimentary speeches were afterwards made. In responding to a toast on the occasion of a visit by the Liverpool Engineering Society, Mr. Brown, engineer of the Runcorn section, gave the following list of steam navvies, etc., used on the canal, with the amount of work done daily by them :—

	Best day. Cubic Yards.	Average.	Number of Hands.
The German Navvy excavated	2,400	1,900	22
The Ruston & Proctor Navvy excavated .	1,910	750	17
The Whittaker Excavator	718	320	7
The Priestman's Grab	362	324	7

On the whole length there were 221 miles of temporary railway, 90 excavators, 171 locomotives, 6,300 waggons, 16,000 workmen, and the value of the plant was nearly £1,000,000.

Addressing the Steam Shipowners, Alderman Bailey said he was sure his visitors understood the economy of force connected with their business :—

As they well knew, a ton of goods was like water going down a hill. If it had to be sent from one part of the world to another, it (like water) took the line of least resistance, and he believed that in the Ship Canal they had that line between Manchester and foreign ports; and if they could take a ton of goods from any part, from the vineyards, gardens, quarries or mines of any country, right on to Manchester, they would not stay to land that ton of goods anywhere *en route* if it could be demonstrated that it would cost more money to send it a part of the journey by rail.

Speaking of the quality of the Irwell water, he raised a laugh by saying that no parasites inimical to ships could live in it, and suggested that a few days in the canal would be as good as a special outside cleaning down of the vessel.

When the Bankers went down the canal, Mr. Oulton, of the Liverpool Adelphi Bank, said :—

As far as he could judge, his fellow-townsmen never had but one feeling that could be regarded as inimical to the canal, and that was an apprehension lest, in dealing with their magnificent river, something would be done which might injure it as a channel of commerce. He had no hesitation in saying that if the canal had dammed the river, the Liverpool people would have damned the canal. But he was glad to say there was a consensus of scientific opinion which went far to remove any apprehension of that kind.

ARBITRATIONS.

The Salford Corporation claimed for 9 acres and 6 perches, of which 7 acres and 6 perches were available land, forming part of their sewage works, and 2 acres were portions of the bank and bed of the river Irwell. Mr. J. H. Clutton, of London, was the umpire :—

Claim made by the Corporation	£10,800
Statutory offer made by the Ship Canal	7,000
Award of the umpire	4,785

Whilst the witnesses of the claimants valued the property at £13,753 those for the company estimated it at £3,719. The award being less than the offer, the Corporation had to pay their own costs.

The Mode Wheel Mills Company claimed for their flour mills and water rights. Mr. D. Watney, of London, umpire. Area of land 12 acres 2 roods 34 poles :—

Valuation by owners' witnesses	£39,044
Valuation by Ship Canal witnesses	13,145
Statutory offer made by the Ship Canal Company	21,000
Award of umpire	21,174

The Boulinikon Floor Cloth Company claimed for their works covering an area of 5 acres 1 rood 3 poles required for docks. Mr. Thomas Fair, umpire :—

Valuation of claimants' witnesses	£20,240
Valuation of Ship Canal witnesses	5,513
Statutory offer by Ship Canal	10,000
Award of umpire	11,143

CHAPTER XXI.

1891.

GLOOMY PROSPECTS—SLOW PROGRESS OF THE WORKS—
DIFFICULTIES WITH MR. WALKER'S EXECUTORS—THE
SHIP CANAL BOARD TAKE OVER THE CONTRACT—
MONEY RUNS SHORT—FINANCIAL NEGOTIATIONS—THE
MANCHESTER CORPORATION COME TO THE RESCUE—
SALFORD AND OLDHAM OFFER HELP—DECISION OF
MANCHESTER TO FIND THREE MILLIONS—PARLIA-
MENTARY BILL—ATTEMPT TO BLOCK BY MR. LEES
KNOWLES, M.P., AND MR. P. STANHOPE, M.P.—DRY DOCKS
—ERRORS IN ESTIMATES CAUSE UNEASINESS—DISCUS-
SION IN CITY COUNCIL—DISASTERS—CANAL OPENED
TO THE WEAVER—CONDITION OF THE WORKS—ACCI-
DENTS—VISITORS—ARBITRATIONS.

Buffalo Creek Harbour was begun, carried on, and completed principally by three private individuals, who mortgaged the whole of their estate in its behalf. Over the grave of Samuel Wilkinson, which faces the harbour, is chiselled, "Urbem condidit". He built the city by building its harbour.—*Harper's Monthly*, July, 1885.

GLOOMY, indeed, were the prospects of the Ship Canal at the beginning of 1891. The very elements seemed to have conspired to prevent its success. Following on three years of satisfactory progress, torrential rains had come twice in rapid succession, drowning the works from Warrington to Manchester, and destroying all hope of the canal being finished in time. To make matters worse, after the rain came one of the most severe winters on record. Ice and snow prevented lost ground being regained, and even the Bridgewater Canal, the one source of income, was paralysed by an abnormal frost that blocked the waterway with ice, and temporarily stopped the traffic. In time it leaked out that the trustees

(80)

of Mr. Walker and the directors had parted company. A slackness in carrying out the work had become noticeable, and indefinable rumours of all kinds were in the air, industriously magnified by articles in the Liverpool Press. It was said that the executors of Mr. Walker were making large claims for extra work, just as they had made a claim of £204,000 under similar circumstances against the Barry Docks Company.

The Ship Canal directors issued reassuring bulletins, stating that the flood water was being emptied out of the cuttings at the rate of 2 feet per day, and that the locks had been so far cleared as to enable a resumption of work.

During the first part of January satisfactory progress on the canal was next to impossible. The ice on the Bridgewater Canal had been broken up by the ice-boat, but it soon froze again into massive blocks, which prevented steam-tugs from forcing a way through. Strings of flats and steam-tugs were helplessly waiting to continue their journey. It was years since the traffic had been so impeded, and one night the thermometer registered 34° of frost.

Early in January the Deputy Chairman, Sir Joseph Lee, asked the engineer to give him a rough estimate of the cost of completing the canal. This came to £2,700,000, and convinced him there would not be sufficient capital to carry out the work. On the 15th January he sent for me, as a shareholders' auditor and a councillor of the city, to talk over his idea of asking the Corporation of Manchester to find or guarantee £1,000,000 for the completion of the canal, and he requested me to sound some of the leading members of the Council. The next day I went to London with Sir John Harwood, and found him more favourably disposed than I could have expected.

On the succeeding Tuesday, 20th January, at the invitation of the Mayor, Alderman Mark, the leading members of the Council met Sir Joseph Lee, who explained the difficulties of the Ship Canal, and how they arose, suggesting that the city should advance £1,000,000 to £1,500,000, and become debenture holders, ranking before the ordinary and preference shareholders. There was a general assent, for it was felt the construction of the canal must not stop. Even Alderman King said that though he had been a staunch opponent of Corporation interference, he should sink his own feelings, and help to sustain the credit of Manchester.

Sir Joseph Lee was asked to report more definitely on the position and prepare estimates of the cost of completion, to be submitted to an adjourned meeting.

The approaching meeting of the shareholders on the 3rd February was looked forward to with much interest; people wanted some explanation of current rumours which were affecting the value of the shares.

When the meeting took place the Chairman, Lord Egerton, made a most important statement. He said that in the summer of 1890 serious complications had arisen with the contractors as regarded extras, which had rapidly increased in amount. By compromise all claims had been amicably settled, and the existing contracts with the executors of Mr. Walker had been determined, and the Ship Canal Company had taken over the works from the 24th November, 1890. The executors also agreed to carry on the works on behalf of the company so long as they were desired to do so. The only interest they, as trustees, would have in the works, would be a contingent share in the value of the plant when it was sold. The plant had cost £948,600, and that sum had been advanced by the company to the contractor. The impossibility of realising the plant at present, the increased cost of the works, the necessity of providing extra works and extra land, and the arrangement with the contractors, would render necessary an application to Parliament for further capital powers.

In moving the adoption of the report, the Chairman gave further details, and went on to say that he preferred to be an optimist. But in the Ship Canal, as in other great undertakings, things did not always run smoothly, and he must confess that the report was not as favourable as he could wish, but there was no reason to be despondent. The directors had a report from Mr. Abernethy, the eminent engineer, and notwithstanding the serious drawbacks they had had, that gentleman was of opinion that the canal would be completed by the summer of 1892. Alluding to the position with the contractor, he did not wish to say anything unfavourable to the executors of the late Mr. Walker. Unfortunately, difficulties had arisen with them as to their claims, and as to the sufficiency of the rates in the contract. They had claimed large extras on the ground of delays at Runcorn and elsewhere, caused by legal proceedings. These claims the directors disputed, and Mr. Walker's executors thereupon stated they were neither empowered nor justified in going further without a settlement, and they declined to place further plant on the works, thus delaying completion.

As the result of negotiations an agreement had been amicably come to whereby the contract with the late Mr. Walker was to be voided, and the company was to take over the work from 24th November 1890. The executors and their staff, under the management of Mr. Topham, were, however, to carry on the works on behalf of the company. If the plant, which had roundly cost £1,000,000, realised more than £460,000, then 40 per cent. of the excess was to belong to the executors. Of course the new arrangement entailed greater responsibility on the company, but the directors felt it would prevent friction and settle disputes that had become acute, avoid a dead-lock, and result in more economical working. Any other course would probably have ended in a stoppage of the works, as the executors were unable to act like a principal. The company had taken over both Mr. Walker's assets and liabilities in the under-taking. No doubt labour was now 15 to 20 per cent. higher than at the beginning of the contract, and this, coupled with the fact that they had bought more land, and had been obliged by public bodies and railways to do extra works, would compel them to raise more money. According to Mr. Abernethy, £1,700,000 extra capital would be wanted to finish the canal. Against this must be placed the value of the land and plant, and the fact that the change in the site of the docks at old Trafford and their increased size would make them much more convenient for the trade of the district.

Lord Egerton further stated that on that very morning he had had an interview with Alderman Mark, the Mayor, and Alderman Harwood, the Deputy Mayor, and had explained to them the position of the Ship Canal. He had pointed out that this was not a speculative venture to enrich one man, or a body of men, but one con-ceived by public spirited men in the interest of the city itself. Further, that in most other countries it would have been considered essentially a work to be carried out by national or municipal funds. He had met with a kindly reception from the gentle-men before-mentioned, who had promised to lay the whole matter before the City Council, and to see what assistance (if any) that body would be prepared to give to the directors. In his opinion the Corporation ought to do what other seaport Cor-porations had done under similar circumstances. He concluded by appealing to the 40,000 shareholders for support and assistance.

Sir Joseph Lee said it was no unusual thing for estimates to be exceeded; indeed it was often the case in big undertakings. The directors' position was due to so many extra works and responsibilities being thrown upon them. The settlement with the contractors would enable the canal to be pushed forward and completed. That very day the issue of £453,000 balance of the first debentures had been effected. The good feeling of the London ordinary and preference share-holders was shown by their consenting to these debentures being placed in front of their securities.

The executors of Mr. Walker, the contractor, were very sensitive, and so many statements were current as to the reason for their relinquishing the contract that Mr. R. W. Perks, their legal adviser, wrote to the London *Standard* that, at the death of Mr. Walker, the executors had to take in hand a large contract in Argentina, as well as the one for the Ship Canal. All had gone well with the former, and it would soon be completed. As regarded the latter, out of a contract for £5,750,000, to be executed in four years, work to the extent of £1,914,000 had been done at the time of the contractor's death, and, going on at the rate of £142,000 per month, they had since done £1,558,000 of work up to 31st October, 1890. At that date they had £2,278,000 worth more to do, and fourteen months in which to do it. Difficulties had arisen, but the contracts were eventually terminated by agreement to the satisfaction of both parties. Had the executors continued the works he believed they would have been finished by July, 1892.

Next came a formal application from Lord Egerton, Chairman of the Ship Canal, to the Mayor, asking if the Council would assist in raising the required £1,700,000. This application was laid before the General Purposes Committee, and by them remitted to a special Committee. During the debate the Mayor said they had to decide whether they would come to the assistance of the Ship Canal, and so maintain the reputation of Manchester, which was at stake. Unforeseen events had placed the canal in an unfortunate position, and he went on to point out that Liverpool, Bristol, Newcastle, Hull and other places had, under similar circumstances, rendered financial assistance to undertakings in order to benefit their citizens.

Sir John Harwood said that the Corporation would not be justified in coming to the rescue, unless every other avenue was blocked, and there was no way of escape. He was satisfied that if they did not come to the rescue, in August next, the works would be standing idle, and that was why he supported the proposal. He hoped posterity would not point to the men of to-day as men wanting in loyalty and fidelity to the important trust that had been placed in them. Other speakers followed, and the only objectors were Alderman King and Councillor Clay, on the ground that Salford, Oldham, etc., ought to join in a guarantee, Manchester taking the lead.

The calm way in which the shareholders, and the public generally, received the Chairman's communication was most astonishing, and here is a remarkable trait in the character of a Lancashire man. If he feels no advantage has been taken of him, and that he has been treated fairly and honestly, he will accept misfortune or loss with the greatest equanimity; but if he thinks those whom he has trusted have deceived

SIR JOHN HARWOOD, CHAIRMAN OF THE EXECUTIVE COMMITTEE, 1891-94; DEPUTY-CHAIRMAN OF THE SHIP CANAL, 1893-94.

Brown, Barnes & Bell. *To face page* 84.

him, then he would rather lose every penny than let the delinquent escape the meshes of the law. The shareholders in the Ship Canal have always been a wonderful example of this. Though from the commencement they have not received any dividend, they have met their directors in the kindest possible way; and though punished, as many of them have been, they have shown no irritation; indeed a stranger attending a Ship Canal shareholders' meeting would think by the attitude of the members that a dividend was being declared.

Astonishment was expressed in the Press that information of a change made in November had not been given earlier than February. This, as already explained, arose from the necessity of keeping the change secret. It was intended that the outside public should know nothing of the new arrangement until the directors and their officials had obtained a good grip of the work. It is well known, too, that workmen will often do better for a contractor than they will for a private individual.

Strange to say, directly the men knew it was no longer a contractor's job they struck for advanced wages. The change was announced on the 3rd February, and on the 4th a dispute, that had been simmering some time, ended in an open rupture. An advance of $\frac{1}{2}$d. per hour had been given to the concreters on the Barton section on the Monday, and the agitation seemed to be over. But on the Tuesday the other hands demanded the same advance, and just at a time when all speed was necessary, the whole of the men left their work. Fortunately, fresh hands were available, but the strikers became riotous and attacked with stones, etc., the new-comers and a few of the old hands who remained; indeed they drove them from the works and damaged the plant. By altering the switches they nearly sent a train over the embankment. About two thousand of the strikers then proceeded to the Pomona section. Here they were met by the police, who by skilful tactics prevented a riot, and induced the crowd to disperse, or rather to return to Eccles Cross, where they held an open air meeting, which was addressed by Mr. Leonard Hall, an official of the Navvies' Union. It seemed there had been a misunderstanding between the navvies and a working ganger, who, it was stated, had used language which the men regarded as an insult. Mr. Hall told them that they had made a mistake and were ill-advised in going on strike before they had made sure of the support of the Union; and besides this he feared they would not have the public with them. The Ship Canal Directors had just had to go on their knees to the Corporation for help, and would view with indifference the men going out for a few weeks. They ought to organise, and by joining the

Union become one compact, strong body, and then they could call the men out on the whole length from Eastham to Manchester, and be irresistible. He advised them to give up one point now that they might gain twenty hereafter. They might ultimately demand a maximum of 6d. per hour instead of 4½d. to 5d. He also advised them to send a deputation to try and get an apology for the insult, but anyway to return to their work.

Next day another mass meeting was held, and Mr. Hall told the men that the masters backed the ganger and were obdurate ; indeed they had informed the deputation that the men could work or go away, just as they pleased. He advised them, as this was not a Union job, to bide their time and strike hereafter. He further said that "When they did have a general strike it would not be for a paltry farthing, but for at least 6d. an hour". Anyway they had better take a show of hands. On one being taken, the result was a decision to continue the strike. Mr. Hall considered they were wrong in policy, but he admired their pluck.

The plentiful supply of labour in the market and the firmness of the directors were too much for the men, and they gradually drifted back into their places, but to the end there was unrest, and a gradual rise in wages followed.

Shortly after the directors had taken over the works, and whilst labour was in a disturbed state, the Deputy Chairman, Sir Joseph Lee, tried to get at the root of the dissatisfaction about wages. One day he approached a burly navvy and asked him several questions about the hours and work generally, to which he received rather gruff answers. When he came to ask what wages the man was getting and where he was working, he turned round savagely, and evidently thinking it an improper question told him to go to h——; what business was it of his? and Sir Joseph retired disconcerted. Shortly afterwards a ganger asked the navvy what the gentleman had been saying. On hearing the impolite speech the ganger told him he had been speaking to the Deputy Chairman, and he certainly would be sacked for his impudence. He had better go and say he was sorry, and that he did not know to whom he was speaking. Shortly afterwards Sir Joseph Lee returned the same way, and the navvy went up to him in a sheepish way, but was uncertain what to say. At last he blurted out, "Mester, I didn't know who you wur, and I told you to go to h——, and I'm sorry for it, but I didn't mean you to go". The Deputy Chairman was much amused at the man's apology.

Outside critics were very busy discussing the position of the Ship Canal. They could not understand why there was not a state of panic and collapse. The possibility that the ordinary and preference shareholders would allow Corporation debentures to take precedence, quite amazed them.   One Liverpool paper wrote :—

The wealth of Manchester is great and its public spirit strong.   Its common-sense, let us hope, is not altogether imaginary.   There has been a shrinkage of £2,250,000 in the value of the property, and yet the Directors have unshaken faith in the ultimate success of the enterprise.

Another article said :—

It is scarcely probable that the shareholders will consent to be sacrificed in order that the Manchester Corporation may be rendered more safe.   There will not unlikely be a renewal of the strife of lawyers.

The Manchester papers were unanimous in the opinion, that rather than let the canal be sacrificed, the Corporation must come to the rescue.   The directors were blamed for not speaking before, and were plainly told that if the Corporation were going to assist, they must be adequately represented.   The *City News* urged that the time had arrived when the canal should become a public trust, and said there had been a resort to secrecy which was inexplicable, both as regarded the contractor's claims and the paying £1,000,000 for plant; further, that the Ship Canal had no right to be its own contractor.   The latter statement brought a letter from Mr. Walker's solicitors, who said it never was his business to find the plant.

Mr. Stanhope, M.P., the well-known champion of public trusts, on the 10th February asked Parliament for a Special Committee to consider the position of the Ship Canal in respect to the Canal Developments Bill, but was told by Sir M. Hicks-Beach that he did not intend to consider the question of municipalising canals.

Ward meetings were held beginning with Ardwick, when it was unanimously declared that the canal must be finished, and if necessary by municipal assistance.

The Special Committee appointed by the Council to report on the advisability of rendering monetary assistance consisted of the Chairmen of the various Committees of the Corporation.

At their first meeting they appointed the Mayor as Chairman, and decided to seek the assistance of Mr. Moulton, Q.C., the eminent counsel, also of the water-works engineer, Mr. Hill, C.E.   Prior to coming to a decision they had asked the Ship Canal Company to supply them with information concerning—

1. Any statutory provisions as regards finance, and the obligations imposed by statute as to the construction, completion, and opening of the canal.

2. Any contracts with the late contractor or his executors, and as to the present position and obligations of the company in relation to the works.

3. Any obligations with which the Corporation should be made acquainted.

When they had consulted Mr. Moulton on the legal position, and had had a report from Mr. Hill on certain engineering questions, also an explanation from Mr. Saxon, solicitor, of the agreement with Mr. Walker's executors and of the legal and statutory obligations, as well as a statement from Mr. Thackray, the City Treasurer, of the financial position of the Ship Canal Company, they made a report, of which the following is an epitome, which was signed by the Mayor as Chairman. This report was dated 4th March, 1891 :—

A. As to the Completion of the Work and its Cost.

1. That no works could be postponed without affecting the efficiency of the canal.

2. That the directors' estimate for completion was £351,000 too low.

3. That it would take two years to complete the canal, and that the estimate of expenditure should be made up to 1st March, 1893, instead of 24th June, 1892, thus adding £97,000 for debenture interest and expenses.

4. That £100,000 would be required for Corporation interest to March, 1893.

The effect of these corrections and 15 per cent. added for contingencies would be that even if the statutory obligation to pay interest out of capital were cancelled, it would take roundly £2,500,000 to complete the canal, and it would be prudent that £3,000,000 should be provided. Any sum received for plant would be required for equipment.

B. Necessity of Assistance from Public Sources.

Inasmuch as the public had failed to subscribe for £400,000 of first debentures, it was manifestly impossible for the Canal Company to raise £3,000,000 within the next few months, and any stoppage of the works would be disastrous, and cripple the usefulness of the undertaking.

C. Causes of the Present Difficulty.

The Committee did not think the present difficulties arose from any innate defect in the project, or that the directors had founded their hopes on an unsound basis. Nothing had occurred to affect the future earning power, and the works as projected were quite practicable. The works were far advanced and could be completed at the cost above-named. The work so far had been substantially done, and there had been no waste.

The causes of the present unfortunate position were an estimated extra expenditure for works of £2,227,500, and for land of £360,300. The latter was likely to be a remunerative expenditure, as the land would be available for resale. The extra cost of the works was further due—

1. To the unforeseen cost of the works at Ellesmere Port.

2. To the protective clauses put upon the Ship Canal by Parliament at the instance of various public and private bodies.

3. To the death of Mr. Walker, and the financial disputes with his executors.

4. To the great rise in labour and materials.

5. To the disastrous floods of last November.

D. Mode in which Assistance could be Given.

The Committee believed it would be unwise to ask for the position of salvors and to be placed before the debenture holders. It would damage the credit of the company. The assistance was imperative, and the Corporation ought to come before the preference and ordinary shareholders. The advance should not take the form of a permanent loan, but debentures should be issued and taken up by the Corporation, and these would be a marketable security for the investing public. They could be sold when confidence was restored. In this way the Corporation could give the required assistance without imposing an unnecessary burden on either the Ship Canal or the ratepayers.

E. Terms on which the New Debentures should be Issued.

The existing debentures were carrying 4 per cent. interest. Second debentures ought to carry $4\frac{1}{2}$ per cent., and be redeemable at par on certain dates, if the Canal Company could borrow at a lower rate. The new debentures should be issued as the Ship Canal required money, so as to save interest.

F. Disposal of Interest Paid to the Corporation.

The Corporation would borrow on the most favourable terms, and receive $4\frac{1}{2}$ per cent. debentures as security. The Committee considered the balance of interest should be carried to a suspense account out of which to pay expenses, and as a guarantee against loss. A strong feeling existed that no profit should accrue to the Corporation from this interest fund, but that eventually the Canal Company should receive the benefit, when the debentures were redeemed, so that the Canal Company would have an incentive to redeem any that were unsold and in the hands of the Corporation. On these terms it would be wise and right to give the required assistance. If other Corporations desired to co-operate, Manchester would be willing to receive an intimation to that effect.

G. Mode of Carrying out the Proposed Scheme.

As neither the Canal Company nor the Corporation possessed the requisite powers, a joint Bill would have to be promoted—

1. To get the necessary powers to issue debentures with priority over all other shares.

2. To give the Corporation power to lend in return for debentures.

3. To enable the Corporation to sell, hold or redeem debentures.

4. To extend the time for the construction of the canal.

As the Corporation were providing the funds, they should have representatives on the Ship Canal Board till the loan was discharged. Also in order to have an effective check on the expenditure they should nominate an engineer to work in conjunction with Mr. Leader Williams, and certify the expenditure. Further no more interest should be paid out of capital during construction, and this would require the assent of the shareholders. The Ship Canal Company should be required to pay Parliamentary and all other necessary expenses, and immediate application should be made to suspend Standing Orders.

The report, of which I have given a *résumé*, came up for consideration by the City Council at a special meeting held on the 9th March. The Mayor (Alderman Mark) moved its adoption, remarking that the Committee felt that the £10,000,000 already spent on the Ship Canal ought not to be wasted, and that he believed it was absolutely necessary that assistance should be given to complete it. They had had the assistance of, and reports from most eminent men, and had spared no trouble in probing the position to the bottom. They were unanimous in their report.

Alderman Sir John Harwood seconded the adoption. He referred to the $4\frac{1}{2}$ per cent. to be charged, which some people thought too high, when the value of money was $3\frac{1}{4}$ to $3\frac{1}{2}$ per cent., and said if the Canal Company could have got the money themselves even at $44\frac{1}{2}$ instead of $4\frac{1}{2}$ per cent., the Corporation would have had no business to interfere in the matter at all. But they could not get it at any price. He wished to say the Manchester Corporation would be glad to hear of the Salford Corporation or other Corporations desiring to help in giving financial assistance. If the Council wished to maintain the character of their city, he believed they would never suffer so much money subscribed by so many needy people, to lie there as a monument of a want of fidelity to their trust as public servants, and of loyalty to the cause, the trade and the interests of this great city.

Alderman Chesters Thompson said if any railway member of Parliament voted against the scheme he would not be a member long. He would not support any man, whatever his politics, who opposed this great Manchester scheme.

The report was carried unanimously, and a copy ordered to be sent to Salford and other Corporations.

The decision of the Manchester Corporation to rescue the canal at all costs caused astonishment, not only in England but all over the world. An impression prevailed that the "big ditch," as it was jeeringly called, was the outcome of a few enthusiasts whose work must in the end miserably fail. In turns it had been said, "They won't get the Bill," "They won't get the money," "They can't make the

SIR JOHN MARK, DIRECTOR OF THE MANCHESTER SHIP CANAL
COMPANY, 1891-97.

Lafayette, Ltd.

To face page 90.

canal, and if they do, there won't be water to work it," and lastly, " They won't get the trade, for ships won't come up the canal ".

It was now clearly established that local patriotism was the keystone of the fidelity of Manchester to the canal.   Many citizens who had been opponents were willing, in the day of adversity, to give a helping hand for the credit of the city.

The Manchester Press was full of praise of the masterly and comprehensive report of the Special Committee.   They admitted it was absolutely necessary that the Corporation should save the canal, but insisted that there should be adequate representation; they held, however, that financial assistance from outside Corporations would be dearly bought if it led to divided counsels.   The *City News* was of opinion that the Corporation ought to convert the canal into a public trust.

Much soreness was shown by correspondents that Manchester should make her help contingent on the cessation of payment of interest on capital during construction; it was said this would be a distinct breach of faith.   The reply was, that if the city did not find money for the canal, the shareholders would lose both capital and interest.

Before going to Parliament for power to lend money to the Canal Company, it was necessary to call a Borough Funds Meeting, in order to obtain the consent of the Manchester ratepayers.   This was held on the 23rd March.   The resolution consenting to an application for the requisite power was moved by the Mayor, seconded by Alderman Sir John Harwood, and supported by Alderman King.   The latter said that whatever the cost, Manchester must finish the canal.   She could not afford to stand before the world face to face with a great failure.   Mr. Higgins, a gentleman who made a point of attending Ship Canal meetings as an opponent, moved an amendment, but could not find a seconder, and in a large gathering the resolution was carried with only two dissentients.   A curious incident occurred at this meeting. An elderly gentleman from Harpurhey, with an idea that he had a great financial plan to elaborate, essayed to speak, but burst into tears, and had to retire, explaining that he was not used to stand before such an audience.

Salford also showed herself desirious of helping in this emergency.   The Mayor, however, felt some delicacy in going to Manchester, cap in hand, and asking to be allowed to assist.   He preferred making an offer to the Ship Canal directors themselves.   As the chief docks were in Salford, and as that borough had already subscribed 2d. in the £ to the Parliamentary expenses (£5,000), and moreover would have subscribed £250,000, if Parliament would have given permission, there was some pique shown that she had not been applied to as well as Manchester.   On the other hand, the action of Salford in asking for her £5,000 back was commented upon, and it was urged that Manchester was better without such a fickle companion. At a meeting of the Salford Council it was decided they could not give any assistance till they were asked directly by the Ship Canal directors, but there was a feeling that it would be unfortunate, in the interests of Salford, if only Manchester had representatives on the Ship Canal Board. So lukewarm were the Salford Council that the inhabitants determined to force their hands, and memorialise the Mayor to call a public meeting. It was pointed out that by the terms of the resolution at the Borough Funds Meeting, Manchester had invited both Salford and other towns to join in promoting a Bill. It was further stated that though Councillor Phillips had moved for a Special Committee *re* Ship Canal, it had never met in the seven weeks of its existence.

On the 4th April at the public meeting, Mr. G. C. Haworth moved, and Alderman Snape seconded, that Salford should assist to the extent of £1,000,000, and this was carried by a large majority. Subsequently a meeting was held between the Mayor of Manchester and a Salford deputation, when the Mayor said that Manchester would welcome any help from Salford, but that as the offer of it had been so long delayed, Manchester had gone on by herself, and had a Bill in Parliament. It would therefore be necessary now that Salford should present one on the same lines, to enable her to subscribe the million proposed. This would be the proper course to take. The dilatory action of the Salford Town Council, which seemed entirely out of accord with the general feeling of the inhabitants, caused some very strong language to be used by the Aldermen and Councillors themselves, who were not satisfied with the position.

At a meeting of the Salford Council held in the first week in May, it was decided to do nothing further in the matter, but "to leave to Manchester all the honour, glory and possible loss in the transaction". A chief reason for this course was stated to be that Salford found she could not borrow money at the same rate as Manchester, and that the Council themselves had not the same enthusiasm for the canal that was manifested outside. On their part the Corporation of Salford alleged that they had not been well treated by the Ship Canal directors, who ought to have asked their assistance at the time they approached Manchester.

Before the deposit of the Bill it was necessary to get the sanction of the ordinary and preference shareholders to the placing of the Corporation debentures before their own securities. A meeting for this purpose was held in the beginning of May. Lord Egerton congratulated the Manchester Corporation on the public spirit they had displayed in becoming the saviours of the canal, and he welcomed Mr. Hill, the engineer, as the safeguard of Corporation interests.   The directors could not help themselves when the Corporation insisted that payment of interest during construction should cease, and he assured the shareholders that in the awkward position of the concern the directors had done all in their power to protect their interests.

Sir Joseph Lee said that they had money to last them until July only, and it had often been a difficult matter to provide the £40,000 required weekly for wages.   Only last March, when, through lack of confidence, the first debentures were not taken up by the public, he had been obliged to appeal to his colleagues, and in an hour he received from them subscriptions for £135,000 of these debentures, which enabled them to carry on the works.   Afterwards on this fact becoming known in London, he was able to issue the balance of £250,000 amongst the shareholders there.

Questioned as to why the directors had not applied for help to Salford as well as to Manchester, Sir Joseph said he believed he was the culprit.   Manchester had received him so sympathetically, and had taken such prompt action, that it never occurred to him that it would be at all desirable to apply to any other Corporation. It seemed to him to be a one Corporation job.   Manchester had suggested others should join, but the Salford Corporation did not apply until the Manchester Bill was drafted and before Parliament.   He assured the meeting there was no intention of slighting Salford.

On the 26th May Manchester applied to Parliament for authority to lend the money to the Canal Company.   As the Bill was nearly six months after time, application had to be made to the Standing Orders Committee for permission to enter the Bill on the list.   After hearing the applicants, the Committee agreed to give leave to deposit the Bill.   It then had to go before the Examiner on Standing Orders.   As it was too late he reported it to the Standing Orders Committee, who agreed to overlook the flaw.   On the second reading Mr. Lees Knowles, M.P. for Salford, moved a resolution blocking the Bill, unless Salford had one or more directors placed on the Board.   This opposition was supposed to have arisen in consequence of Manchester requiring Salford to have a Bill of her own, and it had been openly stated that Salford would lodge a petition against the Bill of the Manchester Corporation unless they were parties to it.

At a subsequent meeting of the Salford Council attention was drawn to the action of Mr. Knowles, which had given the impression that the authorities of Salford were behind him in an attempt to damage or block the Ship Canal Capital Bill.   The Mayor of Salford denied this, and said he was satisfied that Mr. Knowles had no feeling of hostility to the Ship Canal, and that if that gentleman found his motion would embarass the Manchester Corporation and the Ship Canal Company, he would withdraw it. Aldermen Husband and Dickins justified Mr. Knowles in seeking representation even without the instructions of the Council, but the feeling of that body being strongly against anything that would impede the Bill, it was announced that Mr. Knowles was willing to withdraw the instruction of which he had given notice.

No sooner was this done than Mr. Philip Stanhope, M.P. for Burnley, announced that he intended to move an instruction that under the enabling powers of the 1885 Act, the Ship Canal should be converted into a public trust, prior to municipal aid being granted. On the motion for a second reading, Mr. Stanhope moved an instruction to this effect, but it was declared to be premature, and was postponed. At the proper time Mr. Stanhope again moved his resolution, and it was seconded by Sir Joseph Pease, who said that while the Manchester Corporation were in the present stress of weather, finding the crew to man the ship, they were in a minority on the Ship Canal Board of Directors. The canal could only be of service to Manchester and the neighbourhood by having representatives of the local public bodies on its management.

Sir William Houldsworth deprecated the House discouraging private enterprise. As to altering the whole working and converting the canal into a trust it would be the dangerous process of "swopping horses while crossing a stream". Mr. Ritchie thought the unanimity prevailing among every class in Manchester, and the support from the Corporation, showed such an exceptional state of things as justified the Local Government Board in offering no opposition to going to Committee. The instruction asked for would cause delay, and be fatal to the canal. Eventually Mr. Stanhope, in deference to the feeling of the House, withdrew his instruction. On the motion of Mr. Ritchie the Ship Canal Loan Bill was referred to a Select Committee, with Sir Selwyn Ibbetson as its Chairman.

While the Bill was in Parliament a strong feeling was growing in and out of the Corporation that the Board of the Ship Canal should be reconstituted, and that if the Corporation advanced £3,000,000, the proposed representation by three Corporation directors was inadequate. The matter was discussed at the June meeting of the Council, and it was decided that a request should be made for a Board of fifteen members, of whom five should be members of the City Council. The Mayor promised to try and carry out the wish of the Council, though he thought it rather ungracious to alter a Bill which had already had their assent.

When the Bill came for consideration before the Select Committee it was opposed by the railway companies, who pleaded that it was unjust for them to pay rates in support of a scheme that would be competitive with themselves. The Committee in one day's sitting passed the preamble.

On clauses the Committee struck out the proposal to return to the Canal Company any profit the Corporation might make on the interest charged for the loan. They also decided that the sinking fund of 1 per cent. must commence in 1896 instead of 1901.

Bills are seldom opposed on the third reading, but in this case Mr. Stanhope, M.P., and Sir Joseph Pease renewed their opposition on the grounds they had previously urged, but they got little support and the Bill was passed without a division. It should be mentioned that the Ship Canal directors agreed to the Corporation having five members on the Board, and altered the clause accordingly. In order to make room for new directors, it was arranged that Mr. Jacob Bright, M.P., and Mr. Henry Boddington should retire from the Board of directors. Much regret was expressed at the necessity for this step, as both had been stalwart, energetic and liberal supporters of the canal almost from its inception.

When the Canal Bill went before the House of Lords the Examiner on Standing Orders allowed it to proceed. It was found that the railway companies had again petitioned against it. Hence, after being read a second time, it went before a Committee of the House of Lords, with Lord Basing as Chairman. For some reason this Committee was altered and the Bill was postponed. The new Committee consisted of the Earls of Romney and Mayo, Lords Digby, Leconfield and Brougham, the last being Chairman. They sat on the 10th July, and despite the railway opposition passed the Bill the same day, inserting a clause to charge the railway companies only one-fourth of any rate that might be raised for canal purposes.

In an Omnibus Bill promoted this year by the Manchester Corporation, power was sought to buy land near Mode Wheel for markets and slaughter houses with a view to establish lairages. To this Salford raised objection on the ground that if land were acquired at Mode Wheel it would seriously affect their existing cattle market. The Committee, of which Lord Limerick was Chairman, eventually passed the Bill.

At a special Council meeting on 30th July to receive the report of the Finance

Committee *re* Ship Canal, the Mayor moved the report be approved, and said that the money was wanted quickly if they wished to benefit the undertaking. Sir John Harwood desired to emphasise the fact that the Bill was simply "an enabling one," and that they were not *obliged* to pay a penny for finishing the Ship Canal. They must keep absolute control, and he would advance nothing till they had a clear balance sheet showing the position of the company; there was a rumour that the Canal Company had anticipated the loan and had already spent a quarter of a million of it; of this they must have an explanation. The Corporation directors must oppose unnecessary expense, and press forward the finishing of the waterway.

Councillor Clay held that they must not part with any of the £3,000,000 until certificates were presented and the permission of the Council obtained. Councillor Hoy thought that having decided to help the canal they ought not to put such a stumbling-block in the way as might possibly lead to catastrophe and delay progress. The report of the Finance Committee was approved by a large majority. It was also decided that the Special Committee should nominate the five directors.

That Committee having met presented a report recommending as directors the Mayor (Alderman Mark), Aldermen Sir John Harwood, Walton Smith, Chesters Thompson and Councillor Southern. They also proposed that Mr. George Hill, C.E., should be appointed consulting engineer at about £2,000 per year, inclusive of all extras.

At a meeting of the Corporation Special Committee it was decided to raise at once £1,500,000 by an issue of debenture stock at 3 per cent. through the Bank of England, the minimum price to be £94 per cent. A resolution proposed by Sir John Harwood was passed that till the canal was finished and opened, the five Corporation directors should receive no remuneration.

In the Press the recommendation that the directors' services should be gratuitous did not meet with much favour. It was pointed out that gratuitous services were apt to become unsatisfactory, and that it would be unfair to pay the Ship Canal directors and not the Corporation directors. One of the papers went on to say that the proposition would have emanated with better taste from some other quarter.

At a subsequent Council meeting there was a good deal of discussion as to the mode of electing directors. Eventually the ballot was resorted to, and the recommendation of the Committee was adopted, except that the name of Councillor Boddington was substituted for that of Alderman Walton Smith. It was announced that the £1,500,000 3 per cent. redeemable stock which it had been decided to issue through

JAMES W. SOUTHERN, DEPUTY-CHAIRMAN OF THE MANCHESTER SHIP
CANAL, 1897 *SEQ.*

Lafayette, Ltd.   *To face page 96.*

the Bank of England had been tendered for nearly twice over, and that the average price obtained for the stock was £96 0s. 9d., the minimum price of issue being £94.

On the 21st August the Special Finance Committee issued the report on the finances of the Ship Canal, moved for by Sir John Harwood. They explained that the Canal Company had borrowed from their bankers £546,000 on £659,600 of first and second unissued debentures. Further, that the directors proposed to apply £300,000 to redeem £369,600 stock which had been pledged to keep the works going. They recommended that £200,000 out of the loan should be paid to a special account, whence £40,000 weekly would be handed to the Canal Company for wages and expenses.

At a meeting of the General Purposes Committee the above report was approved. In moving it the Mayor said the Canal Company had a bank overdraft of £135,000 which ought to be paid off.

At the beginning of August the shareholders' ordinary meeting was held to sanction the borrowing arrangements with the Corporation. The Chairman explained the delays caused by floods and bad weather, and held out hopes of speedily completing the canal to Weston Point. At the extraordinary meeting which followed, the Chairman asked for sanction to borrow from the Corporation, on the security of the undertaking £3,000,000, at a rate of interest not exceeding $4\frac{1}{2}$ per cent., with the proviso that this loan should take priority of the ordinary and preference shares. This was agreed to. The directors also obtained authority to issue debentures for £65,420, being the amount remaining unborrowed under the Act of 1885.

As soon as the system on which the works were carried on by the trustees of Mr. Walker came under the purview of the auditors, they called the attention of the Deputy Chairman to certain defects, especially as regarded contracts; but he decided to defer taking action or instituting any changes that might in any way interfere with, or delay the progress of the works, believing it wiser to let things go on than to disturb an existing system under which 16,000 men were working at full speed. He said that he had been given to understand it was the same system on which Mr. Walker had managed his business for the last twenty years.

In October Mr. C. P. Scott, of the *Manchester Guardian*, contested North-East Manchester, and he came in conflict with Councillor Chesters Thompson, leader of the Conservative party. At a meeting in the Islington Hall, Mr. Scott called attention to what he called a blackguard picture that his opponents had circulated. It represented Daniel Adamson carrying in his hand £8,000,000 of money in a bag. Behind him came creeping up a scoundrel in a mask, who garrotted the capitalist and seized the bag, appropriating the money to his own purpose.   On the garrotter's cap was the word "Scott".   In good round terms the candidate denounced the attack on him as black-guardly.   The bill had been printed in London and was distinguished by its smallness, its grossness, its virulence and its imbecility.   Had Daniel Adamson lived he would have supported him in that contest.   To this Mr. Chesters Thompson replied denying the *Guardian's* advocacy of the canal, and saying it had always been a thorn in the side of the promoters.   It was not until the pluck, the courage, and the resolution of better men had carried the project through Parliament that the *Guardian* trimmed its sails and changed its tactics.   Further, he said that when Mr. Scott, at the instigation of Mr. Beith, hauled down the flag of opposition and went in for the canal, it was because he was anxious to be on the winning side.   These insinuations were treated by Mr. Scott with silent contempt.

So many visitors were anxious to see the canal that a company was formed under the name of "The Ship Canal Passenger Steamer Company, Limited".   This Company was inaugurated in October and at once put steamers on the canal to visit points of interest.   They were a great convenience and did well, and also acted as ferry boats.   In a short time, however, another company started with new landing places and more modern boats, and the old owners had to retire from the scene, poorer than when they started.

During October the plucky efforts that had been made by a few Newcastle gentlemen connected with shipping interests to get a footing on the Ship Canal ended in the formation of the Manchester Ship Canal Pontoons and Dry Dock Company, Limited.   The first directors were Mr. George Renwick, J.P., Mr. Thomas Bell, J.P., Mr. J. T. Eltringham, J.P., and Mr. Alexander Taylor.   To these were added subsequently two members of the Ship Canal Board.   The promoters, whilst others were doubting, showed their faith in the success of the canal by purchasing 17 acres of land at Mode Wheel, having a frontage of 463 yards to the canal.   They were about the first purchasers and got a good plot at a most reasonable price.   They then floated a company and had conceded to them the right to construct dry docks, shipbuilding berths, workshops and jetties alongside the water-way, and to moor a pontoon for repairing purposes.   The same rights were accorded them near Ellesmere Port where they had also purchased land.   The company was registered with a capital of £300,000, the first issue being £120,000, of which the promoters themselves took £42,000.   There can be no doubt that the establishment of the company has been a great convenience to shipowners, and promises to become a financial success, in which case the Canal Company will receive 10 per cent. of the profits.

As the contemptuous view entertained of the canal in Liverpool, even when it had been opened to the Weaver, may be interesting in future years, I quote the following passage from the *Liverpool Daily Post* of 17th October:—

It is true that Liverpool has not looked with favour upon the Ship Canal, but as years roll on we shall eventually realise that the canal instead of acting detrimentally to the interests of the port will attract trade to Liverpool which otherwise would not have come. Of course no one expects that the owners of large sea-going vessels will ever be induced to use the canal; a sail up the new waterway on one of the small steamers should convince the most ardent enthusiast from Manchester that such a state of things is impossible. But that the canal will be largely used by smaller craft, such as coasters, there is every probability, and that the passage up and down the canal of vessels of even 300 or 400 tons will become quite common may be taken for granted. As is well known the promoters have entertained more ambitious ideas and have at Manchester provided sufficient dock accommodation to put seventy large ships a day through the locks, whereas, as they point out, the average of large ships using the port of Liverpool daily is only thirty.

When another paper doubted the correctness of this statement, the writer of it became indignant, and in a letter signed "Your Correspondent" wrote:—

The general opinion of the shipping community is as I stated. Possibly a few steamers of 800 or 1,000 tons may occasionally use it, but when we see steamers of 3,000 to 4,000 tons pursuing their wild course past Liverpool up to Throstle Nest in Manchester the time will have arrived when we may see "pigs begin to fly".

It is to be hoped the writer has lived to see ships ten times the size he first named navigating the canal.

Various depressing rumours being afloat about the Canal Company's finances, and it being stated that the £3,000,000 borrowed would not complete the canal, Sir Joseph Lee wrote the following letter to the *Guardian*:—

SIR,

Will you afford me an opportunity, by the publication of this letter in your journal of denying the reports that have been circulated to the effect that the Ship Canal Company will have to apply to Parliament for additional powers to borrow money to finish the canal. The resources of the Company are ample to complete all the authorised works and leave a substantial surplus.

I am, etc.,

JOSEPH C. LEE.

13th November, 1891.

Notices were deposited in due course for a General Purposes Bill in the session of 1892. It asked for power to acquire land for flattening slopes and to define the rights and limits of the Ship Canal in the Mersey estuary.

In the reconstitution of committees on the advent of the five Corporation directors, a Special Works Committee was formed on which two of them were placed, the other three being shareholders' directors. The Corporation also appointed a Consultative Ship Canal Committee to whom the Corporation directors could apply for advice and instruction.

When, in November, the Special Works Committee came to make a report as to the progress of the works, there was a wide divergence between the shareholders' directors and those appointed by the Corporation, and it became necessary to issue separate reports. The former, consisting of Sir Joseph Lee and Messrs. Bythell and Crossley, said they had come to the conclusion that the estimates must be increased, and a revised estimate, exceeding by £512,614 that of the Corporation engineer of 28th February, was put forward. Savings, however, of £36,000 might be effected, the balance required being £476,614. This balance comprised extra work not hitherto contemplated, extra land and miscellaneous expenses. Amongst the extras was 7,936,000 yards of additional excavation. It was possible, however, to postpone work to the extent of £364,000, and they believed that the canal might be opened for traffic in the spring of 1893. They were of opinion also that the available assets would not only be more than sufficient to provide for the opening of the canal, but that if the 24,159 ordinary shares in hand were taken at £5 each there would be a surplus of £468,000. The surplus plant was taken at £400,000 and the surplus land at £1,000,000. Assuming the available assets were realised there would be an eventual surplus of £1,400,000. They recommended the introduction of piecework and tendering, and that stores should be bought by public tender instead of private contract. They advocated the reduction of the staff and the introduction of all possible economies.

Sir John Harwood and Mr. Southern, on the other hand, thought the present position demanded from them a separate pronouncement. They had desired to give prompt and substantial aid to the canal, but expected the amended estimates placed before them would be reliable. To their surprise, the amended estimate of September was £863,595 in excess of the Ship Canal estimate of February last. They were not prepared to assert the available funds were insufficient to complete the canal, but this could only be effected by immediate and vigorous retrenchment.

When the Corporation consented to lend up to £3,000,000, it was distinctly understood there would be a large margin above all possible requirements.   During their short term of office they had come to the conclusion that the resources were not economically administered, and they urged—

   1. That wherever possible, work should be let by contract to suitable contractors.

   2. That the workshops and purchasing of stores should be reorganised, and that for the latter there should be public competition.

   3. That there should be a revision of salaries, and that superfluous service should be dispensed with.

   4. That extraneous charges should be curtailed.   Further that a Committee should be forthwith appointed to carry out the foregoing suggestions.

As was to be expected, the warning sounded by the Corporation directors as opposed to the optimism of the shareholders' directors caused a considerable sensation in the city and among the shareholders.   Further written explanations were made by the Chairman and Officials of the Board.   These communications were submitted to the Consultative Committee at a meeting, when four members of the Ship Canal directorate, Messrs. Platt, Bythell, Crossley and Galloway, together with Mr. Hill, the engineer, attended and gave explanations.   Mr. Hill said that his estimate was given on figures supplied to him by Mr. Leader Williams.   He had no time to get them out for himself.   Had he endeavoured to do so, the works would have had to be stopped.   Since he gave his estimate he had exercised no control in the carrying out of the works, the expenditure of the money, or the letting by piecework.   After a long discussion, the general feeling of the Consultative Committee was that though they were of opinion that no adequate effect could be given to the views of the Corporation directors whilst they were in a minority, yet there was no reason to shake their confidence in the ultimate success of the Ship Canal enterprise.

It should be stated here that in the month of October Mr. Boddington's resignation as a Corporation director was received, also that though Messrs. Chesters Thompson and Mark did not sign the minority report, they were entirely in sympathy with it.

When the special report on the Ship Canal Company's affairs came before the December meeting of the Council, Alderman Chesters Thompson contrasted the estimates of those put before the Council by the engineer of the Ship Canal eight months ago when a loan was asked for with those of the present time, which were nearly a million in excess.   True, Mr. Hill, for the Corporation, had added

£350,000 to put them on the safe side, but he could not understand that gentleman taking the quantities of Mr. Leader Williams as correct, and when an error of 8,000,000 cubic yards was found in the excavations, laying the whole blame on the engineer of the canal.

Sir John Harwood did not wish to say anything that would disturb the mind of the people or damage the undertaking, especially as Lord Egerton had declared the willingness of the directors to accept the Committee's report, but he must discharge the duty cast on him.   When he first joined the Mayor in an effort to assist the canal, he was told that a million would suffice : then one and a quarter millions, and at last it grew to three millions.   It now appeared that even this was not sufficient.   He charged the management with extravagance.   Salaries had been doubled since Mr. Walker died.   A coal merchant was paid 2d. per ton for buying coal, and purchased some of it from himself.   Buying goods and stores was delegated to the managers, who did not even get tenders.   A complete reconstruction of the management, accompanied by strict economy, was absolutely necessary.

Alderman Mark was hopeful that now they knew the worst.   The canal, with care and economy, might be finished with the £3,000,000 from the Corporation. Only the engineers could be blamed for the increased estimates.   There was no cause for panic.   All that was wanted was energy, vigilance, and common sense, and though it was not easy for the five Corporation directors to bring their opinions home to their colleagues, he had hopes that in future a better state of things would exist, especially as they had a promise that the shareholders' directors would be amenable to recommendations by the Corporation directors.

Mr. Southern felt sure that when £3,000,000 was fixed upon, the Corporation believed there would be sufficient to finish the canal, and leave a balance of nearly half a million ; and besides there would be an extra asset in the plant and land.   Even if extra cost and insufficient quantities had driven up the estimates by £836,000, there ought to be money to cover half a million of the excess.   He declined to sign the report of the shareholders' directors because it was conceived in a spirit of sanguine optimism, and he would sign nothing he did not conscientiously believe to be true.   In September the Works Committee passed a resolution that timber, stone, coal and cement should be bought by tender, but it had never been acted upon ; and in other matters there had been much laxity.   He believed henceforward that as a result of the action taken, a better state of things would prevail.

Councillor Andrews severely criticised the action of the shareholders' directors and the engineers, and said they stood condemned in the minds of all practical men. He said "they were simply automatons in the hands of others ".

At an adjourned meeting, the Mayor read a letter from Lord Egerton, written with the intention of removing some misunderstandings.   He explained that the Ship Canal Board were obliged to take over the contract on 24th November, 1890, or stop the works.   The works came into their hands of necessity.   They carried out the contractor's system till February, 1891, when they began to reduce the staff.   Admittedly, the salaries paid to the leading officials were large, but they had been fixed by Mr. Walker.   At his death great additional responsibility fell on some of the chiefs, who had to manage the works and 15,000 men, hence the increased salaries.

Another letter, signed by the shareholders' directors, dealt with the impression that they had stood in the way of economy.   They regretted the increased estimates, but could not be responsible for what was not unusual in large works, and they had acted to the best of their ability.   The clauses granted in the 1885 Act had been the cause of endless worry and heavy unforeseen additional expense.   "Time is money," and inasmuch as each working day cost £1,000 for fixed charges and interest on capital, they desired to secure the prompt and economical completion of the canal. They were willing, if the Council wished, that a greater control of the spending departments should be given to the representatives of the Corporation.

Councillor Clay, who could not be present, wrote urging that the report presented by the Special Consultative Committee, which included the report of the Board of Directors, should be referred back for further consideration.   He could not approve of it unless fuller information was obtained; the time might come when further calls for help would be made, and the time had arrived when the Council should know the true position.

Alderman Holland thought the report moderate in its language, and statesmanlike in its suggestions.   He wanted the Board to be strengthened, and suggested an immediate appointment be made in place of Mr. Boddington.

Councillor Gunson urged that there should be eight Corporation directors instead of five, but was told this change could only be made by Act of Parliament. He deprecated both the Corporation and the contractor appointing engineers.   He objected also to the minimum salary fixed for the solicitor of the company.

Alderman Mark said that they were making it impossible for the directors to work harmoniously.   He could not understand why half a million increased estimate should create something like a panic in Manchester.   They ought to pass the report.

After several other members had spoken, Alderman Harwood, in reply, said there was no reason why men should sit with their fingers in their mouths when the money was running away. Was there any justification in continuing a line of extravagance? The management of the canal had degenerated into a state of utter incapacity, and the Corporation representatives were powerless to help it. There never was a contract for the canal worth calling one. Had they sat down and said nothing till the £3,000,000 had been spent, they would have been blamed for not doing their duty.

The report of the Consultative Committee, which included the minority report of the directors, was carried by a majority of eleven.

Naturally the debate, and the strong language used, created a great sensation. The Press teemed with articles and letters on the question. Some few shareholders stigmatised the directors as extravagant and incompetent; others, headed by Mr. Belisha, were disturbed by the shrinkage in the value of the shares from £8,000,000 to £3,500,000, and were anxious to form an association of Ship Canal shareholders, who should aid the directors in preventing further disaster. The great bulk of the shareholders, however, followed the lead of Mr. Reuben Spencer, and were not disposed to blame the old directors for all the ills that had befallen the undertaking. Mr. Spencer's letter was so apropos to the occasion that I give it in full:—

To the Editor of the "Manchester Guardian".

Sir,

I have been closely associated with the canal movement from its first inception, have watched the varying currents of feeling in reference to it, and in my humble way have tried to assist the enterprise. My warm attachment to the work and the workers has brought me into somewhat intimate relationship with many who have been supporters of the scheme, some of whom have ultimately become directors of the company. I am not allowed to see the list of shareholders' or directors' holdings, and therefore cannot give an authenticated list; but from various conversations and the voluntary statements of some, I can venture so far as to say that I can place my hand upon twenty people who hold shares to the extent of £800,000. Some time ago a special effort was made to place a number of ordinary shares, and at another time to issue debentures, which gave me an opportunity for asking what the members of the Board had done or were proposing to do.

I was told pretty nearly what they were doing, and, as is common amongst business men, I added two and two together, and I shall not be far wrong if I say that out of the amount named the directors will hold amongst them upwards of £400,000, Lord Egerton heading the list. This alone, one would think, would lead to great watchfulness and care. With all their capacity and their untiring industry the directors are human, and to err is human.   There can be no great work like this conducted without errors, and it is to be hoped that the further help on the Board will be of great service in the difficult task of watching and wisely directing every department of labour.   I for one rejoice to see such an addition to the Board as has been appointed by the Corporation.   The experience of the new directors must be of value, and I think will be appreciated by the older members of the Ship Canal Board.   Friction, however, has evidently been felt in the beginning, and I can understand that gentlemen who have been fighting a hard and terrible battle for several years, and who have again and again been called upon to put their hands into their pockets and find £10,000 or £20,000 each, as circumstances demanded, are sure to feel sensitive when new minds and thoughts are brought to bear in their deliberations and discussions ; but men of the character and worth of the old directors, together with those engaged in municipal and other public works, tried and trusted men of judgment and business experience, can and must exercise forbearance, and by this means bring to completion a work that will be the pioneer of many great schemes.   They will thus make a mark in the annals of our city which the historian of another day will make memorable to our children and to our children's children in future years.   Meantime, they have the high enjoyment of those who have done a noble work, giving employment to the toiling artisans, and opening the way to further enterprise and the upbuilding of our great city.

> Yours, etc.,
>
> REUBEN SPENCER.

MANCHESTER, 8th December, 1891.

Other shareholders went further, and not only justified the old Ship Canal directors but deprecated the attacks made on them.   They said that Aldermen and Councillors whose estimates had been so absolutely wrong in regard to the Victoria Hotel, the Town Hall, and Thirlmere, should be very chary in using hard language to honest and patriotic citizens who were risking their own money, and they charged Sir John Harwood with wishing to wrest the Chairmanship of the canal from its present occupant.

It is singular, but true, that from beginning to end, those who have been the most bitter in attacking the management of the canal have had the least at stake, whilst the large shareholders as a rule have placed implicit confidence in the directors, feeling sure that they were doing their best for the concern.

The Manchester Press as a rule did not commit themselves.   The *Courier* and *Examiner* were apologetic in tone, but the *City News* lashed out furiously :—

For the scandalous mode of working, which has brought upon the enterprise its present state of danger and disgrace, the directors are to blame, and cannot be acquitted of culpable and contributory negligence.   They had been paying 2d. a ton for buying coal, and this alone amounted to £4,333 a year.   They had been paying eight or nine medical men £400 per annum each to look after the hospitals, etc., with a man over them who got £1,000 per year. They had been paying 25s. 3d. per thousand for bricks that could be bought for 18s. 6d., and when Sir John Harwood carried a resolution to buy certain articles by contract the Chairman had used his position to prevent it being carried out.

The article ended :—

Lord Egerton and Sir Joseph Lee must go. They must take with them two others at least. The Corporation must fill the vacancies.

After the Council meeting the adoption of the minority report was communicated to Lord Egerton, and in reply he asked for an interview between a Ship Canal deputation and the Special Committee. At that meeting an arrangement was come to, whereby in future the executive or spending department of the Ship Canal should be composed of four directors from the Manchester Corporation and three shareholders' directors, who should have power to carry out all works, order all materials and do everything that was necessary to complete the canal. The Chairman of this Committee to be a Corporation director.

This practically settled all differences and placed the spending of future money under control of practically a Corporation Committee. At the first meeting of the new Committee Sir John Harwood was elected Chairman.

The idea of forming a Shareholders' Association, as proposed by Mr. Belisha, was carried out, but when a meeting of those favourable was called, there was a curious scene. It was moved and seconded that Mr. Reuben Spencer should take the chair, but this could not be, for Mr. Belisha had anticipated him, taken possession of the chair and refused to give it up. This caused an uproar in the room, and one speaker said they were making a very bad beginning if they were going to guide Ship Canal affairs and could not manage their own. Mr. Spencer's tact got over the difficulty; he begged the meeting to hear Mr. Belisha's statement, and when that gentleman had ended a long speech he vacated the chair which Mr. Spencer took. Mr. Belisha explained he had called the meeting and was paying for the room, and that was the reason he claimed to take the chair. The meeting discussed the action of the Manchester Corporation in stopping the payment of interest during construction and in seeking to obtain additional directors on the Board, and a Committee was appointed to watch the interests of the shareholders.

Referring to a previous monetary fix, when the public did not respond freely and when it became necessary that a large amount of unissued shares should be taken up in order that the remaining debentures could be issued, a meeting of directors and leading shareholders was held, and though the shares could have been bought much under par in the open market, yet there were found public spirited men who vied with one another in taking up the shares at par, and thus getting the directors out of a difficulty. This was a fine instance of local patriotism, and deserves special recognition.

Amongst the helpers was the late Mr. Hilton Greaves, who piled money and antiquated cheques on the table from all his pockets till they amounted to some thousands of pounds. This was quite characteristic of the man. Having once decided the canal was on right lines and would benefit the country, he rather rejoiced in a hard fight that would need some personal sacrifice. The canal never had a truer or more courageous friend than Mr. Hilton Greaves. I here insert a letter I had from him during the year, addressed to me as Mayor of the City :—

DERKER MILLS,
OLDHAM, 1st *December*, 1891.

DEAR MR. MAYOR,

　　　　I was quite gratified by observing that you spoke at the Scotch dinner last evening of your continual confidence in the Ship Canal, that you think of it as you did nine years ago. I have entirely the same feelings, and so have very very many people. I see that your Corporation had a long meeting yesterday and that it was adjourned unto to-day. I do not in the least regret taking the large number of shares I hold. I would do it again even if I had foreseen these somewhat untoward events, which enterprises of this kind are, under even the wisest, subject to.

Nor do I regret having induced many to support and take shares in this *glorious* under-taking—a noble work putting altogether in the shade gifts of public parks and endowments for even this or that charitable institution of science or art. In the future it will be thought of with admiration, like the Roman Appian Way and the Roman aqueducts.

There is one view, I fear, lost sight of, *viz.*, the interests of the original shareholders. The Corporation will protect themselves in their first charge (after debentures), which is not difficult, and then will come the preference shareholders and *lastly* the shareholders, very many of whom put their little *all* from the best of motives into the scheme.

Now I certainly do feel for these, and what I am driving at is, that inasmuch as they will be the *last* at banquets of dividends, these said shareholders are fairly entitled to the *principal* share of the management.

It seems to be drifting into the Corporation having much management ; this should hardly be so.

The Corporation should certainly have the right to *see* that the money they advance is spent in a substantial and useful manner on the undertaking.

What has come of the members of the Consultative Committee who sat on the under-taking in the early part—would it not be well they were called together informally?

Yours truly,
HILTON GREAVES.

Use this letter as you think fit.

PROGRESS OF THE WORKS.

Returning to the progress of the works, the clearing out of flooded cuttings was a tedious process notwithstanding the powerful pumps used—the one at Latchford raised 32,000 gallons per minute. The railway bridges made rapid progress and became prominent features in the landscape, as also did the mounds raised for railway deviations.

The first work of any magnitude to be finished was the Weston Marsh Lock, which connected the river Weaver with the Ship Canal; it was 315 feet long 42 feet 6 inches wide and 16 feet deep, and had gates made of greenheart. This lock, after being tested by the Weaver Trustees, was opened in the middle of April, Mr. Platt, Chairman of the Works Committee, and other directors being present. The opening was quite a success.

During May and the earlier part of June tremendous efforts had been made to clear out the Eastham section and get it ready for filling. Nearly 3,000 men were engaged in relays night and day. And there was great reason for this haste, because, till the section was open and access given to Ellesmere Port, the temporary entrance through the Ellesmere Port embankment could not be closed. By Wednesday 18th June all the plant had been removed and the slopes were practically finished. Great care was needed to prevent a rush of water doing damage. Next day the Works Committee met at the Eastham end, and their Chairman, Mr. Platt, set the last coping stone of this important section. To prevent a crowd the opening was kept a secret. Mr. Leader Williams had some hundreds of men at work making final pre-parations and cutting a hole in the dam (with boards to regulate the aperture) that separated the canal from the estuary. This dam was at the Ellesmere Port end and had been built to keep the tide out of the cutting. When the work was finished there was a time of great excitement. The navvies stood with their tools in hand watching the rising tide come nearer and nearer the opening. Then it gently lapped over and sent a sheet of muddy water into the vast opening. In an instant the men and visitors joined in a rousing cheer; hats were thrown in the air and for a quarter of an hour the locomotives never ceased sounding their whistles. Each day the water

S. R. PLATT, DIRECTOR OF THE SHIP CANAL; CHAIRMAN OF THE
WORKS COMMITTEE, 1887-91.

To face page 108.

Birtles, Warrington.

LETTING IN THE WATER AT ELLESMERE PORT. WITH ENGINEER ON THE BANK.

To face page 108.

was admitted through the cross dam for about an hour before and after high tide at the rate of about 3 feet per day. In this cautious way it took more than a week to fill the section—a rush of water might have damaged the banks. The next step was to open the canal to Ince, all ships for Ellesmere Port would then have to come up the Ship Canal. The first vessel to actually use the canal was a small launch that was lifted by a crane over the bank at Eastham for the use of the resident engineers.

Many alarmist statements crept into the Press to the effect that the slopes of the canal had given way, but they were at once contradicted by Mr. Leader Williams. It was only to be expected that some slips of earth would occur, but beyond these, and one case where some heavy pitching sank, all went on well, and the damages were repaired at a small cost.

The banks had a further trial, for no sooner had the section been filled than heavy rains set in and cut gullies in any soft earth on the banks, but no serious damage was done. It was amidst this downpour that the Ribble Commissioners, including Admiral Sir G. S. Nares, and accompanied by the engineer, Mr. Leader Williams, visited the works. During the blinding rainstorm the engine drawing the party crashed into a waggon near Ellesmere Port and nearly caused a serious accident. The visitors escaped with a shaking, but the inspection came to a summary end.

During the progress of the works, and in addition to official visits, I often made inspections on my own account, taking a length of 10 or 12 miles at a time, and thus I made myself thoroughly acquainted with the work. One day, accompanied by some ladies and children, I started from Eastham to walk to Runcorn. When we reached the embankment extending across Ellesmere Port bay, 2,033 yards long, I knew great care must be taken, because it was only wide enough for one line of rails; so selecting the dinner hour when work was suspended, I arranged with the lad at the points that we were to have time to cross before the next train of ballast was sent on. To our dismay, when we had walked nearly half the distance, a train came thundering along. Somehow the lad had forgotten us, and there seemed no way of escape. However, I spied in the distance a small wooden projection and some upright poles that had been used for pile-driving. To these we rushed and had just time to get the ladies on to the timber which overhung the deep water, and the gentlemen to the poles, when the train overtook us. Our frail structure quaked so much that it was difficult to hold on, but fortunately we got nothing worse than a severe fright. One and all, however, determined never to trust our lives again to the memory or judgment of a boy at the points.   It was indeed a very close shave.

Travelling the length of the canal in June one noticed the efforts being made to secure stability in the banks.   A number of Dutchmen had been imported to teach the best way to plant and plait the banks with willows, which now seemed to be growing vigorously.   On lengths where dredgers could be employed, a system of filling from the buckets into iron receptacles on rafts had been introduced.   When full, these receptacles were taken to lifts and tipped over the banks on to low land adjoining.   At Barton, Bardsley's old corn-mill had gone; so had the road bridge; only Brindley's aqueduct remained.   A large pier was just finished on which both the new bridge and the new aqueduct were to swing on a circle of rollers.   One noticed that the red sand-stone in Trafford Park was a disappointment when exposed to the weather; veins or layers of soft sand had developed, and what appeared at first a bold face of rock required in time a great deal of patching up with bricks to fill in the soft places.

But the chief effort of the directors was to open the canal to the Weaver Docks, because until this was done they were prevented by Act of Parliament from dealing with the length from Weston Point to Runcorn.

The *Manchester* dredger was working energetically to open the passage from the Sloyne to the Eastham dockgates, and by the first week in June the way was clear from the sea to the outer lock.   It was a proud moment therefore when Mr. Samuel Platt, in his yacht the *Norseman*, entered the canal through the Eastham Locks.   She was the first vessel of any size to navigate the canal, and there was great excitement, when, with her owner and Mr. Leader Williams aboard, she quietly passed through the locks and anchored inside.

The next step was one of great risk and magnitude, *viz.*, to close the passage through the Ellesmere Port embankment left open to give access to that port, and then, after filling the section with water, to dredge away the dam at the head of the Eastham section.   In order not to obstruct the Shropshire Union traffic it was arranged this should be started on a Saturday and completed on Monday.   The tide for months had established a set route through the 250 feet opening in the embankment, and to close this aperture in a few hours was no ordinary task.   An army of men stood ready with many train loads of material at hand to be tipped into the aperture at low water, in hope that before the returning tide the new bank would be able to resist it.   It was a fight against time.   An effort was to be made to have the opening closed and the dam cleared away early on the Monday afternoon, so that the

INTERIOR LOCK GATES, EASTHAM.

Barningham.

To face page 110.

Barningham.

Eastham Lock and Sluice Gates at High Water.

Norseman at the head of a string of Shropshire Union barges might sail up to Ellesmere Port. But the engineers and directors had to learn that human hopes are frequently but "vanity and vexation of spirit".

On Saturday, 11th July, as soon as the Shropshire Union traffic had passed out and the tide was low, train loads of ballast were emptied as quickly as possible into the chasm from both sides. The noise and smoke were almost unbearable; all the men were working at high pressure, and slowly the heap rose above the level of the water. People walked across what had been a deep gulf. There was great elation that the deed had been done so well and so quickly. But the test had to come, and the engineers, though confident, were anxiously awaiting Neptune's attack. Grievous was their disappointment when, as an old salt said, "she (the tide) rubbed her nose against the bank and down it came" At two o'clock on Sunday morning, to the dismay of everybody, the tide forced a small hole, which quickly became a large one, and then played havoc with all the new work, sweeping it away like a pack of cards and repossessing itself of the old passage just as if nothing had happened. I well remember receiving a telegram of the failure and hurrying down on the Sunday morning to the scene of the disaster. I had to cross to Birkenhead, take the tram to New Ferry and walk thence to Ellesmere Port. When I got there despair sat on every face. The new route to Ellesmere Port must be opened by Monday afternoon under a very heavy penalty, and the sea had set at nought the calculations of the engineers. A council of war was held to determine the best way of coping with the tidal invader. It was decided to repeat the attempt, but if possible on a more solid basis. This time, instead of tipping soil and clay, it was arranged to start with heavy boulders, and these were searched for in neighbouring spoil heaps by hundreds of men. Thousands of tons were cast into the gap and heavy piles were driven to prevent them moving about. On the top of the boulders thousands of tons of clay and soil were tipped. From early morning on Sunday till 10 o'clock at night a grand spectacle of perseverance, energy and skill was exhibited. By the time the breach was again made good, every one was thoroughly worn out. All day long engines had been running on lines that were scarcely fastened down after being placed, and if, in consequence, a waggon tipped over along with its contents into the chasm it had to be buried, no time could be wasted in recovering it. When the next high tide was due the bank had been restored, every effort had been made to strengthen it, and there were good hopes of the structure standing the strain. But all was of no avail. "Vain was the help of man." The sea would not be denied and again forced its way up its usual channel, defying all attempts to stop it.   It tossed the bank over and washed away the huge boulders as if they had been marbles, and after a stiff duel the engineers had to admit themselves thoroughly beaten.

This second mishap prevented the usual packet (which had never missed since 1837) and its attendant tugs getting to Ellesmere Port, and the penalty for stoppage was £300 per day.   By the kindness of the Shropshire Union Canal Company it was arranged that the goods should be delivered at Eastham and taken by rail to their destinations.

At a subsequent meeting of the engineers it was determined, as the place could not be taken by storm, to lay regular siege to it, and to stop up the channel in a more gradual way.   Hitherto the empty canal had been a kind of inner vacuum and the pressure outside had met with no resistance.   It was determined now to proceed by stages and to build up the embankment with layers of concrete and suitable material, and then by raising the water in the canal to secure a uniform pressure on both sides.

The final effort commenced at 4.30 on Tuesday, the 14th July, and went on unceasingly for some thirty hours when success was achieved.   A gantry with a railroad on it had been placed over the gap, and this materially assisted operations, as from it the concrete was dropped into the aperture.   To break the force of the tide, dredgers and tugs had been moored outside in the estuary.

These mishaps succeeding one another in rapid succession created much consternation in Manchester, and it was a relief when the Mayor assured the Council that the difficulties were only temporary and would soon be overcome.   On the night of the 16th was issued the bulletin :—

First flotilla of traffic from Ellesmere Port passed down the Ship Canal into the river Mersey at 8.45 this morning.   Time occupied in passing through the Eastham Lock, seven minutes.   Ellesmere Port traffic will henceforward pass down the Ship Canal.

When the dam that separated the Eastham and Ellesmere sections had been cut through sufficiently to allow the *Earl Powis* to pass up stream there was a roar of welcome from all sides.   Mr. Leader Williams received quite an ovation.   The opening was a great event, but unfortunately it did not bring much grist to the mill, inasmuch as ships up to 400 tons are entitled to the free use of the canal as far as Ellesmere Port.   Previously they could only get there when the tide served.   The next move was to complete the canal to Weston Point, and to that end men were working day and night near Ince Hall, where there is a deep rock cutting.   And here occurred a disaster which will never be forgotten in the neighbourhood, which swept ten men

THE GAP, ELLESMERE PORT.

Birtles, Warrington.

To face page 112.

Birtles, Warrington.

INCE.

To face page 113.

SLIP AT INCE, 24TH APRIL, 1891.

Birtles, Warrington.

To face page 113.

into eternity and injured as many more.   The cutting is perpendicular and about 60 feet deep.   Ballast waggons were constantly bringing the rock by a circuitous route from the bottom to spoil tips on the top.   Along the top of the cutting was a siding for empty trucks, with a dead end, and not far off were points where a lad stood, who, by altering the lever, sent the trucks in the direction required.   It so happened that on the morning of the 18th July, night gangs were in the cutting, drilling and chipping the rock by the aid of the lucigen light, and one gang of twenty men was directly under the truck siding.   By accident, or through carelessness, the lad at the points (17 years of age) turned a train of twenty-three trucks, drawn by two engines, into the empty truck siding instead of on to the line to the tip at Ellesmere Port.   These came crashing along, and charging the dead end of the siding fell over into the hollow below, right on top of the gang at work immediately beneath.   Engines, trucks, stone and men were in one almost inextricable mass, lit up by the lucigen light. The scene was appalling, and the shrieks of the injured and dying were awful.   Men rushed to the rescue, and by the aid of steam cranes released those still alive, who were promptly conveyed to various hospitals or attended to on the spot by medical men.   Many who were not killed were maimed for life.   Fortunately the six men on the engines jumped for their lives and escaped with a few bruises.   The lad at the points who caused the disaster was arrested and charged with manslaughter, but was eventually discharged.

Another misadventure which delayed the work occurred on the 25th July.   A slip of about 150 yards of bank took place between Ellesmere Port and Ince.   It brought with it a mixture of clay and peat like a travelling bog and completely submerged the contractor's line in the cutting, besides severing the overland route on the top of the bank.   Hereabouts the slimy blue marsh clay had been most difficult to deal with, and this was the third slip in the locality.

Gradually length after length of finished canal was filled with water.   On 1st August, by means of shoots, the Irwell was turned into the Little Bolton cutting, and then any one standing on Brindley's aqueduct could see the completed Ship Canal on either side of him as far as the eye could reach.

The next important work was the filling of the Ince section, commenced on the 14th September, which took five days to fill.   The dams were then cut away by dredgers, and a direct course for the salt trade was thus opened from the Weaver to the sea.   The filling was effected by means of a pipe 90 feet long and 4 feet wide, which was run through the dam of the full Eastham section, the flow of water being regulated by sluices.   In consequence of the wretched ground about the Gowy the banks of this section had in places been left very flat, and there was no further giving way during filling.

On the 27th the steamer *Fanny*, with some of the directors on board, sailed from Liverpool through the length of open canal, amidst many signs of rejoicing, and entered the Weaver through the Weston Marsh Locks.   Eleven consecutive miles were now in use, but it was not expected they would yield much revenue, in consequence of the nominal toll and other onerous conditions that had to be submitted to in order to get the 1885 Act passed.   Vessels up to 500 tons, carrying salt only, were exempt from dues and tolls, as were also all vessels light or in ballast.   Similar ships carrying other merchandise, and ships bound for the Potteries, paid half the authorised Ship Canal tolls.   Thus shipowners got a canal open every hour of the year and a safe navigation, in lieu of an uncertain and, at times, dangerous river.   The canal has certainly been a godsend to the Cheshire salt trade.

On the 1st November a mishap occurred at the dam dividing the canal already opened from the Runcorn section.   This was magnified into " A great disaster" by the Liverpool papers, and it might have been so if prompt steps had not been taken to repair the mischief.   The dam had been cut down to pass some water into the upper section for sluicing purposes, when it was discovered that the top of it was being carried along with the water.   It became necessary at once to open the Weaver sluices and so reduce the pressure.   This quickly lowered the water in the open canal 3 or 4 feet.   Fortunately no big ships were in the way, and the dam was quickly repaired at a cost of about £100.

The first foreign ship to come up the canal to Saltport, at the mouth of the Weaver, was the Norwegian timber ship *Deodata* from Shediac, N.B.   She arrived on the 10th November, the consignees being Messrs. Pierce, Watts & Co., Liverpool. Her draught was 16 feet 6 inches.   She was towed up by the Canal Company's tug. The *Deodata* took back a cargo of salt.   Before she left, the directors of the Ship Canal Company presented to her owners a handsome clock and barometer bearing the inscription that she was the first foreign vessel to load and unload in the canal.

No sooner had the new Executive Committee been appointed and harmony restored, than the works were visited by another disastrous flood.   During the first fortnight of December there was continuous and heavy rain, and on the 14th December the combined Mersey and Irwell broke into the canal cutting at Irlam, filling it for three-quarters of a mile, to a depth of 20 feet, and submerging a steam navvy and

EASTHAM LOCKS COMPLETED.

Birtles, Warrington.

To face page 114.

other rolling stock that could not be rescued in time. Fortunately the greater part of the plant had already been removed.

The Bollin also broke into the canal at Latchford, doing much damage. The total length flooded was only 2⅓ miles, which compared favourably with the 18½ miles flooded in the previous year. Of course this caused delay and expense. Part of the water was speedily pumped out and the remainder was left in till the spring.

The end of the year saw the canal opened to the Weaver and a new Port (Salt-port) created, and doing a fair amount of trade in iron, timber, resin, salt, etc. The Runcorn section was the most behindhand, but vigorous efforts were being made to complete the locks and the long embankment in front of Runcorn. All the other locks were practically finished, and the railway deviations and approaches had made rapid progress. About 5 miles of the canal between Manchester and Warrington were filled with water, and it might be said that half the length of the canal was completed.

Whilst the work was in the contractor's hands there were practically no floods to contend with, but directly the Canal Company were responsible, disaster followed disaster in rapid succession.

ACCIDENTS.

Many and various were the accidents of this year, and no doubt the liberality of Mr. Walker in providing commodious hospitals and a good staff of surgeons ameliorated much suffering.

At Ellesmere Port a man threw some lucigen oil on an open fire and caused the death of a workman sleeping close by. At Weston Point the chain of a ponderous hammer snapped, and the hammer swinging round killed a man who was not known to any one. Near Barton a man was crushed to death between a steam navvy and the rock, and when the driver of the former was questioned, he coolly told the coroner that all the workmen were supposed to look after themselves, and that no warnings were given.

It was evident that most of the accidents on the canal were caused through drink, and the police were directed to keep a strict eye on shebeening. The result was that the owners of huts were often brought before the magistrates. It was necessary to adopt all kinds of expedients to catch the transgressors and obtain a conviction. In one case a fine of £60 and costs was inflicted on an old offender, and even then the trade was too profitable to be given up.

VISITORS.

On the 9th June the Tyne Commissioners went over the works. They travelled by the saloon to Runcorn, and at luncheon Alderman Cail, a veteran of 81, expressed on behalf of the visitors their pleasure and astonishment at the progress of the works, and ended by proposing success to the canal. The party afterwards went in workmen's trucks to Ellesmere Port and then by boat *via* Eastham to Liverpool.

On the 31st July the Society of Mechanical Engineers visited the canal. They started from Pomona, and when they got to Eastham had only little time in which to see the locks before the last boat left for Liverpool. But they were anxious to see the locks, so armed with permits from Mr. Leader Williams they presented themselves to the navvy on duty at the lock gate. The engineers thought they would only have to present the permit and write their names, etc., and all would be right. But they reckoned without their host. The sun-tanned fellow on duty, who had a wooden leg and a squint in his eye, had never heard of the Mechanical Engineers or their Conference, so he stretched his wooden leg across the bridge and bade them begone. The engineers were indignant and thought to influence him by producing cards and credentials, but all in vain. The navvy stood like a modern Horatius Cocles holding the bridge alone, and no one dared to force a passage. "It's no use bringing books and cards to me," he said. "Anybody could do that, and I have seen them before. You cannot pass here. I've got my orders." The engineers tried kind words and explained who they were, but it was like talking to a stone wall. Then they got angry and stormed, but he would not budge. "I've got my orders." Engineers might be able to control a mechanical navvy but they could not deflect a living one from what he thought the path of duty, so they had to give up in disgust and make for the boat.

ARBITRATIONS.

The last day in July Mr. A. Garrard, as umpire, opened an inquiry in London as to the amount of compensation to be paid to the Chilworth Gunpowder Company for their property at Ince. Messrs. Vigers and Dunlop were arbitrators. The claim originally was for £100,000, but the valuers for the vendors placed it at £57,993. From the evidence it appeared that in December, 1887, the Pulver Fabrik Company in Germany, the then owners, were served with notice to treat and they asked £10,000. Subsequently the present claimants bought the property for £1,500, subject to the vendors continuing to store some powder there, and a second notice to treat was given to them.   It was urged now, that in consequence of Government restrictions, it was next to impossible to get storage in the neighbourhood.   Hence the increased value. This was a continuing licence to store 470 tons of gunpowder, and it was impossible to obtain a similar licence for new areas.   There was no room for additional powder hulks in the river Mersey.

On 10th August the Queen's Bench were asked by the Ship Canal Company to revoke the arbitration on the ground that improper evidence was being accepted as to the value of the interest covered by the first notice to treat, and that the owners were aware of this notice having been given to the former owners.   The request was refused and the arbitration ordered to proceed.   Eventually the Chilworth Company were awarded £15,860, this to include the stock of powder, valued at £1,100.   The Ship Canal Company offered to provide a new site for magazines near Pool Hall, but the offer was not accepted by the claimants.

CHAPTER XXII.

1892.

NEW EXECUTIVE COMMITTEE AT WORK—DEBATES IN THE CITY COUNCIL—SHAREHOLDERS' ASSOCIATION—ISSUES OF 3 PER CENT. STOCK—PARLIAMENTARY BILL—DISTURBING RUMOURS OF MONEY RUNNING SHORT—ENGINEER'S REPORT ON ESTIMATES—COUNCIL INDIGNANT—MR. MOULTON'S OPINION TAKEN—SALFORD AND OTHER TOWNS READY TO PROVIDE CAPITAL—PROPOSED LAIRAGES—MR. GLADSTONE'S VIEW OF THE CANAL—DESCRIPTION OF WORKS—FLOODS—SALTPORT.

EXTRACTS.

It is impossible not to follow the developments of the Ship Canal with interest. The great undertaking has a very exciting history. If ever its story comes to be written by an impartial historian it will be found to have had as many twists and turns as the plot of a sensational drama.—*Manchester Evening News.*

THE new Executive Committee lost no time in overhauling the affairs of the Ship Canal Company with the view of deferring work not absolutely necessary and of effecting all possible economies.

Originally there were two sets of engineers on each length, one to represent the contractor and another to check and supervise on behalf of the directors, and this arrangement had continued since Mr. Walker's death. One set was now dispensed with, and notice to expire in March was served on the whole of the chief officials (except Mr. Leader Williams). This gave the directors a free hand and enabled them to re-engage the best men.

At the January meeting of the City Council Mr. Clay opposed the issue of the remaining moiety of the £3,000,000 loan on the ground that a report giving the whole cost of the work had not been presented. Mr. Gunson wanted to insert in the new Bill a clause providing for a Corporation majority on the Board.   Sir John Harwood explained that the issuing of the loan did not stand in the way of the proposed check on expenditure, and said that the Corporation could not add clauses to a Bill not their own, but that they intended to petition against the Bill with the view of obtaining voting power for the debentures issued, and a larger representation on the Board.   He assured the Council he knew of nothing that ought to disturb the public mind.   The Ship Canal directors had met him in a very kind spirit and given every possible assistance in the arduous work in hand.   It was difficult to revise a settled policy; they were making economies, and had to complete the canal whatever the cost.   Some people wanted to wind up the company, but he assured them that meant to write on all goods shipped, " Manchester has fallen," instead of the proper watchword " Excelsior".   Their commercial character would be ruined if the enterprise were to fail.   He had nothing to conceal, and he believed in time everything would turn out satisfactorily.   When the General Purposes Committee met at the end of the month a Sub-Committee was appointed to take charge of the petition against the Bill.   Mr. Gunson carried a resolution that the Town Clerk be instructed to insert a clause to secure eight Corporation directors out of the fifteen to be appointed, and said they might yet have to find £6,000,000 or £7,000,000 for the canal, and that so far the Council had been hoodwinked.   He could not, however, give any authoritative ground for this statement when asked to do so.

On the following Wednesday the minutes of the Special Ship Canal Committee and the General Purposes Committee came up for confirmation.   Singularly, whilst the former contained the opinion of Mr. Moulton, Q.C., that Parliament would certainly rule out of order a new clause for additional Corporation directors, the General Purposes Committee had instructed that such a clause should be asked for.

The Mayor, on opening the proceedings, pointed out the difficulties created by violent and random speeches, especially while negotiations for a loan were going on with the Bank of England.   He showed how irresponsible members had damaged the credit of the canal, and how veiled hints had already depreciated the shares.   He urged the members of the Council not to indulge in them but to be discreet in their utterances.

Sir John Harwood believed everything was now being done at the proper time and in the proper way, and he asked the Council to adopt the proceedings.   They had not been parties to the Ship Canal Bill, and decency forbade them having any-thing to do with it.   He defended the action of the Special Committee in seeking Mr.

Moulton's advice.   Personally he was in favour of increased and permanent representation on the Board, inasmuch as they were going to be the saviours of the concern, but he deprecated trying to force the hands of the Canal Company when they were advised that Parliament would not permit it.   They ought to get at the cost of completing the works.   For seven-tenths of the work yet remaining they were going to get estimates, for valid reasons they were doing the rest themselves, and the Council ought to await the completion of their estimates and entrust them with the task of trying to get favourable clauses in the new Act.   Councillors Williams and Gunson defended the position they had taken up, but felt disposed to accept Sir John Harwood's explanation.

Councillor Southern regretted the random statements of Mr. Gunson when he said the Council had been misled and that a debt of £6,000,000 or £7,000,000 would be placed on the city.   Any one who made a statement like that ought to consider the damage it might do.   When the Finance Committee wanted to negotiate a moiety of the loan with the Bank of England they found the Bank had heard of it.   It had disturbed the public mind and had had a grievous effect upon the credit of the city and the property of the Ship Canal.   The Corporation directors and their colleagues were doing all in their power to put affairs on a right basis, and so far there was nothing to justify the conclusion that half a million or a million more would be required.

Alderman Chesters Thompson said if the former extravagant management had been allowed to continue it would have entailed a cost of £5,000,000 or £6,000,000 on the Corporation, and he recounted some of the savings that had been effected.   Eventually the proceedings were passed, and it was arranged that Mr. Moulton should visit Manchester and give both Committees an opportunity of hearing his views.

On the 6th February Mr. Moulton, Q.C., attended a joint meeting of the Consultative and Parliamentary Sub-Committees.   Some of the Councillors became very offensive in their language towards the Ship Canal directors and had to be called to order.   Great soreness was displayed because of the increased estimates.   Mr. Moulton then spoke and said that hard words and hostility would never secure increased representation on the directorate, which must be obtained by conciliation and reasonableness.   His advice met with general approval.   Sir John Harwood, though present, took no part in the proceedings.   At a subsequent meeting it was arranged to follow Mr. Moulton's advice and not oppose the Ship Canal Bill.

The newly appointed Ship Canal Shareholders' Association, under the presidency of Mr. Reuben Spencer, was soon on the war-path to defend the shareholders' interests. A correspondent expressed their feelings when he wrote :—

It is all very well to stop leakage and waste and to put financial matters on a good footing; that was what the Corporation directors were chosen for, and they have done their work well and nobly so far, but it was quite another thing to bodily take the whole concern. True, the Corporation lent the £3,000,000 generously, but they made good and safe terms, and they said so at the time. They had £8,000,000 to fall back upon, and good interest in the meantime. Why do they claim supreme control? that belongs to the shareholders.

Another correspondent objected strongly to Sir John Harwood saying that if it had not been for the Manchester Corporation the Ship Canal Company would have been in liquidation. He called attention to the fact that Salford and other Corporations had expressed themselves willing to assist, but their help had been declined.

The courage, the patriotism, the generosity and self-sacrifice of the original promoters and shareholders of the Manchester Ship Canal have no parallel in modern commercial history, and although at present stricken down by misfortune, the same indomitable spirit which inspired Daniel Adamson will help us onward with at least a determination and satisfaction to see the work done.

Sir John Harwood had credited the shareholders with "splendid loyalty and patriotism to the city of Manchester," the correspondent continued, and it was to be hoped "that whilst he (Sir John) dutifully guards the interests of the ratepayers, he will at no period of an honourable municipal career stain his reputation by being a party to dishonouring our noble city at the expense of those without whom we would still have had 'the old river' and not the great Manchester Ship Canal".

The Shareholders' Association at their next meeting asked the canal directors for information as to past and future expenditure, and desired to have a progress plan of the works placed in the Town Hall. A public trust not being part of the policy of the Association, the *City News* quickly told them they were "backboneless, and in the air, and as they had nothing better to offer, the sooner they committed the happy dispatch the better". They also incurred the wrath of Mr. Joseph Scott (author of *Leaves from the Diary of a City Auditor*), because they would not support him in his endeavour to become a Ship Canal shareholders' auditor. He prophesied "they would be defunct in a very short space of time".

In order to judge for themselves, the Shareholders' Association made an inspection of the works from Eastham to Manchester, and expressed themselves satisfied that every effort was now being made to complete the canal speedily and with efficiency.   Mr. Leader Williams, who accompanied the Association, in responding at luncheon to the toast of his health, said he had £6,000 of his own money invested in the canal, and he thought that ought to be a guarantee that he was personally interested in its success.

When the canal got into financial trouble, unpleasant remarks were made about the contractor and the Ship Canal directors.   Hints were thrown out as to the reason why the former did not complete the work, and as to the incapacity displayed by the latter in first entering into a vague contract, and then allowing it to be broken.   Consequently, on 26th February, Mr. Perks, solicitor to Mr. Walker's trustees, wrote to the papers to make the matter clear.   He said the financial troubles of Messrs. Baring did not affect Mr. Walker and his contracts.   It was not true Mr. Walker was unable to make his repayments for plant.   He had paid the directors £194,895, and only refused to make further repayments when they repudiated certain contractor's claims which they afterwards admitted.   Mr. Walker had paid cash for the £500,000 in shares he had promised to take up.   Half of these he sold at a great loss, and the remainder were transferred to the directors as part of the bargain when, by mutual consent, the contract was cancelled.   There was no ground for the statement that Mr. Walker's executors were unable to carry on the works.   Mr. Walker and his executors had been charged with mal-administration.   During Mr. Walker's lifetime, work to the value of £1,914,864 had been done in two years.   In the eleven months after his death, work which cost £1,558,077 had been done up to the time of transfer, and at a rate of progress unequalled since.   No serious accident had happened whilst Mr. Walker or his executors, were in control.   They had not to design the work but to execute it, and had nothing to do with quantities or prices : indeed they were of opinion the latter were inadequate, and time had proved it.   The only interest Mr. Walker's executors now had was in the realisation of the plant, and they had reason to complain this had been placed in the hands of other contractors without due provision for repair and maintenance.   Mr. Walker's executors had rendered the company all the aid in their power, and were satisfied with the way in which they had been treated by the directors, but as erroneous statements had been made, they wished to place themselves right with the public.

At the end of February the directors issued their report, explaining the arrangements made with the Corporation as to an Executive Committee, and showing how the progress of the works had been impeded by various catastrophes.

A report was also issued by the Ship Canal Stores Committee demonstrating the great savings made by the tender system they had inaugurated, also stating that they were abolishing, as unnecessary, many of the store departments, thus saving £2,000 per year.   They claimed, too, that they had put an end to a buying commission on coal and were buying it more cheaply direct, also that they were saving 20 to 25 per cent. on many articles purchased under the new system.

The Finance Committee now decided to ask the Council for power to advance the remaining moiety (£1,599,400) to the company, and to negotiate with the Bank of England for the issue of a 3 per cent. redeemable stock.

When the shareholders' half-yearly meeting was held, it was stated that Sir Joseph Lee, the Deputy Chairman, was prostrated through anxiety and overwork, and had gone (under doctor's orders) to the Canary Islands.   There was a wide-spread expression of sympathy, and of hope for his speedy recovery.

Lord Egerton explained the delays and difficulties that had impeded progress in the canal.   There had been delay in the Weaver section and this had stopped the Runcorn works; then they had had a dispute as to foreshore ownership to contend with.   The Mersey Conservators had both blocked the works and added to their cost, as also had the Dock Board; and the Canal Company had been obliged to go to Parliament for relief.   They were also being delayed by the onerous conditions imposed on them by the London and North-Western Railway.   To crown all this, disastrous floods and storms had come upon them, such as had never occurred in Mr. Walker's time.   Last year they had fifty days of solid wet weather that had stopped the works and inflicted severe loss.   These and other unforeseen causes had increased the estimates by £863,000.

Mr. Bythell followed Lord Egerton, and in an able speech palliated or explained away charges of extravagant administration.   He was demonstrating to an impatient audience how almost all big railways had far exceeded the engineer's estimates, when in a stentorian voice, Mr. Mark Price (once a well-known and popular orator), speaking from the body of the hall, said, " Mr. Chairman, we have something else to do than to stay here all day listening to a discouraging history of past failures ".   Mr. Bythell pulled up at once, and turned to another subject.

After some remarks from Mr. Forrest, a member of the Shareholders' Association, the report was passed unanimously, and what was expected to be a stormy meeting ended harmoniously.   No doubt the felicitous speech of Lord Egerton under very trying circumstances was mainly instrumental in securing this end.   An impression had been prevalent that the Chairman was antagonistic to his Corporation colleagues, instead of which he held out to them the right hand of fellowship, and welcomed their advent. As he said, anything in the nature of recrimination or quarrel could only be injurious to the interests of all concerned.

When the reports of the Special and Consultative Committees came before the March meeting of the Council, they were passed, Mr. Gunson alone protesting.

On the tenders being opened for the million and a half of Ship Canal 3 per cent. stock it was found to have been applied for three times over. Tenders at £95 10s. received 94 per cent., and higher tenders the full amount applied for, the average price obtained being £95 14s. 4d. The issue was regarded as a great success. To show the intention as regarded any profit to the Corporation at the time the loan was made, I quote from the *City News* of August, 1892 :—

The Corporation magnanimously engages to hand back to the Canal Company, on the final closing of accounts, any profit that may have been made on the transaction, as between the costs of the loan to the Corporation and the interest paid by the Canal Company.

It was Parliament that afterwards prevented the Manchester Corporation from carrying out this generous intention.

At the adjourned meeting of the City Council, Alderman Lloyd moved that the resolution previously passed to prevent Corporation directors from sharing in the Ship Canal fees be rescinded. On its being seconded, Sir John Harwood said he did not wish to pose as a superior person. He objected at the time to the continued payment of interest out of capital whilst the canal was in difficulties, he was also opposed to the directors taking fees until the concern was remunerative. It was in the shareholders' hands to fix if any (and what) payment should be made to the directors, but seeing they had voted an honorarium, it would be monstrous for him and his colleagues to say they were not entitled to share in it. After Alderman Mark had stated that though he would work gratuitously for a philanthropic or charitable object, he was not disposed to do so for a public enterprise, the motion was carried unanimously.

When the Council met in May a resolution was unanimously passed that the Mayor (Alderman Leech) be requested to fill the vacancy in the directorate caused by the resignation of Councillor Boddington.

It was hoped to bring the foreign cattle trade to Manchester Docks as a source of revenue, but directly this was mooted, Salford became uneasy lest her market might be injured. The Board of Agriculture, under Mr. Chaplin, was also strongly opposed to adding to the ports of entry, on the ground of the risk to English cattle if infected cattle were allowed to pass up the country. In May an influential deputation from Manchester, accompanied by the Mayor, waited on the Board of Agriculture and asked for the same privileges for foreign cattle that were already granted to Liverpool, Glasgow and other ports. Mr. Chaplin promised to consider the matter, and confessed he had no idea of the size and importance of the Ship Canal, or of the meat trade round Manchester. His idea had been that cattle would have to be transhipped at Liverpool and conveyed up the canal in barges.

At one time it was hoped to avoid a Parliamentary Bill in the session of 1892, but this was found to be impossible. An Omnibus Bill was therefore lodged in due course. It sought power—

To construct roads at Barton and elsewhere for canal purposes.
To alter the position of a dam, locks and sluices at Warrington.
To abandon a bridge at Fairbrother Street.
To buy up to 100 acres of land for flattening slopes and other purposes.
To vary railway deviations.
To make a charge on barges moved or propelled otherwise than by animal power. Also on tugboats towing vessels on the canals.

It also asked for exemption from the payment of tolls to the Mersey Commissioners on vessels passing through the Eastham entrance.

After passing the Standing Orders Committee the Bill came on the 24th May before the Lords Committee with the Earl of Ducie as chairman. There were twenty-one petitions against the Bill, the Rochdale Canal Company being the chief opponents.

After Mr. Pember had opened the case, and Mr. Bythell and Mr. Leader Williams had given evidence, Mr. Arthur Sinclair, on behalf of the Mersey Commissioners, opposed the Bill on the ground that it would affect their revenue. A compromise was eventually effected, whereby all vessels going to Ellesmere Port, Weston Point, and Runcorn, should pay toll, and a minimum revenue of £2,800 per annum was guaranteed. When the clause—to charge for barges on the Bridge-water Canal towed otherwise than by animal power—came on for consideration, Mr. Balfour Browne said the Ship Canal Company had spent £112,000 on improvements to make propulsion by steam possible, and that the maintenance of the banks in the future would be very heavy in consequence of the increased wash from the steam-tugs. The Leeds and Liverpool and other canals did not object to the clause, but the Rochdale Canal, though it would cheapen their carriage materially, opposed it on the ground that the question was one which the Railway and Canal

Commissioners or the Board of Trade ought to decide. Here a curious incident occurred. At a critical moment when it was necessary that either Mr. Pember, Mr. Balfour Browne or Mr. Cripps should reply to an able speech made by Mr. Worsley-Taylor for the opponents, not one of the Ship Canal's counsel was available —they were all addressing other Committees. There was a deadlock. Mr. Coates, the Ship Canal Parliamentary agent, made the best apology he could to appease the irate Committee, and asked them to adjourn till one of the counsel was at liberty. But the Chairman said angrily, " We cannot allow this sort of thing ". After a few minutes they permitted Mr. Coates to make a few remarks. When the room was cleared the Committee were in the worst possible humour, and they quickly decided to strike the clause out of the Bill. The Rochdale Canal counsel sought to take advantage of the position, and to get costs, on the ground that the proposal of the promoters was "unreasonable and vexatious," but this was refused. The Committee passed all the other clauses of the Bill.

In consequence of the dissolution of Parliament the Bill did not reach the House of Commons, but was carried over to be resumed next session.

The directors of the Ship Canal Company were very much annoyed at the way they had been left in the lurch by their counsel at a critical moment, and especially as they believed the proposed charge to be just and equitable. The result might largely affect the revenue of the Canal Company, and it was certain that a sore had been created likely to disturb the amicable relations of the two companies.

Discussing the incident in a leader, the Liverpool *Daily Post* said :—

Three Parliamentary counsel of high distinction were retained to plead the cause of the Bill, having been paid of course very substantial fees and " refreshers ". Yet at a critical point in the progress of the Bill, when a clause of the first importance was being weighed in the balance, not one of the three was present to throw the weight of his eloquence into the scale in favour of his clients, and the clause was consequently struck out. The scandal is no novelty, though it is rarely exemplified in quite so glaring a form. Even allowing every possible latitude, and recognising in the fullest measure the right of each of the learned gentlemen to pocket two fees while doing work only for one, it might reasonably have been expected that they would have arranged between themselves that one, at least, should have held himself free to attend the business for which they had all been highly paid.

For some time disquieting rumours had been current as to the growing cost of the canal, which the Chairman of the Executive had hinted would exceed the estimates. In the middle of July he was asked how soon the promised report would be ready. In reply, Sir John Harwood apologised for the delay. It had been very difficult to get the necessary information from the engineers. Instead of letting the Weaver to Runcorn section, the directors had found it better to do it themselves ; and the dredging had turned out both troublesome and expensive. He could, however, say that the work was being done much cheaper under the new dispensation than under the old, though the canal could not be completed before January, 1894, also that they would require a sum in excess of the money already lent, but he did not think it would be very much. It was a serious outlook, but they must meet it calmly ; and he hoped so to deal with it that in the end the canal would be an accomplished fact. He promised that a comprehensive report should shortly be issued.

This pronouncement caused great disappointment. It was commented upon by an adverse Press almost in a tone of jubilation, and a heavy decline in canal shares immediately followed. Those who, like the *City News*, had persistently kept in view the ultimate conversion of the canal into a public trust, came to the front, pointing out that the necessary powers existed, and that where such trusts had been established, they were, as a rule, successful.

And now must be recounted an incident of a most extraordinary and annoying kind, but which fortunately is exceptional in municipal history. A member of the Committee betrayed his trust and handed over to the *Manchester Guardian* an immature document marked *"private and confidential"* Thus a report still under consideration by the members of the Special Sub-Committee, and subject to alteration or correction, suddenly appeared *in extenso* in the papers. Sir John Harwood and his Corporation colleagues on the Canal Executive had prepared a statement on the Ship Canal works, and on the financial position, which after revision by the Special Committee was to have been presented to the Council at the usual meeting on the first Wednesday in August, and it ought not to have been made public before it had been thus presented and discussed. This breach of faith caused great indignation among the members of the Corporation.

When the Council received the report they arranged a special date for its discussion.

The report commenced by an expression of regret that previous figures submitted by the engineers had proved unreliable, and proceeded to describe the course pursued by the Executive Committee. They had reversed the previous policy, and had decided as far as possible to do the work by tender. But in consequence of the involved and unfinished state of the work, difficulties had arisen in the measuring up necessary for tendering, and they had not been able to go as far as they could have wished. The canal had been divided into six sections. Four of these had been let for £915,817, and the other two would be done under supervision. The cost of dredging and widening between Eastham and Weston Point would be £30,000. Half the work had been done. The engineers had been busy revising the estimates, and as the canal would not be ready before 1894, provision for interest and management expenses would have to be made up to that date. After deducting the cost of works which might be deferred, it appeared that in round figures an additional £1,250,000 would be wanted to complete the canal.

The chief London and provincial papers had long leaders on the report. Pity for the poor shareholders was expressed on all sides. Liquidation was counselled, as was also the dismissal of the directors, who had spent the money. It was predicted that the £1,250,000 wanted meant £3,000,000. The *City News* was averse to the city taking further responsibilities as the ratepayers had been grossly misled. The directors, however, had their apologists. A local paper was of opinion that the cost of the canal had been swollen by difficulties that could not possibly have been foreseen; for instance the pressure of water in adjoining lands had pushed in the bank; also large areas of soil, to the extent of tens of thousands of tons, had been precipitated into the canal bed. Further, the railway companies had insisted on all works done for them being carried out in the most expensive way, and they were now claiming £250,000 for the cost of extra haulage on increased inclines, and giving no credit for the gain on descending gradients. Similar drawbacks had been met with in the construction of the Suez Canal, and yet it was a success in the end. Many shareholders urged that a cheeseparing policy was suicidal; the undertaking must be well carried out whatever the cost. Universal interest was felt in the difficulties of the Manchester Canal, and the pluck and perseverance of the shareholders and citizens was so much admired that it was felt the struggle could not be abandoned with the world anxiously looking on.

On the 12th August the Corporation met to discuss the report, which stated that on 29th July the Sub-Committee had asked the Ship Canal auditors, Messrs. Thomas Wade & Guthrie, to furnish them with a financial report up to date, and to certify the statement of estimated expenditure. The auditors' report was dated the 9th August, and showed that the sum required to finish the canal by the 31st December, 1893, was £1,264,282, or with contingencies, £1,489,282. Towards providing this sum the following resources were available :—

Estimated revenue from the Bridgewater Canal £330,235
Surplus lands taken as worth 1,000,000
Unsold plant taken as worth 400,000
Unpaid calls. 19,651

and 24,159 ordinary £10 shares taken over from the executors of the late contractor.

In the previous estimate of September, 1891, these shares, together with a small balance of mortgage debentures (in all £186,215), were regarded as available resources, but it was thought safer to omit them now. On the other hand, no account had been taken of incompleted works, such as the Warrington Dock, the Dry Dock, Runcorn, additional dredging, plant and land, road connections, etc., amounting in all to £333,400, which could well be postponed. The auditors' report was founded on figures provided by Messrs. Leader Williams and G. H. Hill, the engineers; the latter in a note stated that his estimate did not provide for contingencies like floods, etc., other than the sum of £85,000 added for further possible works. In his detailed report, Mr. Hill explained that the excess on his report of January, 1891, had occurred mainly in consequence of the incorrect quantities and particulars supplied to him, and the disastrous floods that had since taken place. He also pointed out now that four contracts amounting to £909,118 had been let, it was possible to form a more accurate estimate of the cost of completing the canal. Opening it by the end of 1893 depended chiefly on the weather, and in a lesser degree on the arrangements made with the railway companies. After this followed Sir John Harwood's report, the substance of which has been previously given.

The following list gives the successive estimates for the purchase of the Bridgewater undertaking and for the making of the canal:—

Parliamentary (1885) £9,812,000
Canal Company (January, 1891) 12,651,300
Corporation (January, 1891) 13,199,491
Joint amended (September, 1891) 13,769,444
Report (1st June, 1892) 15,176,172

The excess in actual cost over the original estimates largely arose from additional works prescribed by later Acts of Parliament.

On the 30th August the general meeting of the shareholders was held. The report showed an increase of £1,391,284 over the estimates of 1st September, 1891, and the directors regretted it would be necessary to obtain further Parliamentary powers to raise at least £1,500,000 in addition to the capital already authorised. It recorded the nomination of Mr. Edward Guthrie as auditor in the place of Alderman Leech. The Chairman, Lord Egerton, in moving the adoption of the report, said he had to tell them some unpleasant truths, but in spite of gloomy prognostications he was full of confidence in the future of the canal. Their increased expenditure had been unavoidable and was beyond the control of the directors. Difficulties must always beset a large undertaking, and unforeseen expense was common to all huge engineering schemes. Their aim was to make Manchester a seaport, and though the works were directed by an Executive Committee, the majority being Corporation directors who had worked energetically and harmoniously, still they had found it necessary to increase the estimates. They had let the straightforward work in four contracts, and in them there was an increase of £91,000 on the engineers' estimates. The Weaver portion they were compelled to do themselves, because of the intricacies of the work, and because at all costs they must push on with it in order to open the Runcorn navigation, which was temporarily diverted. After explaining the position caused by the death of Mr. Walker, the cost of extra dredging at Messrs. Wiggs' works, landslips and heavy floods, he alluded to the delays caused by the action of the London and North-Western Railway Company and their extreme demands. Until their claims were settled, they had declined to allow any relaxation of the time required, *viz.*, six to nine months' working before passenger traffic started. They had put off sending in their claims, and now when a portion of them had come in, they were asking for the cost of doubling their main line to Warrington. It was right that the shareholders of that railway should know the position, and how every attempt was being made to delay the opening of the canal by dilatory tactics. He admitted that, when fully equipped, the canal might cost £15,000,000, but even then he believed it would pay. He thought it possible Manchester, with some neighbouring towns, might assist in finding the additional capital required, and he paid a compliment to the great capitalists in London, without whose help the canal could not have been begun, who had found as much capital as Manchester and Salford put together, and yet who were willing to let Corporation debentures take priority over their own securities. He hoped the Corporation would help them in completing the work by arranging for the additional capital. After Mr. Bythell had seconded the report in an explanatory speech and had suggested a further loan of £2,000,000, Mr. Alexander Forrest asked if the Chairman thought £1,500,000 would be ample to complete the work. Expressing his unshaken faith in the success of the canal, and his detestation of croakers and pessimists, he proceeded to make an onslaught on the London and North-Western

Railway Company. "He did not like to advocate a policy of reprisal or boycotting, but too little had been said of the malevolent, spiteful, and unjustifiable position that company had taken up towards the Manchester Ship Canal." Some personal and influential representation ought to be made to the directors of that great monopoly.

Lord Egerton, in reply, said that the greatest care had been taken in the estimates, but he would not pledge himself to an exact sum whilst so many contingencies might arise.

Lord Egerton's tone of undaunted confidence and unwavering resolution told well with the meeting. All passed off smoothly, almost pleasantly, and again by their faith and trust the brave shareholders encouraged the directors to try and pull the canal through. The way in which they trusted their leaders and uncomplainingly waited for better times was the admiration of the whole of England. It was an instance of local patriotism, unique of its kind. How it was appreciated is touchingly shown by the following letter which I received :—

TO THE MAYOR OF MANCHESTER (ALDERMAN LEECH).

BELFAST, *September*, 1892.

DEAR SIR,

I always admire pluck, energy and enterprise, and this the people of Lancashire have shown in making the Manchester Canal. A work of such magnitude is not easily made or estimated for, and the shareholders are entitled to the kindest sympathy of all high-spirited citizens. I therefore enclose £1 as a present to the company, and hope all working men will act in the same way (say 5s. to 20s.), and this will enable the present shareholders to get four or five per cent. reward for their pluck and courage.

Yours truly,
" A BELFAST MECHANIC."

I believe in the canal and the people of Lancashire, and that it will be a success. You are at liberty to publish this if you wish, so I refuse to give my name.

Belfast early showed her desire to trade with Manchester, and it is a tribute to the importance of the undertaking that in January, 1891, a vessel was launched there, specially designed for the canal traffic.

On 31st August the Mayor received from Lord Egerton a copy of the report, and a letter calling his attention to the fact that at least £1,500,000 would be required to complete the canal.

The adverse Press issued the usual jibes at the increasing estimates, and reminded their readers of the sneering utterance of Mr. Pope, Q.C., who, before a

Parliamentary Committee had said, "They talk of millions in Manchester as if they were threepenny bits".

On the 7th September the Ship Canal report again came before the City Council, together with Lord Egerton's letter to the Mayor. Sir John Harwood moved that they should be referred to the Special Committee for report. He said that when the Corporation came on the scene they had either to help, or the whole concern would have gone into liquidation. The Ship Canal directors trusted the engineers' estimates, but they soon found them unreliable. A reinvestigation then took place, and the result was an increase of £863,000; the reason for this had not yet been cleared up. In the estimate of January, 1891, the cost of dredging was put down as £30,000, and in September of the same year at £369,688. In January, 1891, the estimate for excavation in the dry was 2s. 3d. per yard, and in September 6s. 6½d. Could there be any wonder at a disparity of estimates? He then dealt with the difficulties the railways were putting in their way, and said in spite of them all the Executive Committee were striving to do their duty conscientiously and economically. It would never do to have the fair name of Manchester associated with a failure. More money was necessary, but he trembled to have to go to Parliament again. He would not dare to say that a further £2,000,000 would be sufficient; with economy he hoped it would, but with the London and North-Western case unsettled, no one could be certain.

Councillor Hoy gave great credit to the Executive who were doing the work, and he considered the Corporation was pledged to finish it.

Councillor Gunson said the Corporation had been badly treated, and had been misled from the first. They had had blind leaders of the blind in the directors, the Council, and the Press. They had been deceived; also the shareholders. He urged that the number of directors should be increased, with a Corporation majority, and that this should continue after the canal was completed. He estimated the minimum cost to complete the work at £2,321,000.

Councillor Southern deprecated alarmist speeches like Mr. Gunson's, which unnecessarily damaged the Ship Canal, and would prevent the company from issuing further debentures.

Councillor Williams characterised the Ship Canal policy as one of expediency. The question with them seemed to be "How much can we conceal, and how little can we tell, in order to get over the present emergency?" There had been systematic deception, and hitherto the estimates had proved unreliable. He predicted a rise of 1s. 6d. in the rates, and said a further loan would impose a burden of £10 on every man, woman, and child in the city.   He asked that some competent and outside engineer should be called upon to give an independent estimate of the amount necessary to complete the work.   He hoped also that help proferred by Salford or other towns would not be rejected.

Councillor Trevor did not think the tone of the speeches calculated to mend matters.   If it were their own private concern, they would put the best and not the worst face on it.   He did not believe the money was irretrievably lost.

Alderman Chesters Thompson was convinced of the earnestness and honesty of the old Board.   They had put large amounts of their own money into the concern, and that was a guarantee to the shareholders of their fidelity to the trust placed in their hands.   An attempt was made to prevent him from giving individual amounts, but he insisted on doing so, and gave the following figures, showing the holding of the shareholders' directors :—

Lord Egerton of Tatton	£69,000
Sir Joseph Lee	53,150
Alderman Bailey	50,000
Mr. Alexander Henderson	59,000
Mr. S. R. Platt	38,700
Mr. W. J. Crossley	38,700
Mr. C. J. Galloway	28,500
Mr. Joseph Leigh, M.P.	27,250
Mr. Henry Boddington	27,000
Mr. J. K. Bythell	8,830
Sir E. G. Jenkinson	5,660

This much abused body had a heavy stake in the concern, and people must not try to persuade him that these gentlemen had not the best interests of Manchester at heart.   His advice was to give up quibbling about past mistakes, and trust the Board, who were themselves deeply interested in the success of the canal.

Councillor Clay objected to Manchester having the honour of finding all the money, if such might be called an honour.   The continued drain on their funds had put them in the position that "there was not a petty-fogging town of 100,000 inhabitants whose stock did not stand better than that of the Manchester Corporation".   He did not object to Manchester lending the money, but they must have ample security.   In the end the report was adopted without opposition.

The discussion in the Council afforded plenty of food for the local Press.   The

Guardian blamed the Council for not finding out that before making the first loan the Canal Company were overdrawn at their bankers, and had borrowed on debentures, but thought the city must see the canal through. The *Courier* saw a rate of 8s. or 9s. in the £ looming in the distance, but deplored Sir John Harwood's rhetorical bombshells, which had helped to bring down the value of the shares by one-half; he had posed as the great friend of the ratepayers, and the heaven-born saviour of the scheme, but it was certain his so-called courage and sagacity had weakened the confidence of the public. The *City News* took Sir Joseph Lee severely to task for saying a few months before the first loan, "This statement will dispose of the rumours that we are getting short of money". It hailed with pleasure the prospect of aid from Salford and other towns, but twitted Salford with having hitherto given very meagre support, in view of the great advantages to be obtained by her. With half the population, Salford held £26,430 ordinary shares, whilst Manchester held £232,834.

Reverting to the proposal of the Corporation to establish lairages at Mode Wheel, where it was intended cattle should be slaughtered within a few days of their arrival, Sir John Harwood had prepared and sent to the Board of Agriculture a plan showing the convenience and safety of the intended landing-place, and a report from the medical officer of health, stating there could be no fear of the spread of contagious disease, and that healthy, nutritious and cheap food would be provided for the working classes. Also, a certificate from the owners of cattle-ships that they were prepared to come up the canal, accompanied the report. Sir John Harwood also explained, in order to stay Salford's threatened opposition, that a market for Canadian and American cattle, a class not before brought into the district, could not possibly interfere with Salford's established market for English and Irish cattle. Despite a remonstrance from the Mayor of Salford, however, the Council of that borough passed a resolution that the establishment of such a market would have a most detrimental effect on the interests of Salford.

A second deputation from the Manchester Corporation, accompanied by the Mayor, subsequently had an interview with Mr. Chaplin at the Board of Agriculture. They explained the conveniences for lairages, and stated that within 50 miles of the site there was a larger population than within 50 miles of the centre of London.

Mr. Chaplin feared the risk of diseased cattle passing up the canal and communicating the disease to cattle grazing on the banks. In reply he was told that cattle now passed 20 miles up the narrower Clyde without any evil result. He again promised to consider the matter.

Meanwhile, the Special Committee of the Corporation were taking counsel with Mr. Moulton as to the wisest and safest way in which the necessary assistance could be rendered to the Ship Canal, now that Lord Egerton had declared there were no capitalists to whom they could go for assistance. In Salford there was a growing feeling, in response to a letter from Lord Egerton, to advance the canal £1,000,000, provided a proportionate representation could be secured on the directorate. The Oldham Council said they were quite ready to find £250,000, should it be wanted. Warrington also promised to subscribe £150,000, provided the Canal Company would make the dock there at once instead of waiting three years, as stipulated in the Act. This, however, was quite inadmissible.

The report of the Special Ship Canal Committee to be presented to the Council at the November meeting was made public a few days previously. It commenced with some correspondence between Lord Egerton and the Mayor, the last letter stating that the directors found it impossible to get the requisite financial help in the open market. Then followed the names of the special Sub-Committee, consisting of the Mayor (Alderman Leech), the Deputy Mayor (Alderman Mark), Sir John Harwood, Alderman Chesters Thompson and Councillors Clay, Southern, and Andrews, appointed to confer with Mr. J. F. Moulton, Q.C.

Councillors Andrews and Clay had been persistently attacking the action of the Sub-Committee, and great was their surprise when they were added to it. The latter soon got into trouble with his colleagues: and he and another member of the Committee went so far as to interview Mr. Moulton on their own account, a proceeding which was severely criticised.

After an interview with the Committee, Mr. Moulton stated his opinion in writing. Given in brief, it was that in case of a further loan, the Corporation should insist on having a majority on the Board until such time as the loan was paid off, and that as holders of $4\frac{1}{2}$ per cent. debentures, they ought to have a vetoing power at shareholders' meetings. He would fix the limit of the advance at £2,000,000. As regarded the co-operation of Salford and other boroughs, he was not in favour of a joint Bill. Let Manchester act by herself, with an enabling clause for others to join on equal terms and to take proportionate risks. He advised that each other contributing borough should apply in its own way to raise money and assist the canal. The expenses of the Bill to be borne by the Ship Canal Company.

While the Corporation Sub-Committee were in London, an effort was made to come to terms with the London and North-Western Railway Company, so as to get possession of the disused lines, and on 3rd November, 1892, a deputation, introduced by Lord Egerton, waited on the directors of that railway at the Euston Hotel. Sir John Harwood pointed out the loss and inconvenience caused by the Ship Canal Company in not being in possession of certain land, and volunteered that the Corporation of Manchester would guarantee any amount that might be decided on by an arbitrator. He ventured to say that the citizens of Manchester would feel incensed if their offer were not accepted, and it would be doubtful policy for any railway company to outrage the feelings of their customers. Lord Stalbridge, the Chairman, on behalf of the directors, repudiated the idea that they were wantonly delaying and damaging the Ship Canal, and promised to consider Sir John Harwood's offer.

Mr. Pritt, the Parliamentary agent, acting on Mr. Moulton's report, suggested Parliamentary powers should be sought—

1. To advance a further £2,000,000 to complete the canal, on 4½ per cent. debentures, as before, and ranking *pari passu*.
2. To appoint five extra Corporation directors.
3. To give the Corporation voting power on their total advances.
4. To provide for other towns joining, who should be entitled to a director for each £500,000 advanced.
5. To empower the Corporation to take up any unissued first and second debentures, if thought advisable.

The Special Committee recommended the Council to apply for a Bill giving effect to the above recommendations.

A letter was sent by the Mayor to Lord Egerton, with a copy of the report, saying the Corporation would be glad of the co-operation of Salford, or of any other towns that would seek Parliamentary powers to aid the canal.

This report was favourably received by the public. Shares went up when it was seen that help was certain, and that Manchester was not going to jeopardise the £3,000,000 already advanced. Salford, too, was better satisfied when she found Manchester was ready either to welcome assistance or to find the whole of the money herself.

At a Special Council meeting on the 28th October, after the report of the Sub-Committee had been moved and seconded, Alderman Joseph Thompson said he thought it was greedy and unfair to ask for ten directors for £5,000,000 of capital, whilst the remaining £10,000,000 would only have five directors. This view was shared by Alderman Mark and other members, who felt that the Council ought to be

MARSHALL STEVENS, FIRST MANAGER OF THE MANCHESTER SHIP
CANAL.

Warwick Brooks.   *To face page* 136.

satisfied with a simple majority on the Board. They ought not to displace men who had helped the canal financially, and had hitherto borne the burden and heat of the day. There was considerable discussion, and after it had been explained that the Ship Canal Board would have to be consulted, both as to the number of directors and the amount of the loan, the report was carried unanimously.

The Salford Council on the same day decided to render financial assistance to the extent of £1,000,000. The Sub-Committee, however, were to continue their investigations and report to a future meeting. During the discussion, Alderman Dickins said things indicated that—

Manchester desired to be master of the situation, and simply desired Salford to play the part of monkey, and help it to take the chestnuts out of the fire.

Speaking of the revenue prospects of the canal at a meeting of shareholders, Mr. Marshall Stevens, the manager, made the statement that two general reductions had already been made in railway rates as a result of the canal; one on the 1st January, 1883, after the deposit of the first Bill, and the other on 1st November, 1885. On raw cotton and cotton manufactures alone, the saving was £120,000 per annum, sufficient to pay 4 per cent. on the £3,000,000 already advanced by the Manchester Corporation. During the Parliamentary fight for the original Bill, the railways made their calculations on all goods being carted from the docks at Manchester, and none going by railway; now the Lancashire and Yorkshire, the London and North-Western, and other railways, were all arranging for their lines to have direct access to the docks.

On the 3rd November Lord Egerton wrote to the Mayor that he fully appreciated the liberality of the Corporation in offering to find a further £2,000,000, but whilst admitting the justice of the expenditure being controlled by Corporation directors, he suggested the following alterations, which would achieve the result, and be more acceptable to the shareholders:—

1. The number of directors to be twenty-one.
2. The lenders of the £5,000,000 to have eleven directors, or a majority of one, on the Board.
3. That from the date of opening, the fixing of the rates and tolls be under the control of the shareholders.

He found that the shareholders did not favour the condition that the lenders should have voting power at the shareholders' meeting. The Canal Board would have preferred to join the Corporation in a Bill, but in view of Mr. Moulton's advice they were willing to promote one of their own.

The old directors smarted under the following paragraph in Mr. Moulton's opinion :—

But it is clear that there was much in the condition of the canal which was unknown to the Corporation. The conduct of the Manchester Ship Canal Company in conceding to the Corporation the practical control of the expenditure is a proof of this

They wrote a letter to the Mayor, signed by all the directors. Against Mr. Moulton's statement they entered an emphatic protest, and also denied the following assertion made in the *Times :—*

The work had been conducted wastefully, and work which had been reported as done, proved, on investigation, to be no more than half done.

They said they had not fully assumed direct control of the works when they first applied to the Corporation. A Committee of that body, with their own engineer, had made inquiries into the engineering and financial departments, and nothing was withheld from them by the company.

They recapitulated the unforeseen difficulties they had had to encounter, and said the works had been carried on by a Works Committee on which the Corporation directors had a majority. They did not wish now to raise any controversy, but could not allow their silence to be construed into an admission that the allegations against them were true.

Under the advice of Mr. Moulton, the Manchester Corporation were firm in their determination to have a preponderating influence if they advanced more money. It was also decided that Sir Benjamin Baker should be asked to give advice as to the state of the works, and on 10th December that gentleman met the Committee, and made an inspection, giving his special attention to the Runcorn Bridge piers, which the railway company asked should be underpinned.

By a majority of 40 to 3 the Salford Corporation decided to promote a Bill in Parliament to enable them to lend the Ship Canal £1,000,000. When the Oldham Corporation found they would have to promote a private Bill in order to lend £250,000, they decided not to proceed. As a dock could not now be constructed, Warrington also withdrew her offer of £150,000.

On the 3rd December, 1892, Messrs. Fisher, Renwick & Co., of Newcastle,

UNDER RUNCORN RAILWAY BRIDGE.

Killon.

To face page 138.

commenced their line to London, which has been running ever since. The first ships sailed from Saltport.

The City Council met on the 16th November, and decided to promote a Bill for the provision of lairages for foreign cattle at Mode Wheel, and also a Bill to enable financial assistance to be given to the Ship Canal Company. The same evening a Borough Funds meeting was held in the Town Hall. The first Bill was approved of unanimously, but when the Ship Canal Finance Bill came on, a Mr. Joseph Waddington made a rambling speech, and opposed the Bill on the ground that "it was illegal" and because "the Council had displayed an indifference to human life". Though he had practically no support he insisted on demanding a poll.

At the Borough Funds meeting in Salford the socialistic element turned up in force to oppose the Bill. Mr. Horrocks said he thought the nation, as a whole, should undertake such work. There had been 20,000 men at one time on the job, but not 500 were Salford men. The canal had brought a great many men into the local labour market from all parts of the world, and when these men could not get work, they underbid in other employments. This argument carried the day, and the majority was more than 2 to 1 against the Bill. A poll was demanded by the Mayor.

Then came a time of great excitement. Many public meetings were held in Salford, at which the Socialists made themselves heard. When the poll was announced it was found that 13,385 ratepayers had voted for advancing £1,000,000 to the Ship Canal, and 3,052 against; showing a majority in favour of 10,353.

So enthusiastic were the people of Oldham in support of the canal that when the Council decided not to subscribe, a public meeting of the ratepayers was held, at which a resolution was passed almost unanimously, asking the Council to contribute £250,000 towards the extra capital required.

During the Assizes the judges expressed a desire to see the canal, and on the 6th of December, with deep snow on the ground, Justices Grantham and Gainsford Bruce, with their marshals and Miss Grantham, visited the Manchester end to view the docks.

When Mr. Gladstone was receiving the freedom of Liverpool, he, in his usual happy way, thus referred to the Ship Canal :—

There is something grand in the rivalry of Manchester, the great sister city of Lancashire and undoubtedly the spirit of enterprise which first prompted and has now carried forward to an advanced point the Manchester Ship Canal, is a spirit of enterprise so strong, so resolute and so determined to surmount every obstacle, be it great or small, that you Liverpool men, and I (for I was born in Liverpool) along with others, cannot but admire it.   But when I come to speak of its consequences, there have been those who believed that Liverpool in some degree, forsooth, may lose its palmy state, and will have to yield the honours of the second commercial city of England, competing very sharply with the first, London. Well, gentlemen, these ideas of danger to Liverpool are utterly visionary.   There may be some reduction of charge.   There may be reductions in railway rates.   I do not wish to tread on sacred ground, but I can conceive it possible that questions may arise with regard to dock dues.   Nay, more, that even in warehouse rents and other matters of that description, should the Manchester Ship Canal become an efficient engine of commerce, as I believe some people think it will, certain modifications will find their way into the arrangements which have hitherto dominated in Liverpool.   But yet, although there may be some inconvenience to particular persons and particular bodies attending some of these processes, yet, for Liverpool at large, do not conceal it from yourselves, the result must be an enormous gain.   The Ship Canal, if it adds largely to the commercial means of the district, will draw to the Mersey more than ever the commerce of the world.   And if that commerce of the world liberally bestows itself upon the Mersey, depend upon it that not only in your days, but in the days of your children, your grandchildren and your great-grandchildren, and I know not how far down the course of generations, it will be Liverpool that will have the first, best, and the most of it.

Unfortunately the broad spirit of Mr. Gladstone was not shown in the attitude of the timber trade towards the Mersey new port, for when the Pitch Pine and Export Timber Company and another firm, found that Saltport was saving them money (up-wards of five shillings per ton), and began to increase their business, a dead set was made by the Liverpool buyers, who established a boycott on all timbers brought to Saltport.   When a broker at Liverpool offered by public auction some timber lying there, the buyers assembled about fifty yards away and would not go near the auctioneer, and so made the sale a fiasco.   At the same time, Liverpool merchants were telling customers on the Manchester Exchange that though they were willing to bring cargoes up the canal, no one would give them tonnage to Saltport.

Soon, however, this little port, at the mouth of the Weaver, became so busy with the import of timber, grain, etc., and the export of salt, that it was necessary to in-crease the accommodation, and a long landing stage was erected.   This accommoda-tion proved most useful.   The directors of the Salt Union, accompanied by the principal Liverpool Salt Shippers, made their annual inspection in July, when they expressed themselves surprised at the facilities for despatching salt.   The Chairman, the Hon. L. Ashley, formally named the place "Saltport," throwing a bottle of wine against the new jetty and wishing success to the Port.

WRECK OF THE BRIDGEWATER LOCKS.

Birtles, Warrington.

To face page 140.

When the opening of the Weaver navigation cleared the way, every effort was made by the agent, Mr. Jones, to ensure rapid progress on the Runcorn section. Several locks had to be constructed.   By far the most important was the Bridgewater Lock, because it gave access to the sea for the vast trade coming down the Bridgewater Canal.   Here the red sandstone seemed to offer an excellent foundation, and the lock wall quickly rose to a considerable height.   The Runcorn Bridge, close by, had stood well, and no one dreamed of disaster.   But on the 2nd February there came an abnormal tide in the Mersey.   Gradually it topped the embankment and flooded the partly finished locks, the chief pressure being from water that had gathered outside the inner canal wall.   Even then no one thought damage possible.   But it so happened that the sandstone on which the embankment was built was laminated with streaks of a mineral silicate.   These, being porous, became softened by the water they absorbed and formed a soft moving substance.   So when at high tide the water pressed on the ponderous dock walls they began to slip, with the result that the whole of the massive brickwork fell down into the excavation made for the locks—excellent concrete and brickwork becoming a mere heap of ruins.

Just when a difficult work had been almost completed, and, as it appeared, in a most substantial way—for the wall was 21 feet wide at its base, tapering to 7 feet at the top—the whole structure collapsed for a length of 200 to 300 feet.   This was perhaps the most disheartening of a long series of misfortunes, and it was lamentable to see such good work so ruthlessly destroyed.   The removal of the wreck was a big business, and the disaster materially delayed the completion of this section.

The temporary closing of the Runcorn Locks, although unavoidable, caused a very strong feeling in the town.   This was allayed by the directors promising to construct a lay-by at once, and possibly a new dock in the future.

About the end of April a disastrous fire at Ellesmere Port destroyed two cement stores, some dwelling huts, four railway trucks and 470 tons of cement.   The fire was brought home to some young incendiaries, who had started it from a love of mischief.

No sooner did it become known that the canal had ceased to be a contractor's job than there was ferment among the men.   When an increase of wages was refused, the dissatisfied workmen took to doing damage.   An attempt was made to burn a new station just completed; waggons were tipped over into the empty canal and smashed; and when a watchman, as it is supposed, tried to stop mischief, he was killed and the miscreants got away.   Disaffection spread even to the nippers or lads. For some slight reason they struck work, marched down the canal bank armed with clubs, and under threat of corporal punishment obliged all the other lads to join them. They destroyed several huts, but at last, possibly afraid of punishment for the havoc committed, they quietly dispersed and started work again.

During the excavations for the Manchester Docks and approaches, several huge oak trees were met with ; they were of considerable size, and so hard that they resisted ordinary tools, and had to be broken up with an axe ; evidently there had once been a forest on the site of the docks. Another feature of the year was the completion of a dredger, built in the dry, near Hollins Green, by Messrs. Ferguson & Co. After floating, she had to cut a way for herself out of the cutting in which she had been built. The dredging department had recently been re-organised and placed in the charge of Mr. A. O. Schenk, an engineer in whom the late Mr. Walker placed great trust. At the end of September the City Council Consultative Committee inspected the whole length of the canal. At Pomona the four docks were excavated to their full depth, and the river walls were being put in. The three Salford Docks were nearly finished. Mode Wheel Locks were completed, but much dredging remained to be done between there and Barton, largely the result of heavy matter from the Salford Sewage Works. At Barton both the swing aqueduct and the swing road bridge were nearly finished. The canal from Barton to Irlam was ready for use. with the exception of the hydraulic apparatus for the locks. The Irlam Locks were also finished. At Partington coaling basin and sidings two-thirds of the work was done. The Works Committee, for certain reasons, retained a small portion of the work to be done by administration ; beyond this the completion of the canal was divided amongst the following contractors:—

> No. 1. Mr. John Jackson, Old Quay to Randles Sluices.
> No. 2.   „   „   Randles Sluices to Latchford Locks.
> No. 3. Mr. C. J. Wills, Mill Bank Weir to Barton Locks.
> No. 4.   „   „   Barton Locks to Pomona Dock.

The work of carrying dredged material out to sea had become so expensive that the engineer arranged to deliver the spoil from the dredger into iron skips on steel barges alongside. These skips were then floated to a crane, and each skip was lifted and emptied at the top of an inclined plane, down which the sludge ran into the low-lying land. This method proved very successful.

A statement having been made by an official of the Navvies' Union, that 1,000 to 1,100 men had been killed during the construction of the canal, 1,700 permanently injured and 250 partially injured, Mr. Bird, of the Master Builders' Association, in

Killon.

PARTINGTON COAL BASIN.　TOP VIEW.

To face page 142.

Killon.

Building Dredger at Partington.

To face page 142.

DREDGER AT WORK, MILL-BANK.

Birtles, Warrington.

To face page 142.

giving evidence before the Labour Commission, gave the correct figures as 130 killed, 165 permanently injured, and 997 slightly injured.   For these ample medical aid and hospital accommodation had been provided.

In the second week of December heavy rain, following a thick fall of snow caused a flood, and the canal cutting near Barton Bridge was inundated and several cranes submerged.   The same week the swing bridge in Trafford Road was success-fully placed in position.   It is 265 feet long, 30 feet deep, 50 feet wide, and the approximate weight of the entire structure is 1,800 tons.   It works on a central hydraulic ram and sixty-four cast-iron free rollers, and is turned by hydraulic power.

Before the end of the year some important works were completed.   Amongst these were the Morris Brook cutting and the railway viaducts over the canal at Latchford and Acton Grange.   For six months goods traffic only was allowed to cross the viaducts ; after which time the line was considered safe for passengers.   On the Runcorn-Weaver section 4,000 men had been employed night and day, and it was hoped that in six months this section would be completed.   In the next section Mr. John Jackson was showing what a methodical contractor and clever administrator could do.   He had excellent plant and a well-ordered staff, and was making progress at a rate never before exceeded, and this with practically no accidents.

CHAPTER XXIII.

1893.

OLD DIRECTORS DEFEND THEIR POLICY—FEARS IN LIVER-
POOL—HOSTILE ACTION OF THE LONDON AND NORTH-
WESTERN RAILWAY COMPANY—OFFERS OF MONETARY
HELP—CATTLE LAIRAGES—ARRANGEMENTS WITH MAN-
CHESTER TO FIND £2,000,000 FURTHER CAPITAL—PARLIA-
MENT RATIFIES THE LOAN AND REJECTS THE SALFORD
AND OLDHAM BILLS—LORD BALFOUR'S RAILWAY ARBI-
TRATION—MANCHESTER TO PROVIDE LAIRAGES—ELLES-
MERE PONTOON DOCK—LIVERPOOL ALARMED—PER-
SONAL ADVENTURES—DIRECTORS' PRIVATE VIEW OF
CANAL—PROGRESS OF WORKS—INCIDENTS AND ACCI-
DENTS—SEWAGE DIFFICULTY—ARBITRATIONS.

*The shipowner had done his part, the men who now held the key to the success or non-
success of the canal were the merchants of Manchester. Unless they determined to support the
canal, ships would not come.*—GEORGE RENWICK, M.P.

STUNG by the charges of feebleness, extravagance, and incapacity levelled at
them, both inside and outside the Council, the shareholders' directors, as
recorded in the last chapter, had sent an indignant remonstrance to the Corporation. They protested that they had been the victims of circumstances, and that
even with the welcome aid of their Corporation colleagues, it had been difficult to
cope with the natural and other drawbacks they had had to encounter. On the other
hand the Corporation resented the non-acceptance of their terms as to a further loan.
The year 1893, therefore, commenced with somewhat strained relations between the
Canal Company and the Corporation.

The Mayor, to smooth matters over, wrote that the Corporation had a satisfactory reply to the statements made by the aggrieved directors, but as it would

Killon

IRLAM LOCKS AND SLUICES.

To face page 144.

only lead to an undesirable controversy the matter had better stand over, and a conference be held between the two bodies. Lord Egerton replied approving the suggestion. On the 4th January the Manchester statutory Borough Funds meeting was held, and both the Financial Aid Bill and Cattle Lairage Bill were passed without opposition. In moving the latter, Sir John Harwood explained there would be a saving of 6s. on each beast in carriage alone, besides other economies, and the meat would arrive in better condition. As a security against disease, the cattle would be medically examined. He gave his experience of seeing American cattle landed at Deptford, and spoke of their excellent condition. Much of Mr. Chaplin's objection would have been removed had he known that the cattle came direct to Manchester without transhipment.

When the Parliamentary Bills were being considered by a Special Committee of the Council, there was a difference of opinion as to the number of directors the Corporation ought to have, but ultimately an agreement was come to. Sir John Harwood said, " The question of representation would have to be settled by Parliament ". On 26th January, 1893, a meeting of the Consultative Committee and the Ship Canal directors took place to discuss the charges that had been made against the latter. Sir John Harwood stated that when the three millions were advanced, Sir Joseph Lee agreed that all land, plant and the Bridgewater surplus should be ear-marked as security. This Sir Joseph Lee promptly denied, and Sir John was so much annoyed that he left the room. Mr. Bythell was indignant against those who had accused the old directors with deception as regarded the estimates. Councillor Williams admitted he had done so, and said Mr. Bythell himself had done likewise, for he had charged the engineers with misleading and misrepresentation.

So alarmed was Liverpool at the progress of the canal that the Liverpool *Courier* published a series of articles, " How to Meet the Competition of the Ship Canal ". Some of these were amusing. It was suggested that cotton mills should be built at Liverpool and Birkenhead, with a view to save carriage. Considering also that Liverpool had the finest flour mills in the country, and that the trade in foreign-made flour was fast disappearing, the writer thought Manchester would be wise if she retained her position as the " Slaughter-house " of the trade, and allowed outsiders to cut one another's throats for the glory of supplying her with flour. Then came a suggestion, " Why not originate in Liverpool a system of conveying by bands?" This would improve the arrangements for dealing with corn, and help the corn merchants of that city in their struggle with other ports. The suggestion was a valuable one, and it has since been adopted in the Manchester Docks.

As if desirous of adding another difficulty to impede the Ship Canal, the London and North-Western Company, notwithstanding the protection afforded them in the 1885 Act, introduced into their Omnibus Bill of this year a clause providing that the Canal Company should not, except with their previous consent, enter upon any part of their system, or that of the Great Western system, or interfere with their railways, till they had deposited in the Bank of England the sum awarded them by arbitration.

The result of the Oldham Corporation poll was not so decisive as that in Salford, the numbers being :—

For the loan	.	.	.	.	.	.	.	.	.	9,805
Against	.	.	.	.	.	.	.	.	.	7,451
Majority for	.	.	.	.	.	.	.	.	.	2,354

Their Bill authorising the loan of £250,000, together with the Ship Canal Company's Bill, came before the Examiner on Standing Orders, and both were declared in order.

On the 1st February, at the Hull Chamber of Commerce, Mr. C. H. Wilson, of the Wilson Liners, made a remarkable speech. He said the high charges at Liverpool were the secret of Hull's success. But now that Liverpool was reducing her charges to compete with Manchester, the trade of Hull was in great peril, and he confessed that when the Ship Canal was opened there would be cause for serious alarm. There would be a great struggle between eastern and western ports for the corn trade, and he feared the result. Then, in the coal trade Manchester would be much nearer the collieries than Hull, and there would be a struggle for that trade also. Hull must put her house in order and cheapen all round if they were to compete successfully with the canal. The rates from Hamburg *via* Saltport were already 1s. 2d. per ton cheaper than *via* Hull.

At this very time, strange to say, the railway companies were creating much ill feeling by raising some of their rates. At a meeting of boiler-makers in Manchester, complaints were made that the railway charges on boilers to London, Southampton, Birmingham and Cardiff had increased 25 to 45 per cent., and on engines 20 to 50 per cent.

At Liverpool there was a conflict between the Dock Board and the cattle importers, who complained of the insufficient accommodation provided for them.   A deputation of the latter visited the proposed site for the lairages at Mode Wheel, and expressed themselves well pleased.   They intimated an intention of coming to Manchester if good prices could be obtained.

The February report of the Ship Canal directors stated that they, along with the Corporations of Manchester, Salford and Oldham had deposited Parliamentary Bills.   The two first-named towns were prepared to provide lairages, and all three offered financial help.   The five deviation railways were now finished, and goods trains were running on some of them.   Two powerful dredgers were being constructed in the dry at the Salford Dock, and the estimated cost for the completion of the canal would not be exceeded.

This report was submitted to a special meeting of the shareholders on 6th February.   In consequence of a recent bereavement the Chairman was absent, and Sir Joseph Lee was also away through ill-health.   Mr. J. K. Bythell presided.   He dealt with the wild rumours that were afloat.   If the shareholders "disbelieved everything they heard as to the Ship Canal, especially as to its finances, unless verified by the Ship Canal office, they would not be far wrong".   It was now ascertained that at least £1,500,000 would be wanted to finish the canal, and they were met to consider a Bill to raise £2,000,000 at not more than $4\frac{1}{2}$ per cent. interest.   They had three offers, Manchester, Salford and Oldham, who were respectively willing to lend £2,000,000, £1,000,000 and £250,000, and who all had deposited Bills taking power to lend.   They acknowledged the ungrudging, full-handed monetary assistance given by Manchester, but as trustees for the shareholders, he regretted they were compelled to criticise the terms asked.   Manchester had neither consulted, nor negotiated with the Ship Canal directors as to the time to be fixed for the completion of the work, or as to the other conditions in the Bill.   This they regretted, as it obliged the other boroughs to deposit separate Bills which would be costly.   Manchester now wanted fifteen directors, and that they should remain in office, not merely till the new loan was repaid, but till the whole £5,000,000 were paid off; also that her debentures should give voting powers at shareholders' meetings.   He was glad to say that since the deposit of the Bill, interviews had taken place which made him hope there was a prospect of the objectionable terms being modified.   The real cause of the error in the estimates, for which they had been so much blamed, was the mistake made by the experts before Parliament.   One thing was certain, viz., that Manchester need never fear about her security.   The ordinary shareholders had shown they were mainly influenced by disinterested and patriotic motives.   When they were asked to allow £4,000,000 preference stock to be placed in front of their investment they need only have said "No" and the canal would have been lost to Manchester, and without cheap access to the sea her commercial supremacy was doomed.

Their London friends, too, in finding money were no doubt influenced by a desire to help a great national undertaking, and by admiration of the indomitable pluck displayed by the ordinary shareholders and promoters.   Assistance from a municipality in aid of dock and harbour works, that would benefit a community, was by no means exceptional, and there were many precedents.   To his mind the Board should consist of twenty-one members, and the municipalities should have a majority, but this would now either have to be the subject of arrangement, or be decided by Parliament.

Mr. Belisha was strongly opposed to receiving aid from Manchester on the terms proposed, and had to be reminded by the chairman that Manchester made a great concession in allowing the £2,000,000 to be borrowed *pari passu* with the previous £3,000,000.   Power to proceed with the Bill was given.

The Chairman further said there was no doubt about completing the canal during the year.   The railway companies were showing a better spirit, and there was no reason why the substituted and unused lines should not be taken over immediately, and the dam of earth on which they stood, and which blocked the canal, be removed.   If this could be done the danger of floods would be lessened.   In reply to a question, the Chairman said that less than one mile of the canal remained to be excavated, and that consisted mainly of the portions over which the railways were running, which could not be dealt with till the trains were turned over the new diversions.   The remainder of the excavating could not be done in the dry, but the soil would be removed by dredgers now being built on the canal.   The meeting was cheerful and the report was adopted without opposition.

Traffic, meanwhile, was pushed on vigorously at Saltport, and during the month of February a consignment of tea (35,000 lb., for D. Melia & Co.), tobacco (B. J. Robinson & Co.), and wine (Mr. Leroy) arrived at Manchester, where special arrangements had been made by the Bonding Warehouse Co. for storage and distribution.

During March the first shipment of cotton came up the canal, *viz.*, 269 bales of Chinese cotton consigned to Messrs. Malcolm, Ross & Co.   It was transhipped in London and came by Messrs. Fisher & Renwick's steamer *Blencowe*.

At the Council meeting on the 1st March, Councillor Andrews advocated the withdrawal of the Bill, and went on to say " he should be glad to see the concern closed and a receiver in to-morrow".

When subsequently the Corporation directors were appointed to take charge of the Bill, an attempt was made to associate with them five members of the Council who had persistently opposed it in Committee. One of them, Alderman Clay, " was sorry to say the present directors do not carry out the views of the Corporation". He further stated that the directors " were delegates paid with the ratepayers' money". Alderman Southern, in a cutting speech, resented these insinuations, and said that violent and rabid Ship Canal haters had no right to go to London.

On 7th March the Ship Canal Bill of 1892, which had passed the Lords, but in consequence of the Dissolution did not get to the Commons, came before a Committee of that House, with Sir Charles Dalrymple as Chairman. The offer of the Ship Canal Company to guarantee to the Upper Mersey Commissioners £2,800 as a minimum for dues was accepted, and the preamble was passed. When the time had expired for petitioning in the Lords against the Corporation and Ship Canal Bills, it was found that ten out of the eleven presented were cross petitions, the exception being one from the late Mr. Walker's trustees. Manchester objected to the advances of other Corporations ranking *pari passu* with the £3,000,000 already advanced by her, and she objected also to Salford appointing an engineer to supervise the works of the Ship Canal Company. The position was described as " a family quarrel".

The Ship Canal directors were in a great difficulty, as they could scarcely move without vexing one or more of the friendly Corporations who had offered to come to their help. The Mayor of Manchester wrote that his city was willing to withdraw the claim for fifteen directors provided she could be assured of a clear majority on the Board, and that the election of Chairman or Deputy should vest in the Corporation till the loan was paid off. Unless this was agreed to, no new loan would be allowed to rank *pari passu* with the first loan. Manchester was willing to let the Ship Canal directors have a majority on the Rates and Tolls Committee.

To this Lord Egerton replied he could not see how his directors could accept the terms, because the negotiations had proceeded so far that they could not, without a serious breach of faith, recede from the arrangements made with Salford and Old-ham. He also reminded the Corporation that in October last, Manchester had expressed her willingness to let other towns come in on *pari passu* terms and be proportionately represented. During February and March much friction existed, especially between Sir John Harwood and the old Ship Canal directors.   No doubt Sir John was in ill-health and much harassed by the difficulties that beset his path. It was discovered also that an engineer, in the service of Mr. Hill, who had strong Liverpool sympathies, was poisoning Sir John's mind by telling him that all his labour would be in vain, that for a variety of reasons large ships would never come up the canal, that railways were useless about docks, that the railways never intended to assist the Ship Canal in getting goods away, and that the enterprise was doomed. This same gentleman so irritated and disgusted me by his dolorous croakings that I told him he was in a wrong position, and with his views he ought not to remain another moment in the service of the Corporation or the Canal Company.   About the same time another Liverpool man, whilst in the Ship Canal service, was found to be writing depreciatory letters to the Press, and actually advising shareholders not to pay up their calls.

On the 13th March the Ship Canal Finance Bill and the Manchester, Salford and Oldham Bills came before a Select Committee of the Lords, Earl Cadogan being Chairman.   Mr. Pember persuaded the Lords to take all the Bills together.   He then gave the history of past negotiations, and said that the Canal Company did not care two straws who lent them the money required, their duty was to safeguard the shareholders and to obtain the best terms possible.   Manchester had now made them a firm offer of £2,000,000 on conditions which would be satisfactory to the Canal Company, and as they held the previous loan of £3,000,000 and would not allow other loans to rank *pari passu* with it, they felt obliged to accept the offer of the Manchester Corporation.   This announcement caused quite a sensation, and Mr. Bidder stigmatised it as " a breach of faith ".

Mr. Pember said he felt very sorry to do anything that might appear a breach of faith, or ungracious to Salford and Oldham, but he could not help it.   Those towns might refuse to accept debentures ranking after the £3,000,000, or Parliament might refuse to let them advance the money at all.   Now that Manchester had met them, they felt it was their plain duty to the shareholders to come to a settlement.   The terms arranged were :—

1. The shareholders to elect the Chairman, and the Corporation directors the Deputy Chairman.

2. Twenty-one directors to form the Board, eleven of them to be appointed by the Corporation, the Board to be reduced to fifteen when the £3,000,000 loan was paid off—of these fifteen the Corporation were to appoint two.

3. The Rates and Tolls Committee to consist of five shareholders' and four Corporation directors.

4. The Corporation to give a written undertaking to provide the extra £2,000,000.

Mr. Pember said no doubt the action of Manchester in insisting on having a majority on the Board and in refusing to let the advances of other Corporations rank *pari passu* with their first loan might seem arbitrary.   But the fact that Oldham and Salford had only offered a portion of the money, and obliged them to go to Manchester for the remainder, placed the Ship Canal Company in a dilemma.   There was only one way out of the difficulty, and that was the one he had indicated—he regretted to have to say it, *viz.*, that Salford and Oldham should take his apology, and he should take none of their money.   Mr. Moulton, for the Manchester Corporation, accepted the terms mentioned by Mr. Pember and withdrew all opposition.

Alderman Marshall, Mayor of Manchester, in cross-examination by Mr. Bidder, was asked: "You and the Ship Canal Company became one in this matter at 10.45 this morning?"   To which he replied: "Earlier than that, on Saturday we had very nearly settled".   In reply to further questions, the Mayor said they were unwilling to act with Oldham and Salford because those towns would not consent that Manchester should have an absolute majority on the Board.   Having a very much larger sum at stake, Manchester ought to have a larger proportionate share of the representation. They would decline to find the money, considering what they had already done for the canal, unless favourable and equitable terms were arranged.   It was not dictation, it was a matter of private arrangement.

Mr. Littler, for Oldham, said the Manchester Corporation had tyrannised over the Ship Canal Company, and had brought them to their knees.   He was instructed on behalf of Oldham to complain of as gross a breach of faith as was ever committed.   This led to a scene between counsel, Mr. Moulton saying: " There is not a fraction of truth in that statement, and I don't believe my learned friend is instructed to make it".   To which Mr. Littler replied: "I am not in the habit of being so accused.   If it were a member of the Bar who generally practised here who made the statement, I should call upon him to withdraw it; but as it is not, I pass it over, assuming it is made in ignorance, which arises from want of knowledge."

Earl Cadogan here interposed and said that there might be a moral status to Oldham and Salford, but he failed to see how they could legally claim a share in the loan.   It was an unusual position to have to determine the relative merits of various claimants to the honour of lending money to a company. The competition had generally been the other way.

After the speeches had been heard, the room was cleared, and the counsel tried, but ineffectually, to come to an arrangement.

Sir John Harwood was then called on behalf of the Manchester Corporation. He denied that a majority on the directorate was an afterthought, or that Manchester had tried to keep Salford and other Corporations out ; they had never negotiated with them.   In cross-examination, he said whoever came in or whatever was done, so long as £3,000,000 was owing to Manchester they must have control.   He denied the ultimatum of the Corporation was "a try on ".   Manchester would advance no more money on any other terms.

The Chairman said the position of Manchester varied entirely from that of the other towns ; she had agreed to lend, and the Ship Canal Company had agreed to borrow.   It would be difficult to force the company to borrow elsewhere, and they must take Salford and Oldham as occupying a different position.   The Committee must confine themselves to the Corporation and Ship Canal Bills, and decide which preamble must be proved.

Mr. Pember suggested that they might be made uniform, as he could not conceive the Committee were going to force Manchester to lend, and the Ship Canal to borrow, on conditions objectionable to both.

The Chairman said that the Committee had decided that the Canal Company and the Corporation had the prior right, as they had come to terms, and the preambles were passed.   Salford and Oldham had no *locus standi* as petitioners against the preamble.   They could only be heard on clauses.

When clauses came on for discussion, Oldham and Salford withdrew their Bills. A clause was inserted in the Bill that when half the £5,000,000 loan was repaid the eleven Corporation directors should be reduced to seven.

Unfortunately, the success of Manchester caused much heartburning and ill-will. There can be little doubt the decision of the Committee was a wise one, inasmuch as divided authority was not likely to work well, and Manchester was so far involved, that in justice to her past sacrifices she was entitled to be the dominant partner.   It was a misfortune, however, that Salford and Oldham had been encouraged to assist, and it looked like ingratitude when their generous offers were rejected after they had been at the trouble and expense of going to Parliament.   In justice to Manchester, it must be said she had always insisted on having a majority on the Board.   The refusal, however, to let outsiders rank *pari passu* with the £3,000,000 loan might well have been declared earlier. It must be borne in mind, too, that Salford and Oldham never showed the same unanimity and heartiness as Manchester. In both places much opposition was encountered, and there can be no doubt that the request of the former to appoint a canal engineer was unreasonable. Further, the opposition of Salford to the Manchester lairages had not contributed to the cordiality of feeling necessary between representatives working on the same Board.

Salford and Oldham will always have the credit of offering to help the Ship Canal in her difficulties. They got the kudos without any cost, for the Ship Canal Company, as some recompense for the disappointment, paid all their expenses.

The same Committee then considered the Manchester Corporation Bill to establish lairages and slaughter-houses for foreign cattle at Mode Wheel. Mr. Fitzgerald explained the objects of the Bill, and referred to the opposition of Salford to the Bill of 1891. He also enlarged upon the great benefit that would arise in a densely populated district from a cheapened supply of meat, and he showed that the Board of Agriculture were making regulations to secure the country from the danger of infection.

After considering the opposition of Salford, the Committee passed the Bill.

An Extraordinary General Meeting of the shareholders was held on 27th March to sanction a Surplus Lands Bill. The Chairman explained that the 1885 Act did not provide for selling land on chief rent. The company could lease for 999 years lands which had been compulsorily purchased, but not such as had been acquired by agreement. They were short, too, of power to reserve rights or easements. The object of the Bill was to repair these omissions. After a protest from two share-holders, one of whom said, "If the directors sold land they *would be* duffers, as it would be worth ten times as much in ten years," the shareholders assented to the Bill.

In the Commons, the Corporation Lairage and Slaughter-House Bill passed without opposition, as did also the Surplus Lands and Additional Capital Bills. At the May meeting of the Council Sir John Harwood stated decisively that the canal would be open for traffic in January, 1894. At the same meeting the Chairman of the Improvement Committee promised to push on the city approaches to the canal.

One of the obligations of the Canal Company was for loss incurred by the rail-way companies through the gradients being increased at the various canal crossings and railway deviations. The amount by the 1885 Act was to be settled by arbitra-tion, and Lord Balfour of Burleigh was appointed arbitrator.

On the 30th May he presided for the first time, at the Surveyors' Institute, in the case of the London and North-Western Railway and Great Western Railway *v.* Manchester Ship Canal, Sir George B. Bruce sitting as assessor. The plaintiffs claimed for the increased cost in the maintenance of works, and in the working of the traffic resulting from the deviations necessitated by the Ship Canal.

The claim of the London and North-Western Company was nearly £500,000, afterwards reduced to £333,000, and that of the Great Western Company £200,000. Most important of all was the question of time. Till the damage had been assessed, the railway barred the way to the completion of the canal. They declined to give up possession of the old railway lines, which ran at a low level across the canal, till they had been settled with. Both sides employed most eminent and costly counsel. Mr. Pember, Q.C., Mr. Moulton, Q.C., Mr. Balfour Browne, Q.C., Mr. J. C. Graham and Mr. W. Compton Smith represented the Manchester Ship Canal Company.

Mr. Pope, Q.C., opened the London and North-Western case. He claimed that under the 1885 Act, compensation was due for raising the lines to 75 feet above water level, and the consequent increased gradients. His company would also claim for increased cost of maintenance. They would, in future, be bound to shorten the goods trains (necessitating new marshalling grounds), to increase the power of their engines, or to use bank engines to help in surmounting steep gradients. His claim was founded on the capitalisation of extra costs. The railways, in order to assist the Ship Canal, had agreed to waive their rights of retaining possession of the old lines for six months, on condition the Ship Canal Company deposited a sufficient sum as security for the compensation that would be awarded.

He then produced his witnesses, who admitted that gradients, equally severe, existed on other lengths of the London and North-Western line, but claimed for additional cost of maintenance, such as extra water troughs, sidings, signal boxes, etc. They estimated that whereas an engine could draw thirty-eight to forty trucks up the old incline, only thirty-two trucks could be drawn up the substituted one. Numerous witnesses proved that the gradients of 1 in 135 caused trains to stick on the incline, unless they had the help of a bank engine, but it was elicited that they often crossed more severe gradients without assistance. Evidence was also given that extra painting would cost £18,496.

Very similar evidence was given on behalf of the Great Western Railway Company. It was admitted that the existing gradients between Chester and War-

rington were not good.   As the result of negotiation, Mr. Cripps, Q.C., on behalf
of the Great Western Railway Company, withdrew or adjusted various claims, and
these were taken out of the list submitted to the arbitrators.

On the 3rd July Mr. Moulton, Q.C., for the Ship Canal, opened his case, and
submitted rebutting evidence.   He said that no injury would be done to justify the
enormous claim put forward.   Whilst the railways declined to give the necessary
information it was exceedingly difficult to check their figures.   If the railway companies claimed for damages, they should, on the other hand, bring into account the
advantages that would accrue to them.   If they used steel, in substitution of iron
viaducts, they must not charge for extra maintenance, and if at Latchford they had
saved the cost of maintaining seven level crossings, this saving should go in reduction
of other extra costs.   If they got a new railway in place of an old one, they should
not charge the whole cost.   It was allowed that the ruling gradient 1 in 135 was not
unusual: indeed Sir George Findlay admitted this, and there already existed one of
1 in 120 not far away.   Mr. Moulton ridiculed the idea of £25,000 for bank engines,
and said such an expenditure was quite unnecessary.

Sir Benjamin Baker said the gradient 1 in 135 was a good one, and with a
powerful engine, a bank engine would not be necessary for an ordinary luggage train.
He considered the estimates for painting, signal boxes, etc., excessive.

After sitting twenty-four days the arbitration proceedings were closed with a
reply from Mr. Pope, who said his speech, like Bob Acres' courage, had been oozing
from his finger-ends all the time.   He maintained the evidence for the Ship Canal
fell far short of the propositions Mr. Moulton had made.   It was one of those
examples of science not standing the practical test.   The London and North-
Western Company did not want to make a profit out of the transaction, and it was
impossible to put them in as good a position as they were before the deviations were
made.   Lord Balfour then adjourned the arbitration in order that he might have
time to consider his decision.

At the beginning of June a special meeting of the City Council was called to
consider the report of the Special Sub-Committee, who, along with Mr. Moulton, had
drawn out a form of procedure for the election of Ship Canal directors, and for a
definition of their duties.   The report also dealt with the powers granted by the
1893 Act, which included the loan of £2,000,000 by the Corporation to the Ship
Canal Company.

When the report came on for consideration, the mode of election of Ship Canal directors was discussed, and it was arranged there should be a succession of ballots, and those who first obtained a majority of the Council should be elected.   The result of the first ballot was that Sir John Harwood, Alderman Leech, Alderman Marshall (Mayor), Councillor Southern, Alderman Clay, Alderman Chesters Thompson, Alderman Mark and Alderman J. Thompson were elected.

In subsequent ballots Alderman Walton Smith, Councillor McDougall and Councillor Pingstone were added, making up the statutory number of eleven.

The Mayor moved the resolutions for the necessary issue of stock.   In seconding, Sir John Harwood said the London and North-Western Railway and the Great Western Railway had sent in a claim to the arbitrators for £533,000, but the unwarrantable feature was, that the companies refused to give up their old lines till the Ship Canal Company had deposited the whole of the money, and they would not even be satisfied by the undertaking of the Corporation.

In London he had told the London and North-Western Company that the Corporation had often trusted them, and they in their turn ought to trust the Corporation. He told them that when the Corporation had issued their stock, he was prepared on their behalf, to pay over the amount fixed by the arbitrator.   But the company were inexorable, and would not part with the lines unless the money were paid down.   If the directors could get possession of the old lines, the canal would be completed in five months.   They had arranged with the bank for a loan, and had informed the railways they were ready to deposit the money, and had asked them to be good enough to hand over the lines that they might get on with their work.   Mr. Pope had said the railways did not want to stop the work.   He (Sir John) could only judge people by what they did, and not by what they said.   Let their actions prove that their words were sincere.   He believed they could hand the line over next week if they liked, and it should not be kept a day longer than was necessary for the public safety.

This strong speech of Sir John Harwood's caused a commotion in the railway world.   It brought all negotiations with the London and North-Western Railway Company to a standstill, and it roused a strong feeling of resentment on the part of the traders of Manchester, who were the best customers of the railways.   One firm, Messrs. Peter Spence & Sons, at once wrote: "As customers of the London and North-Western Railway Company and also of the Ship Canal Company, we have read in to-day's Manchester newspapers the report of the speech of Sir John Harwood referring to the alleged extraordinary persecution to which the Ship Canal

TESTING LATCHFORD VIADUCT.

Birtles, Warrington.

To face page 156.

Company, and to the affront to which the Manchester Corporation have been subjected by the London and North-Western Railway Company.   We hear to-day of various Manchester firms who have been so deeply pained by the statements made that they have decided to transfer all the traffic they can influence from the metals of the London and North-Western Railway Company."   They ended by asking for an explanation before they, themselves, decided on the course they should take.

This letter and the action of the traders in threatening their business, quickly brought a reply from the secretary of the London and North-Western Railway.   He denied obstructing the progress of the canal, and claimed that they had forgone their legal rights so that the old railways might be dealt with at a time earlier than was fixed by Parliament.   They were not to be touched till the new deviations had been opened six months for goods traffic, and approved of by the Board of Trade ; and not until three months after they had been so passed could they be used for passenger traffic.   Nine months after the deviations were completed was the earliest period at which the old lines could be abandoned.   He claimed the railways had never strictly enforced their rights.

The secretary referred to a speech by Lord Stalbridge in London, when that nobleman had assured a Ship Canal deputation they only wished to be properly safeguarded, and not to delay the canal, and stated that an agreement had been made to give up the railways as soon as the arbitrator had made his award.   When the compensation was either paid or secured, they were willing to give up the railways. When Sir John Harwood proposed the Corporation security, they offered to accept it, if it were legally competent for the Corporation to give such a guarantee.   On this Sir John offered to get counsel's opinion, but they had never heard from him since.   He went on to say Lord Stalbridge had promised Lord Egerton to give up the railway lines before the award was made if the Canal Company would deposit a sum to cover their claim.   In consequence, however, of Sir John Harwood's speech, this offer had been withdrawn.

An article in a local paper thus sums up the dispute : " The London and North-Western declare they have not obstructed the canal work, but have met the Canal Company and have forgone legal rights in abandoning old railways, and turning them over at a period earlier than fixed by Act of Parliament.   Sir John Harwood alleged the railway company had declined the guarantee of the Corporation.   The reply is they did not doubt the guarantee, but, doubting the legality of giving it, they wished the point cleared up by counsel, and this Sir John Harwood undertook to get, but failed to do. Finally, the London and North-Western Company state after the now pending arbitration, *viz.*, on 5th May, Lord Egerton offered to deposit the whole of the railway claim if they would give possession before the making and payment of the award; and to this Lord Stalbridge consented, subject to certain conditions, which were under consideration when Sir John Harwood's speech was made."

At the next meeting of the City Council Sir John Harwood gave his version of the negotiation; he denied he had either a personal grievance or had had a quarrel. When the London and North-Western Company doubted if the Corporation could guarantee the payment of a sum to cover the claim of the railway companies, it was not left to him to get counsel's opinion, but to the legal advisers of each side, *viz.*, Mr. Saxon for the Canal Company and Mr. Mason for the railway companies. He had not interfered in legal matters; all he had done was to try and guard the interests of the ratepayers. He understood the railways to say, "Give us the money and we will give you the canal". Knowing the Corporation were going to assist the canal he said, "We will find the money and it could be paid to-morrow" Then the railways wanted an agreement to which it was impossible for him to consent. By this they blocked progress effectually, and kept the shareholders out of the canal. The Cheshire lines did not take nine months after the new line was completed before they let the canal touch the old one. Why should the London and North-Western be more difficult to deal with? He wished for nothing unreasonable or inconsistent with the public safety, or the interests of the railways concerned. He said unhesitatingly there was no justification for keeping the Ship Canal out of possession of the disused railways. There was nothing to prevent the canal being opened at the beginning of 1894 if the railway companies would do what was just and reasonable. Subsequently, conferences were held between representatives of the Corporation, the Ship Canal, and the railway companies, at which explanations were made, the result being that on the deposit of a sum equal to the whole of the claim, *viz.*, £383,713, the railways agreed to waive their rights and give up possession of the unused railways on the 10th of July, instead of on the 27th November, and at the subsequent Council meeting, Sir John Harwood announced "that the agreement relating to the railway company was signed, and that the Ship Canal would get possession of the railways on Monday next".

It was not till the 25th August that Lord Balfour of Burleigh made his award. Originally the claim of the joint companies was £533,000, but when the Great Western withdrew a portion of their demand it was reduced to £450,000. The total amount awarded was £100,661, of which the London and North-Western Railway Company got £63,991, the Great Western, £5,310, and the joint London and North-Western and Great Western Companies, £31,360. In order to obtain possession of the lines £383,713 had been deposited, so that the Ship Canal Company had to receive back £283,052. It is said the costs of this arbitration amounted to nearly £25,000. The result was eminently satisfactory to the Ship Canal Company.

An Extraordinary General Meeting of the canal shareholders was held on the 6th June, when they assented to borrowing a further £2,000,000 from the Corporation at a rate not exceeding 4½ per cent., and agreed that the loan should take priority of the preference and ordinary shares. The same week the Corporation Finance Committee arranged with the Bank of England to issue £1,500,000 3 per cent. Corporation redeemable stock.

Though powers had been granted to the Ship Canal for the erection of cattle wharves, the details were subject to the approval of the Board of Agriculture, and on the 19th July the Hon. Thomas H. W. Pelham and Major Tennant, on their behalf, held an inquiry at the Manchester Assize Courts.

Mr. Pelham explained that the Commissioners had to decide, Whether Manchester as a port was a suitable place for the importation of foreign cattle? Whether there was a risk of introducing disease by bringing cattle up the canal? If there was a prospect of trade sufficient to justify lairages, and whether they should be located in Salford or Manchester? Many witnesses were examined to prove that more cattle vessels than could be well accommodated now came to Birkenhead; that there was no danger of infection by their being brought up the Mersey; that a great saving would be effected in carriage, and that the accommodation offered at Manchester was admirable. Sir John Harwood stated that the offal would afford cheap and nutritious food for poor people.

Counsel for Salford and Birkenhead cross-examined the witnesses, and tried to prove that, because one of the Beaver Line of steamers which came into the canal to winter, had grounded, the canal was therefore unnavigable.

On the second day Salford's case was heard. Witnesses asserted that the royal borough was more accessible by road and rail, and more convenient to the railways than Manchester, and besides a cattle market already existed there. At present one-third of the cattle killed in Birkenhead came to Salford and Manchester. If slaughtered in Salford there would be a saving in carriage of £10,000 per annum. In re-examination, it was admitted the proposed Salford site was now covered with sludge, and that the cattle must be landed up a gangway, or by a winding staircase.

When witnesses for Birkenhead and Liverpool were called, they denied that they had much to complain about in the accommodation at the Birkenhead and Woodside lairages, though they admitted there had been delays. They also feared that if foreign cattle were allowed up a narrow waterway there might be a repetition of the cattle plague, which had saddled a debt of £270,000 on Cheshire. Further, they maintained the navigation was dangerous, and that cheap offal was not a sufficient reason for the creation of lairages up the canal. The chief traffic manager of the Dock Board said that Liverpool required more meat than Manchester, because it contained stronger people and they required stronger meat, and the secretary of the Dock Board made the extraordinary statement that, whilst the dock rates and town dues payable on a 4,000 ton cargo at the ordinary rates were £614, the Dock Board in 1884 decided they should be reduced during their pleasure to £90, thus admitting an immense saving had been effected through fear of the canal. Mr. Pelham's decision was not given till December, when an intimation was received from the Board of Agriculture approving the Manchester Corporation wharf for the landing and slaughter of foreign cattle. The scheme of Salford for a wharf at Weaste was rejected.

Now that the Ship Canal was nearly completed, it became necessary to realise the disused plant, and as Mr. Walker's trustees claimed a contingent interest, it was arranged with them that Messrs. Fuller, Horsey & Co., London auctioneers, should sell it at the various centres where it lay. The first sale of 200 lots took place at the Salford Docks, when fourteen large locomotives, seven portable engines and other machinery were sold. The best price for a locomotive was £540 and for a portable engine £180. A number of similar sales followed at different depôts on the line of the canal.

By the end of August a very considerable foreign trade had been established at Saltport, chiefly in timber, salt, pitch and general merchandise. During this month Messrs. Smith, Wilson & Smith had the honour of importing the first entire cargo of timber into Manchester *via* Saltport, from Bay Verte. A full-rigged sailing ship of 2,400 tons register, the *Fort Stuart*, shipped 4,000 tons of salt at Ellesmere Port for Calcutta.

The August report of the Ship Canal Company conveyed the cheering news that excellent progress had been made in the past six months, and that the work had been carried out under the estimated cost, and that if nothing unforeseen happened the

BUILDING DREDGERS, SALFORD DOCKS.

Birtles, Warrington.

To face page 160.

canal would be opened in January, 1894. The shareholders were congratulated on the very satisfactory award of Lord Balfour. In moving the report Lord Egerton alluded to the favourable weather they had had, and said that similar weather in previous years would have saved two or three millions in the cost of the canal. They had had no floods and no extra dredging.

He wished to recognise the public spirit of Salford and Oldham in offering financial help, and he thanked the citizens of Manchester for the noble way they had taken upon themselves the whole responsibility of finding the £2,000,000. He also felt grateful to the Newcastle Syndicate who were placing dry docks on the canal. After thanking Lord Balfour, and the gentlemen who had so ably placed the arbitration case before him, he mentioned an agreement whereby if the London and North-Western Railway wanted to widen their line they must apply to Parliament within three years; if they did not, they would have to recoup the Canal Company the cost of the foundations now being put in near Warrington. He alluded to various sales of land which would bring business, and to the various arbitrations on hand, and spoke very hopefully of the future prospects of the canal.

Sir John Harwood, in seconding the report, referred to the declaration of the Chairman of the London and North-Western Railway that they were always very anxious to co-operate with the Canal Company. He wished to believe it, but he thought public opinion had done a great deal to induce the railway company to hand over the old railways.

Mr. Reuben Spencer mentioned that Lord Egerton had declined the honour of a seat on the Agricultural Commission lest it should interfere with his Ship Canal work.

On the 30th August the whole of the City Council visited the canal, and were much cheered by the forward progress of the works. Travelling by train to Liverpool, the visitors crossed in the *America* to Eastham and then sailed up the canal to Saltport, passing on the way the newly arrived floating dry dock which was waiting till its new berth, near Ellesmere Port, was excavated. After lunch the party walked along the Runcorn embankment to view it and the Weaver sluices, and then went by the contractor's line all the way to Manchester, going in switchback style, sometimes on the top of the bank and then at the bottom of the cutting, where hundreds of navvies were working their hardest in order to finish in time. On nearing Mode Wheel they examined the site of the proposed lairages, also the equipment of the transit sheds, eventually emerging at the Pomona Docks.

This journey recalled to my mind the many trips of inspection, often fraught with danger of no ordinary kind, which I had made along with my co-directors, sometimes in the directors' car, frequently in the contractor's open trucks, and at times clinging as best we could to an engine. I had travelled over temporary lines on the edge of a steep bank or over shaky trestle bridges, at one time running for some distance on the permanent way and then dipping down to the bottom of the canal cutting, always in risk of a collision with the contractor's trains which flitted about everywhere, often guided by careless lads who manipulated the points. I consider myself fortunate to have passed through this ordeal unscathed, but I had several hairbreadth escapes. Once passing through a narrow gulley called Tom Paine's Bridge, our engine came in contact with a projection and caused a smash. Another time, unknown to us, the flood had washed the ballast from under the permanent way, and we were in great peril; several times also either the engine or trucks got off the line.

My colleague, Sir John Harwood, was less fortunate, for in going to Warrington to inspect the works, the carriage and engine left the lines, and he was so severely shaken that he could not attend to his duties. I have already recounted a similar accident that occurred to Admiral Nares and Sir John Coode.

One other adventure of a more humorous kind was that, as a member of the Buoys and Beacons Committee, I had to assist in arranging the navigation from Eastham to Runcorn, and at the end of the visit I intended to join the London train at Runcorn. The same day Sir John Harwood and the Works Committee preceded my party, also on a visit of inspection. To their care was committed our food supplies for the day, which were to be left for us at a certain point. Unfortunately, we had a breakdown, and were two hours' late. When we arrived at the trysting-place, thoroughly exhausted, we found to our dismay that we had been given up, and that our lunch had been sent back to Manchester. No food was procurable, and I made the best of my way to Runcorn, but even then misfortune followed me, for there was not a seat to be had in the dining-car, and the train did not stop till it was half-way to London, when I got something to eat, but I never before realised what starvation meant.

In view of the early opening of the canal, the Master Cotton Spinners' Association decided to demonstrate to its members the saving likely to be effected by the new waterway, and an appeal, signed by spinners running 2,500,000 spindles, was circulated, asking for general support. Messrs. Pinkney & Company, of the Neptune

Tunnelling Under Knutsford Road.

Birtles, Warrington.

To face page 162.

Navigation Company, agreed to put vessels on the berth at New Orleans for Manchester. Literature, too, was distributed showing the great savings the canal would effect on various articles of consumption and manufacture. Lord Egerton also asked Lord Rosebery to instruct the Foreign Office officials abroad to make known the advantages of the canal, but that nobleman took a peculiar view of the position and replied, "It was contrary to the practice of the Office to employ Her Majesty's representatives in calling attention to new industrial enterprises," and that it would be a bad precedent.

The syndicate of Tyneside men who had bought land at Mode Wheel for a pontoon and graving docks, now sought to promote a company and raise the capital in Manchester, but they met with practically no response. Nothing daunted, they determined themselves to find the capital in Newcastle and build a graving dock at Mode Wheel, and floating dock to the patent of Mr. A. Taylor, himself one of the directors both at Ellesmere and Mode Wheel.

When the company sold land to the syndicate at a very low rate, one condition was that two Ship Canal directors should be placed on the Board, and that 10 per cent. of the profits should be handed over to the funds of the canal. The directors nominated were Messrs. Bythell and Southern.

In the middle of August the pontoon dock for Ellesmere Port was launched and christened by Mrs. George Renwick. Her husband, the Chairman of the company, at the subsequent luncheon said :—

His company was formed to add to the development of that magnificent undertaking—the Manchester Ship Canal. They would all express their sympathy with the almost superhuman exertions of the directors of that great undertaking. He believed they would soon see an end to their difficulties. He was proud the founder, Mr. Adamson, was a Tynesider. They might depend upon it Newcastle would find them ships if there was the trade which he expected. If anything was to be got out of the canal, Tynesiders would not be far away ; and even if there were risks, they were willing to strike a new line and not merely travel in the footprints of others. The pontoon was about to start on her 1,000 mile journey to Liverpool and to find a resting-place at Ellesmere Port.

The huge and unwieldy structure, made more conspicuous by being painted in bright colours, had a successful journey, and on the 18th October it was safely moored in its place. At the luncheon to celebrate this event, the Chairman, Mr. George Renwick, said they were now able to deal with vessels up to 5,500 tons burden. People had told him ships would never go to Manchester, to which he had replied that not only would they go there, but that their owners would be most anxious to send them; and after all, the foul Manchester Docks would be "a paradise" to those who had sailed on the Lagoons, or had been to Santos. He did not believe Liverpool would suffer, but he did believe a vast increase of traffic would follow the opening of the canal.

Liverpool did not view the approaching opening of the canal with equanimity. "How to Scotch the Canal!" was a plank at the Municipal Elections, and the *Liverpool Mercury* showed that the quay attendance, cartage, and master porterage of cotton could be done away with, and the dock and town dues lowered one-half, so that the cost of cotton could be reduced from 13s. 8d. to 5s. 5d. per ton, or to 1s. 7d. less than that proposed by the Ship Canal, and other goods in proportion. Unfortunately for the *Mercury*, it had not got the consent of the master porters or the master carters, who would be seriously affected. Nor had it thought of the feelings of those who had for years been paying 8s. 3d. per ton too much.

During November a new Passenger Steamer Company was floated with £60,000 capital in £1 shares. It bought up the old company, and proposed to increase the fleet and cater for the passenger and ferry business of the canal. For a time, while the canal was a novelty, it flourished. Then a rival company started, but eventually both dwindled away, and the shareholders in the Passenger Steamer Company lost their money.

At the Canal Rates and Charges inquiry held before Mr. Pelham, the question was raised whether the Ship Canal came in the category of Inland Canals and was liable to have its rates revised by the Board of Trade. The decision was that it did, unless special exemption was given by Parliament.

Another inquiry before Sir Courtney Boyle was held in reference to the prohibitory by-laws proposed by the Liverpool Dock Board, dealing with the importation of dynamite and other explosives up the Mersey. These by-laws were opposed by the Ship Canal and other authorities and by private firms. Eventually they were passed, the Committee stating they acted on precedent, and would not decide the legal question as to the meaning of the 1875 Explosives Act.

It was found necessary to deposit an Omnibus Bill for the ensuing session, chiefly for additional powers to buy land, construct railways, fix rates, etc., and this was done in the usual way.

On the 20th November the actual filling of the last section commenced, and one saw the bottom of the Ship Canal for the last time; there was really a melancholy interest in parting with an old friend. Those who in after years view the 35 miles

Birtles, Warrington.

English Navvies and the Iron Horse, Acton Grange.

To face page 164.

of placid water will have little idea of the vast chasm beneath, every mile of it crowded with incidents of one kind or another.   Here was a rock cutting, there was a peat-bog, here a landslip took place that delayed the work, there a sad accident occurred, further on a quicksand was encountered or an engineering difficulty mastered. In another length disastrous floods had inundated the canal and submerged the machinery, and it had taken hundreds of men to repair the damage.   Few will ever realise the miles of contractor's railway lines or the army of navvies that once found occupation in the bottom of the canal.   And here let me say a good word for the navvies who constructed it.   The genuine navvy under a rough exterior hides a kindly heart, and if excesses do sometimes occur, they are generally due to the casuals who, possibly because they are wanted by the police, find their way into remote spots where no character is required, with the hope of getting a few days' work.   With all his belongings in a bundle, the genuine navvy tramps from one job to another; he loves good, fresh air, he enjoys a change of scenery and place, and when he works he does it with right goodwill.

To those who do him a kindness he manifests the greatest gratitude, and I have known a young lady nurse safer from insult or injury in out-of-the-way places among these rough men than she would have been in a large town.   She had tended them when injured or sick, and they would have showed their devotion even at the cost of their lives: no one would have dared to molest her.   On a dark night, or on her way from hospital, she was never short of an escort with a lantern.

No doubt Mr. Walker's care for his men was beyond the average.   He did all in his power to promote their welfare, both in sickness and in health, and if an accident happened to a man in his employ he never lost sight of him, and generally found him some light employment.   The navvies were always very comfortably housed, and they appreciated it.   People who anticipated their advent with terror were sorry when the work was done, for they enriched the whole neighbourhood by earning good wages and spending them freely.   The Hon. Mr. Grimston and the Navvy Mission did excellent work amongst the men, and brightened their stay by many entertainments at which good counsel and advice were given.

The first week in December was a busy time.   Sir John Harwood had declared some weeks before, that on the 7th December he intended to navigate the canal from one end to the other, and it was well known he was a man of his word, and that any one who was not ready would have a bad time of it.   So all submerged dams and unfinished excavations were being dredged away at full speed in order to allow a safe passage.

The completion of the canal caused a flutter in Liverpool. What was to be done to retain the city's trade? Were they meekly to view ships passing the port on their way to Manchester? Had the Dock Board done what ought to have been done years ago, this terrible canal might have been avoided; they had brought the mischief on their own heads. So wrote many correspondents in the papers. Sir Wm. Forwood at the Press dinner said :—

The Manchester men were invariably united in their demands. Let Liverpool learn the lesson and be wise.

Mr. A. B. Forwood suggested a conference of representative men for the promotion of public opinion, and to bring pressure on the Dock Board with a view to reconstruction. A Liverpool correspondent wrote :—

All reformers now recognise that over the portals of the Board Room is written in Roman capitals, " Abandon hope all ye who enter here ".

Again :—

Candour compels admiration of Manchester's enterprise in contrast to Liverpool's apathy.

A generous correspondent to the *Liverpool Courier* wrote :—

Honourable rivalry has hitherto existed between the two cities. Let it still continue; the spur of competition will stimulate and benefit both. Above all avoid all petty, spiteful trade jealousy. There be some amongst us whose object seems to belittle Manchester's enterprise, and churlishly criticise the efforts of our neighbours to obtain cheap water carriage. Beware of such vain babblers. The canal is constructed. The wiseacres prophesied it would not be. The trade will come to it and to us. Liverpool will lose nothing by the rivalry, if we do not lose our heads in our anxiety to overreach our competitors. There all the danger lies.

Belfast's friendly feeling was expressed in a telegram to the " Brither Scots " on St. Andrew's day :—

> Dear Manchester friends ye'll ha'e a great haul
> When ance ye ha'e opened you're great big canaul,
> But tak ye guid care tae leuk tae your locks
> Or the ships 'll gang back tae Liverpool Docks.

On the evening of the 6th December some of the directors of the Ship Canal might have been seen wending their way to the Adelphi Hotel, Liverpool, where they were to stay the night preparatory to an early start next morning. The visit was a business one and strictly private; the intention was to sail along the whole length of the canal.   Neither the Press nor visitors were allowed.   No one could tell what might happen on this the first complete inspection, and if a mishap occurred it was thought best to have no outside spectators.   The greatest vigilance was used to ward off intruders.   The burning anxiety of a youth, however, who wanted to be one of the first to travel on the canal, baffled all efforts.   In the dress of an engineer he smuggled himself on board, and it was not till the trip was nearly over that he was discovered.   He turned out to be a son of Alderman Bailey, who had donned a working engineer's clothing so as to get on board unnoticed and be able to say he was one of the first party to travel on the canal.   At 7 A.M. on 7th December, as it was coming light, the party were tramping down to Princes Dock to join the *Snowdrop*, which was to leave the landing stage at 7.30 A.M.   It had been a stormy night and there was a biting wind, but the excitement kept every one warm.   On board were Sir John Harwood, Mr. S. R. Platt, Aldermen Clay, Leech and Mark, Councillors McDougall, Pingstone and Southern, with the engineers and other officials.   The steamer reached Eastham at eight o'clock, and in ten and a half minutes was safely through the locks.   On leaving, the sun shone brightly on Mount Manisty, and the large floating pontoon at Ellesmere Port was soon sighted, indeed it almost seemed to bar our way.   Passing round her, Saltport was quickly reached, and very busy it looked with a lot of ships loading and unloading.   They had all their bunting hoisted in honour of the occasion, and their crews gave two hearty cheers as the boat passed.   From this point the canal, as a canal, was new ground to all the directors, it had never been traversed by them before :—

> We were the first that ever burst
> Into that silent sea.

After an inspection of the Weaver Sluices, Runcorn was reached at 9.30, and the party remained for twenty minutes at Bridgewater House.   Here the Lord Mayor of Manchester, the Mayor of Salford, Sir Joseph Lee, Sir Edward Jenkinson, Mr. Joseph Leigh, M.P., and Messrs. Bythell, Crossley, Galloway and Henderson joined the steamer.   After inspecting the Bridgewater Locks and noting the working of the various swing bridges *en route*, Latchford Locks were reached at 11.22 A.M. Considerable time was lost through the fact that Armstrong's representative, Mr. Homfray, wished personally to be present at the first swinging of each bridge, and the *Snowdrop* had to slow down till he could come up by land.   Alderman Walton Smith joined us at Latchford, and the ship passed through the locks there in eight minutes.   At Irlam, where there is a lift of 17 feet, the process occupied twelve minutes.   When the steamer passed under the Barton Swing Bridge, Mr. Galloway took the opportunity of proposing three cheers for Mr. Leader Williams for the successful results of his great work, and the compliment was very heartily paid.   At lunch Sir John Harwood proposed " Prosperity to the Ship Canal," and Sir Joseph Lee responded.   The party were so much before time that they dallied in their inspection of the lairages and landed at No. 4 Pomona Docks punctually to time—three o'clock. Here a crowd received them most enthusiastically, and hearty cheers were given for Lord Egerton, Sir John Harwood, the Mayor of Manchester, and the Mayor of Salford.   The whole inspection was a splendid success, and it was evident that the passage could have been made in five hours instead of seven and a half, if it had been necessary.

On the 16th December the trip was repeated in order to give the Press of England an opportunity of viewing the canal, and the journey was accomplished in six hours.   On the way Mr. Leader Williams and Mr. Marshall Stevens explained all details.

The question of the limits of the customs port of Manchester caused representatives from Manchester, Liverpool, and Runcorn, to have an interview with Sir John Hibbert at the Board of Trade.   Runcorn was quite willing to form part of the port of Manchester, provided certain rights were reserved to her.   Eventually a limit was arranged, and on the 22nd December Mr. D. P. Williams, the Liverpool Surveyor General of Customs, formally presented the Mayor with a warrant, issued by the Treasury, constituting Manchester a harbour and port for customs purposes.   Mr. Williams said that the new port included the Manchester and Salford Docks, the whole of the canal from Eastham, and the rivers Mersey and Irwell, from where they began to be navigable down to the point at Ince, where the port of Liverpool terminated.   Manchester and Liverpool thus became adjoining ports.   Runcorn, once a port, now formed a "creek" of Manchester, but still possessed her ancient privileges.

Prior to the opening, revised schedules of canal rates were issued.   The maximum rates granted by Parliament were at the time just one-half the previous cost through Liverpool and by railway.   Afterwards the railways lowered their rates. To meet this, a reduction was made in the canal tariff.   The following is an instance :—

	1883 Cost delivered in Manchester.	Ship Canal maximum fixed by Parliament.	1894 Charge *via* Liverpool.	1894 *via* Ship Canal.	Saving.
Cotton	14s.	7s.	13s. 8d.	6s.	7s. 8d.

In view of the near completion of the canal, all kinds of additions and extensions were going on at the dock terminus.   Banks and shops were opened, also a mission-room for seamen, and a constabulary staff was established.

Now that the canal was proved to be an established fact, and the Pressmen of the country had seen it with their own eyes, the articles upon it were many and various.

The *Liverpool Mercury* now saw clearly that big ships could come up the canal, and pointed out that even two Boston steamers coming to Manchester would be a loss to Liverpool of £5,000 per annum in dock and town dues, and of £20,000 in actual wages to the labouring classes of Liverpool, and would, moreover, effect a saving in carriage of 5s. 8d. on every ton of cotton.

The *Daily Telegraph* complimented Manchester on accomplishing a stupendous task, and acknowledged that the city must become the natural and necessary port for 8,000,000 of traders and wage earners.   Whether the canal paid or not, they would offer a tribute of national recognition to the promoters and Corporation of Manchester for the resolution they had shown, and for the colossal efforts they had put forth.

The *Liverpool Courier* anathematised the Dock Board as being the real cause of the canal.   They had scorned the warnings given them, and it was futile to teach old dogs new tricks.   "The Board was too big and too old to go to school again to study Benjamin Franklin's wise sayings."   They had been weighed in the balance and found wanting in every essential qualification for successful administration.   Of the canal itself they were good enough to say, "Whatever opinion may be entertained as to the commercial success of the Ship Canal, there is no doubt that that waterway, connecting the Mersey with the great seat of the cotton industry, is one of the most remarkable undertakings the world has ever seen".

The *St. James's Gazette* said "Manchester declines to decline.   It is such things as the Manchester Ship Canal and the Forth Bridge which ought somewhat to cheer the pessimist who looks round and tells us of the decadence of Great Britain".

A correspondent in the *Pall Mall Gazette* was most pessimistic.   As an engineering feat the canal was a success, but it would be a white elephant.   The Lancashire and Yorkshire Railway had no connection with the docks, and if it had, the railways had combined to raise their carriage from Manchester to adjacent towns, with the object of diverting traffic back to Liverpool, and there was going to be concerted action between the Dock Board and the railways to swamp the canal.   It would take ocean steamers two days to come up the canal, and the shipowners in consequence would charge 5s. per ton more to Manchester than Liverpool.

This pungent article brought a rejoinder from Mr. Joseph Lawrence, an early pioneer of the canal, who showed the absurdity of the threatened boycott, and pointed out that the railways could not levy heavy charges to punish the canal—their Acts would not allow them—and if they did, they would be cutting their own throats. If the railways carried for nothing, the Liverpool charges would still be more than the Ship Canal tolls. As to cost, the Liverpool Docks had cost twice as much as the Ship Canal, and the Board's assets consisted mainly of dock walls.

The Liverpool City Council felt spurred to do something, and passed a resolution: "That it is desirable that a Joint Committee be appointed, consisting of representatives of the Corporation, the Dock Board and the various railway companies in the city, to make inquiries as to the effect likely to be produced on the trade of Liverpool by the working of the Manchester Ship Canal, and to report thereon".

Following suit, the Bootle Council also resolved: "That at the conclusion of the Council Meeting the Council do sit in Committee for the purpose of considering the relations of the Manchester Ship Canal to the trade of this borough".

The mover said that anything happening to their shipping trade would cause the people to flee as if from the plague.

The Liverpool Dock Board seemed quite calm in the midst of the storm raging around them. The Chairman at their meeting said there was an impression abroad that they were a singularly apathetic body and by no means alive to the changed position caused by the Ship Canal, but it was a mistaken idea. They meant carefully to watch how the Ship Canal would affect Liverpool. He did not fear Manchester being a cheaper port than Liverpool, and he believed the work going on at their new docks would bear comparison even with the Ship Canal. There was no Board in England more devoted and faithful in its work than the Dock Board of Liverpool.

As the day of opening approached, great anxiety was shown in the shipping world to be represented in the procession of ships. About the first applicant was the *Sophia Wilhelmina*, from West Bay, Nova Scotia. She agreed to wait a month, the merchant in Manchester giving the captain a handsome douceur for doing so. Being the first foreign vessel to enter the canal, the directors undertook to bear the cost of towage and of cutting the masts, and the captain was presented with a gold watch.

At Latchford ships were congregating to take part in the opening ceremony. In order to give the 40,000 shareholders an opportunity of seeing the first ship sail into Manchester Port, the dock wharves were allocated to them, and very quickly all the tickets were applied for.   It was arranged that Mr. Sam Platt's steam yacht, *Norseman*, should lead the way with the directors on board.   The Corporations of Manchester and Salford were to follow in separate steamers.   In their wake was to come the *Skirmisher*, specially retained through the exertions of Councillor Batty, for the families and friends of the members of the Manchester Corporation. A number of private steamers (also specially retained) were to follow, and after these, cargo ships of all kinds and from all places were to complete the procession.

The Manchester *City News* ended the year with a review of what it called "An Accomplished Fact," and thus plumed itself on its unwavering allegiance to the cause :—

The campaign for cheap water carriage began with only one friend in the Manchester Press.   The scheme was scoffed at on 'Change, the chief merchants held back.   Our richest merchants, with some splendid exceptions, refused to entertain the idea.   The great majority of the London and Provincial Press derided the scheme and called it " a wild dream ".   Advocates were stigmatised as " ward enthusiasts," and were told their figures were illusory, and their arguments illogical and untenable.   Daniel Adamson was even denounced for asking people to subscribe to a scheme in which they would lose their money.   Our local M.P.'s, with three exceptions, declined to take the responsibility of lending their names to the scheme, as by so doing they would be giving a warranty, as it were, for its success.

The article pointed out the lamentable state of the city prior to the Ship Canal, showing how trades were being driven from the city, and stating that in 1881, 18,632 dwelling-houses, sufficient to shelter 100,000 people, were uninhabited.

Then it was that a few earnest and courageous men determined that the only way to prevent further dwindle and decay was to remove the burden that was pressing on our industries, and bring Manchester in direct communication with the sea.   Without the help of as devoted a body of men as ever unselfishly took part in any cause, even Daniel Adamson, with all his marvellous persistence and strength, might have failed to create and sustain a public opinion and induce Parliament to sanction the scheme.

PROGRESS OF THE WORKS.

On 26th January the new swing road bridge over the Irwell at Barton was temporarily opened for traffic, the Mayor of Eccles being present.   He, and Mr. J. C. Mather were the first to go over it in a cab.   The bridge weighs 800 tons, is 195 feet long, 25 feet wide and is turned on sixty-four steel rollers worked by hydraulic power.

As the railway diversions were now completed, the Board of Trade Inspector certified them. In January Major-General Hutchinson tested Deviation Railway No. 4, on the Liverpool and Stockport line, with ten of the largest engines he could procure, and he found the girders stood the test satisfactorily.

A further step was taken when the Cheshire lines, instead of waiting six months, allowed passenger trains after one month to run between Glazebrook and Flixton. This enabled a dam to be removed, and facilitated the work of the contractor and the company.

On the 27th March the Ship Canal Docks at Salford were filled to the height of 10 feet, or within 7 feet of the keel of the dredgers *Medlock* and *Irk*. These dredgers were constructed by Messrs. Fleming & Ferguson, of Paisley, and brought to the docks in parts, to be put together on a "grid".

When the water rose to 24 feet, the dredger *Irk* was to float away to the scene of its future labours, and the *Medlock* was to eat its way out, by cutting away the dam that divided the docks from the canal proper. The number of Scotch shipwrights busy at the docks gave the works the appearance of a Paisley or Greenock boat-building yard. Besides the dredgers, twelve steel barges had been constructed by Messrs. Gilchrist, of Liverpool. Each barge held twenty boxes, and these boxes when filled by the dredgers were to be tipped on the adjoining land.

In May of this year the results of the dredging by the Dock Board at the Mersey bar were announced. They showed that, so far, 5 feet more water was available than heretofore, and a minimum depth of 17 feet at low water was established.

Every mishap on the canal, however small, was magnified by the Press into a disaster. Thus a second slip at the Bridgewater Lock in May, said to be very serious, turned out to be no more than a slip of the embankment that delayed the work a couple of days.

During the same month the flood water broke the bank at Barton and filled the cutting, submerging the machinery and temporarily stopping the works.

A really serious accident, however, occurred on the 29th May. Perhaps the greatest engineering triumph in connection with the undertaking was the carrying of the Bridgewater Canal over the Ship Canal at Barton, by means of a swinging trough or aqueduct, worked by hydraulic power, and joined at each end with the main canal. The total length was 1,335 feet, *viz.*, 234 feet of aqueduct and an

Killon.

BOARD OF TRADE INSPECTION OF THE IRWELL VIADUCT. LOAD, TEN LOCOMOTIVES = 750 TONS.

To face page 172.

Killon.

BARTON AQUEDUCT APPROACH AFTER THE COLLAPSE, SHOWING WHERE THE WATER ESCAPED.

To face page 172.

average approach on either side of 550 feet. At the end of the approaches were dams, separating the new work from the old course of the canal. Everything was complete, and it only remained to make an aperture in both dams, and fill the new length with water. Amidst the cheering of the people the apertures were made, but when the trough was half full, an alarming accident occurred. Through some fault of construction a considerable portion of the bed of the approach on the Barton side collapsed, and a large body of water rushing through the hole forced its way through three of the arches, which carried the wall and inundated, or washed away the structures below. Fortunately the water found its way into the Ship Canal, and thus limited the damage. The accident postponed the completion of the canal, inasmuch as the old waterway had to be used during repairs and thereby delayed the pulling down of Brindley's aqueduct, which had to be removed before the section could be opened.

It did not take long to repair the damage at the approach to the swing aqueduct. On 14th June pumping operations were commenced to fill the tank, and a few weeks later boats were able to cross the aqueduct. The first barge to enter the new aqueduct was the *Ann*, of Lymm, which passed through on the 21st August with a cargo of vitriol, for Hapton, near Accrington.

When the Barton Swing Bridge and the Aqueduct were both in full action, the contractor commenced to remove Brindley's aqueduct and the rock in the river bed. The latter was, to a great extent, excavated in the dry, a coffer dam was made in the centre of the river, after which first one side was blasted, and then the other.

About the same time the Flixton Weir, at the confluence of the Mersey with the Ship Canal, was completed. It is 8 feet deep, and under the advice of Mr. Hill it was widened to nearly double the original width.

On the 9th June the docks at Runcorn, which had been closed for twelve months, were re-opened, and on the same day the first passenger train ran over the London and North-Western deviations at Walton and Latchford.

When Telford constructed the Weston Canal, he built a massive stone wall fronting the river Mersey. This Mr. Leader Williams proposed to use as his inner wall, the canals being adjacent to each other. But when Mr. Jones, the local agent, put his men to excavate for the new canal, Telford's wall began to crack, huge blocks of stone split in two, and the wall went $3\frac{1}{4}$ inches out of plumb. At once a big weight of clay was placed against it and the necessary propping executed. The clay was dredged out after the canal was filled : the water subsequently acted as a buttress, and the old wall has not moved since.

No doubt Sir John Harwood, as Chairman of the Executive, did the right thing in fixing a day for the canal opening and making the whole staff work to it. High pressure does not adequately express the feeling that pervaded the job, from the contractors downwards. Everybody felt that the opening having been announced for the 1st January, 1894, it would be a disgrace not to be ready. The directors were constantly on the works, exhorting and encouraging the workers. In order to keep them in touch with each other, bulletins were regularly issued to show the position and progress of the work. Glancing at one issued on the 28th September, it would appear that Mr. Wills' contract, the length from Mill Bank to Warrington, was the furthest behind, whilst Mr. Jackson's, from the Old Quay to the river Weaver, was a weak spot. On a portion of the former it was decided not to go to the full width intended, and so for three-quarters of a mile it has remained, and without disadvantage, to this day. One great difficulty was the slipping in of the banks prior to the water being turned in, and so late as the 4th October a large landslip occurred near Tom Paine's Bridge that smothered a man, and caused much delay.

On the 12th November it was announced that the length of the canal—8 miles in all—from the dam at Runcorn to the dam at Latchford was practically completed, and that the engineer was going to fill it by slow degrees by means of the various brooks, and by a pipe carried through the dam. Mr. John Jackson, the contractor, had, perhaps the worst piece on the whole canal to deal with. Highways and byways had had to be kept open, and he had temporarily to divert several streams, and to excavate deviation lines under railways. In addition to this, the soil passed through was most treacherous, being often either bog or quicksand. On this length alone are five swing bridges and one permanent bridge, besides two railway viaducts, Randles Sluices, and the locks and sluices at Latchford. The whole work is a monument of skilful engineering and contracting.

Just when every one was elated at the prospects of early completion, news came of a disaster to the railway viaduct at Irlam, which cast a cloud over the promoters. It was on the direct line to Liverpool, after a severe test had been passed by the Government inspectors and was in full use for passenger traffic. It should be explained that the viaduct was on piers 86 feet long, 20 feet thick and 192 feet high. Borings to the depth of 17 feet had been made in the clay, and the foundations were

CHESHIRE LINES RAILWAY BRIDGE.

Birtles, Warrington.

deemed satisfactory by the engineer of the railway. The work was begun in 1889, and in January, 1893, goods traffic was sent over the viaduct, followed by passenger traffic three months later. Slight signs of movement on the easterly pier were evident in February, but little importance was attached to it, though in May the pier was found to project 1¾ inches at the top. In June the deflection rapidly became worse, and it was then discovered that the bed of clay, 21 feet 6 inch deep, on which the piers were built, had, underneath it, a bed of subterranean water that was not constant, but was affected by the weight of the superstructure and caused the piers to move forward at the top. When the water was admitted into the canal, the overhang increased to 7⅜ inches. Ominous cracks appeared and grew worse, till in October, 1893, the Cheshire Lines Committee propped the structure, and commenced to remove the arches, and to replace them by means of horizontal steel girders in order to eliminate the thrust on the upper parts which was produced by the arches. The work was completed in May, 1895, and has been perfectly secure ever since. The enemies of the canal were jubilant over the mishap, and wondered that the canal directors had ever been allowed to imperil a great highway to Liverpool.

But there was yet another disaster. All seemed to be going well, and the Runcorn and Latchford length was being gradually filled, when floods of a very serious nature occurred on the 25th and 26th November. A few days later they could have done no harm, as the canal would have been full, the inside and outside pressure equalised, and the dam at the Runcorn end, which really caused the mischief, removed. As it was, an immense body of water came down the Irwell and Mersey valleys, greater than the old course of the Mersey could pass. This was passed through the various lock sluices and rushed into the 9 miles of partially empty canal, where it was stopped by the Runcorn dam. If it had risen only to the 26 feet all would have been well, but it kept rising till it reached a height of over 50 feet. Then as it could find no outlet, it burst through a weak spot at Randles Creek, and passed into the adjacent river. In its sweeping rush downwards it carried away the temporary structures by which Bob's Bridge and Tom Paine's Bridge had been replaced. It also did damage to a new wharf in course of erection for Messrs. R. Evans & Company, and loosened the bank in many places, especially at Latchford. Happily, the channel was clear and ready for filling, so that no pumping out had to be done. It was only necessary to clear away the débris and repair the banks, and when the Runcorn dam was lowered, the water passed quietly away. Unfortunately, a young fellow was drowned in trying to navigate a raft and relieve some people whose houses were surrounded by water. The Liverpool papers contained greatly exaggerated accounts of the "catastrophe," and thereby created much alarm. A few hundreds of pounds covered the damage. The water speedily washed away the Runcorn dam to a depth of 12 feet, and on the 27th November the *Falmouth Castle* passed over it, showing there was now a passage through to Manchester.

INCIDENTS AND ACCIDENTS.

On the 1st of October, 1893, after a visit to the works, the directors dined together, and I happened to sit next to Sir Joseph Lee. His health had been failing for some time, but that day he was brighter than usual, and in a communicative mood. We talked about the difficulties of the past, and he said few would ever know how nearly the Ship Canal was financially wrecked in 1886. On my expressing a desire to know the history of the crisis, he went on to say that the first failure to raise the money clearly proved that Mr. Adamson had not the confidence of the financial world. He (Sir Joseph) had long thought it absolutely necessary that some one should head the concern who was well-known in London and the country, and whose name would raise the enterprise to a higher platform.

After consulting with friends, he made up his mind that Lord Egerton was the right man if only he could be got. Accompanied by Sir William Houldsworth he called upon him at his London house. His lordship was out, but they waited for him from 6 to 8 P.M. Then Sir William said he must go to a dinner-party. Just as he was leaving, Lord Egerton came in. They turned back and made the most forcible appeal to him in their power, saying that his lordship had the opportunity of doing his country a great service, and that as a landed magnate he would command respect and support. Lord Egerton at first gave a distinct and positive refusal. Sir William Houldsworth was obliged to go, but he (Sir Joseph Lee) stuck to his task and would not be shaken off. Lady Egerton, who was tired of waiting, at length appeared on the scene, dressed for dinner, and reminded her husband of an engagement. Another forcible, nay even impassioned appeal was made, on the ground that his taking the helm would probably save the canal. At last pertinacity won, and his lordship gave way, saying, "Well, be it so". They then shook hands and parted.

Sir Joseph Lee also spoke of a critical time, when the company had to pay £1,710,000 for the Bridgewater Canal on a certain day. The directors had relied upon a large amount of the calls being paid in advance, but they were disappointed. There were only three days left in which to raise the money, and it was a tight corner.

BOBS BRIDGE, ACTON GRANGE.

Birtles, Warrington.

He took a cab and called upon various English and foreign financiers, but they could not, or would not help.   At last he found Glyn, Mills & Co. were in a position to find the money, and he persuaded the head partner to lend it on the security of the Bridgewater deeds.   But then another difficulty arose.   Had the Ship Canal Company power to raise money on them?   Fortunately, a very old deed turned up containing powers that had not been abrogated.   On this somewhat doubtful title, and through the goodwill of the lender, the money was obtained and punctually paid over.

Sir Joseph Lee attributed the first failure to raise the capital to a "cheeseparing" policy.   The London brokers boycotted the issue because their usual terms were denied to them.   He also gave some amusing incidents, showing how Mr. Hilton Greaves used to help them when they were in a pinch for money, and how he persuaded Mr. James Jardine to invest £5,000 in the canal.

During the London and North-Western arbitration, one of the claims made by the railway company was for the cost of painting the underside of the railway bridges. Mr. Pope said it would be a costly and dangerous work.   Sir Benjamin Baker suggested an ingenious travelling stage which would overcome the difficulty, and by which the men when painting could quickly get out of the way of ships.   Mr. Pope threw doubt upon it, whereupon Mr. Moulton offered to guarantee that it would work to the entire satisfaction of Sir George Bruce, the arbitrator's assessor, and undertook to bind his clients to do the work.   Mr. Pope rejoined: "I doubt your clients.   I don't think ten years hence your clients will be in existence."   The time has elapsed, but it is the counsel, not the canal, that has ceased to exist.

This year was an exceedingly dry one, and as soon as the warm weather commenced, complaints were made of bad smells.   Salford and Stretford were up in arms, along with the other inhabitants of the valley of the Mersey.   Everybody was blaming everybody else, and no one could fix the responsibility.   The Local Board, the Joint Rivers Board, the Ship Canal Company, and the Corporations of Manchester and Salford were by turns invoked.   When Salford was loud in condemnation, a correspondent wrote:—

What will be the outcome of the canal's efforts I cannot say, but if no better result is attained by the Salford Corporation in their so-called purification of sewage, the result will be a downright failure.   I do not hesitate to say that the stuff turned out of the sewage works at Mode Wheel is even more vile than it was before it underwent the process of purification.

Dr. Hewitt, too, at a meeting of the Joint Rivers Board, described the effluent from the Salford sewage works as "abominable". Manchester admitted she was an offender, and asked for time to complete the intercepting sewer and the works at Davyhulme.

At the June meeting of the Salford Council, Alderman Walmsley proposed that an injunction should be applied for to prevent further deposits in the old bed of the river, and eventually it was arranged to inform the Ship Canal Company that the Corporation intended to take action of a strong nature, unless the nuisance were abated. A meeting was suggested between a Sub-Committee of the Corporation and the Ship Canal Board. During the discussion Alderman Bailey blamed both Manchester and Salford. He plainly told the latter that :—

Before this filth was let in from the Salford works, the bottom of the canal was clear and sandy. He had seen the deposits placed, and seen the bank accumulate right opposite the sewage works. It was the necessary removal of the accumulation that caused what was undoubtedly a horrible smell.

As the summer went on, complaints about the canal and docks became fast and furious. The newspapers teemed with letters from persons who believed that they were about to be asphyxiated by pestilential fumes. Manchester was accused of turning the untreated sewage from 20,000 closets into the river, and Salford of acting a sham with their new works. The Ship Canal Company were charged with bottling up vile filth in the docks, and then periodically stirring up the gases by their dredging operations. Public meetings were held in many districts. At Stretford very strong language was used about Manchester pouring liquid filth into the Irwell, and this was alluded to in the City Council by Alderman Joseph Thompson, who gave an assurance that very shortly the whole of the Manchester sewage would be intercepted, and would go to Davyhulme.

The Salford Council went so far as to ask the Local Government Board to institute an inquiry into the pollution of the waters impounded by the Ship Canal, and received a reply that it was the duty of the new Joint Rivers Board to look into such matters, upon which Salford waxed very indignant, and resolved to inform the Local Board that the Joint Rivers Board had failed to exercise due diligence in carrying out the provisions of the 1892 Act. They again asked for a public inquiry.

This was followed by a question asked of the President of the Local Government Board by Mr. Knowles, a Salford M.P. He wanted to know if the President's attention had been called to the Ship Canal, as being dangerous to the health of Salford, and if he would institute a Government inquiry? The President, in reply, said the questions were difficult to answer. His attention had already been called to the position. He would suggest that the borough should take legal advice as to the Ship Canal creating a nuisance, and he promised any assistance in his power if the Corporation came to Parliament for additional powers.

But the most savage attack was made in *The Lancet* of 4th November by the Society for the Abolition of Vivisection.

To the shame of the country this canal will be a sewer, and nothing better than a sewer, when it is opened. . . . But man, proud man, drest in a little brief authority . . . like an angry ape, plays such fantastic tricks. On the one hand, he creates diseases wholesale ; on the other, he unjustly and cruelly tortures his weaker fellow-creatures in the selfish hope of finding cures for those diseases ; thus making the innocent suffer for his own guilt.

Several public meetings of people living near the canal docks had recently been held, and at the Salford Town Hall it was stated that 250 dead dogs and cats were taken out every month. It was predicted the docks would be a cesspool for the cats and dogs of Lancashire.

Strange that the men of Manchester did not foresee as much. Liverpool had better look out, or she may have to dredge Manchester sludge.

These prognostications are only worth quoting to show how unreasonable scientific men may become when their minds are warped by prejudice. Certainly there have been odd times when a strong odour has arisen from the canal, but as a rule the people who work on the canal are perfectly healthy, and I am not aware of any serious epidemic breaking out on land adjacent to it.

The Mersey Commissioners, having decided to spend £10,000 on dredging operations at the Mersey bar, started to work in August, 1890, with two hopper dredgers of 500 tons each. They soon found that though they were on the right track, they must get a dredger capable of containing up to 3,000 tons of sand. They therefore ordered one from a Barrow firm, which could be filled in three-quarters of an hour. By the end of 1891 the depth of the bar had been increased by about 5 feet. The Dock Board got power to spend a further sum of £60,000 in plant. Dredging, however, was slow in 1892, in consequence of having to use old dredgers. Every favour was shown to Liverpool dredging craft, for, whilst they were allowed to deposit two miles from the scene of their operations, the Ship Canal Company had to take their dredgings a distance of four nautical miles from the Bar Lightship. On the 14th June, 1893, the Naval Construction and Armament Company of Barrow delivered their double screw 3,000 ton hopper dredger, the *Brancker*, and she very soon made herself felt at the bar.   By the end of the year there were 20 feet of water at the lowest tides, and twelve months later this had been increased to 23 feet at low water spring tides.

This improvement affects Manchester equally with Liverpool, as it allows almost all vessels to enter the Mersey at any state of the tide.

On the 9th January a singular and fatal accident occurred at Lymm.   It had been usual to bore a row of holes in the rock, to be blasted from off a pontoon, then to move the boat away when the fuse was applied; but through some entanglement the explosion took place before the pontoon got away, and as a result, three out of fourteen men on the boat were injured, and one (Lincoln Jack) was killed.   It was shown at the inquest that by some neglect the charge in one of the holes had not been rammed down, and had exploded prematurely.   A verdict of accidental death was returned.

Later in the month a serious collision occurred on the overland route, near Runcorn, which caused the death of Captain Nicol Bain, of the London steamer *Cragg*.   He had got a lift on an engine, which, through mistaken signalling, came into collision with another engine.   The shock caused him to fall on the ground and he had his neck broken.   The coroner strongly objected to the loose way of signalling, and the practice of allowing people to ride on locomotives.

ARBITRATIONS.

It will be remembered that when the route of the canal was taken away from the side of the race-course, the owners of the latter claimed they were riparian owners, and entitled to use the waterway.   This was disputed, inasmuch as there was a wide towing-path between the race-course and the river, over which the owners of the former had no right.   The Race-course Company claimed £40,000 for the deviation of the canal.   The dispute was about to go into litigation, when Sir John Harwood for the Ship Canal, and Alderman Hinchcliffe for the Race-course Company, met and came to terms.   The latter were to withdraw their claim for riparian ownership, and were to convey certain property previously agreed for, and the Canal Company were to have the first offer of the race-course property, if ever it was to be disposed of.   The Canal Company undertook to admit that the race-course owners were entitled to the privileges of the 1885 Act, if it were ever determined to convert the race-course into docks.   Then they were to give water access from the Ship

JOHN D. WALLIS, LAND VALUER AND AGENT FOR THE MANCHESTER
SHIP CANAL COMPANY.

A. Coupe, Withington.   *To face page* 180.

Canal into such docks, free of wharfage rates. The Canal Company were, at the cost of £7,000, to build a division wall between the race-course and the canal property. The Canal Company were to give the Race-course Company access to the docks by rail or road, and with suitable junctions to the company's railways adjoining the race-course. The idea of the settlement was, that it was better to encourage the conversion of the race-course into docks yielding revenue, than to spend money in litigation, and prevent the race-course ever being utilised for shipping. In the light of after events, there is no doubt the agreement enhanced the price the Ship Canal Company had to pay for the race-course.

Another case was with the Manchester, Sheffield and Lincolnshire, and the Midland Railway Companies in respect to 4½ acres of land near the end of the docks at Woden Street. For this the companies asked £36,805, the Canal Company offered £22,500, and the umpire awarded £22,773, or about £1 per yard.

This arbitration was held at the Queen's Hotel, and damaged several reputations. One valuer gave it as his opinion that land of the kind had gone up materially in the last twenty years, and he valued it at 13s. 4d. per yard. The opposing counsel then proved that the same man had valued the same land in 1875 at 35s. per yard. It was an unhappy half-hour for him. No doubt the excellent evidence given by Mr. Lionel Wells did much to win the case for the Ship Canal Company.

The action of the Chilworth Powder Company v. the Ship Canal Company, to enforce specific performance of an agreement to lease certain land situate at Pool Hall, Cheshire, for twenty-five years, at £120 per annum, came on before Mr. Justice Chitty in the Chancery Court in November. The action against the defendants was dismissed with costs against the plaintiffs.

When all the land had been purchased, Mr. J. D. Wallis, of Dunlop, Lightfoot & Wallis, land agents for the company, read a paper at the Surveyors' Institution full of interesting information, and by his kind permission I give a summary of the work done by his firm.

	Land to be Purchased. Acres.	Estimate of Cost. £	Average per Acre. £	Spoil Land to be Resold. Acres.	Value. £
Parliamentary estimate, 1885 . .	3,430	1,100,008	320	1,766	116,672
Particulars of actual purchase and cost	4,664	1,359,490	291	—	—

The increase in the quantity of land bought arose from alterations made in various Bills, and from purchases made to avoid severance.

Nearly a thousand claims were made, *viz.*, by landowners 425, and by lessees and tenants 529. The total claims came to £2,325,208, and these were settled for £1,359,490, and this sum included—

Redemption of land tax	£2,721
Interest on purchase money	42,036
Expenses	101,121

Excluding these items and tenants' claims, the cost of 4,664 acres was £1,094,463 or £234 per acre.

If 70 acres of building land in Manchester, costing £99,747, and 99 acres of land covered with buildings costing £217,639 were eliminated, the remaining 4,495 acres of semi-agricultural land, including expenses, cost £172 per acre. Prices varied from £97 an acre near Warrington to £5,000 per acre for building land at Manchester.

Among the largest purchases were 358 acres from the Dean and Chapter of Chester for £125 an acre, and 514 acres of marsh land and saltings from the Marquis of Cholmondeley and others for £107 per acre. To the Crown £18 per acre was paid for 334½ acres of foreshore. After this was bought, a Major Orred claimed that he, and not the Crown, owned 100 acres of the foreshore and river bed at Weston, and succeeded in establishing his claim in the Queen's Bench. He proved the existence of an ancient fishery from the year 1086 A.D. The Crown had to refund the money. The following is a rough summary of the average prices paid :—

	Per Acre.
River bed (non-tidal) to its centre	£20
Towing-path	50
Land at Norton and Acton Grange, including Sir Richard Brooke's . .	170
„ „ Moore, no severance	97 to £120
„ „ Lower Walton, including Sir Gilbert Greenall's	263
„ „ Arpley Meadows, including thirty houses	250
„ „ Latchford and Grappenhall, including house property . . .	300
„ „ Thelwall and Woolston, including bad severances	175
„ „ Lymm, no severance	115
„ „ Rixton, no severance	110
„ „ „ severe severance	170 to £200
Partington, Warburton and Carrington	140
Irlam	150
Barton-on-Irwell	290
Trafford Park	700
Pendleton, including river bed and towing-path	400
Egerton estate	1,100

FRENCH NAVVY AT WORK, ARPLEY MEADOWS.

Birtles, Warrington.

To face page 182.

Out of 954 claims, 922 were settled by agreement ; of the remainder, twenty-seven were arbitrated, four went before a jury, one, a tenant's claim, went before justices.

In five cases the award was less than the statutory offer.

	Statutory Offer.	Award.
For land at Ellesmere Port	£5,457	£4,432
„ „ „ Stanlow	1,000	900
„ mill at Runcorn	3,640	2,725
„ land at Salford	7,000	4,785
„ „ „ Flixton	250	150

In many other cases the award only exceeded the statutory offer by a few pounds.

There can be no question that Messrs. Dunlop, Lightfoot & Wallis managed the purchasing of the property for the Ship Canal with the greatest prudence, skill, and ability, and that they deserve the best thanks of the shareholders.

CHAPTER XXIV.

1894.

DR. PANKHURST'S APPLICATION FOR CERTIFICATE, AND
SPEECH—BUSINESS OPENING OF THE CANAL—OPINIONS
OF THE PRESS—LABOUR DIFFICULTIES—BOYCOTTING
THE CANAL—SHIPPING RINGS—APPLICATION TO THE
RAILWAY COMMISSIONERS—DRY DOCKS. COMMITTEE,
RE CANALS—FORMAL OPENING BY THE QUEEN—DIS-
PUTES AT WARRINGTON—DEPRESSING INFLUENCES—
THE CHAIRMAN OF THE EXECUTIVE COMMITTEE DIS-
SATISFIED AND RESIGNS—EXCITING DEBATES IN THE
CITY COUNCIL—LIVERPOOL JUBILANT—MR. J. K. BYTHELL
APPOINTED CHAIRMAN—OPENS OUT WITH BOMBAY—
SUGGESTIONS FOR A COTTON EXCHANGE—DEATH OF
SIR JOSEPH LEE.

ON THE OPENING OF THE MANCHESTER SHIP CANAL.

> When Orpheus tuned his lyre, he played so well,
>   The rocks and trees came down the strain to greet ;
> But thou, Mancunium, with a mightier spell,
>   Hast drawn great Neptune to thy lordly feet.
>   —W. BLAKE ATKINSON.

NEW YEAR'S DAY, 1894.

BEFORE the canal could be opened on Monday, the 1st January, certain
formalities had to be gone through. It was necessary to get a certificate
from the Chairman of the Salford Quarter Sessions that the Manchester and
Salford Docks were completed and fit for the reception of vessels.

To facilitate matters, the Chairman, Mr. Addison, Q.C., held a special sitting
of the Court on Saturday, 30th December, when Dr. Pankhurst, for the Ship Canal

Company, applied for a certificate, and put in the box the engineer, Mr. Leader Williams, who proved that the docks were finished and ready for use, on which the Chairman signed the certificate. The occasion was historical, and evoked from Dr. Pankhurst a speech so full of fire and enthusiasm as to be worth recording. He said :—

They all remembered the obstacles, difficulties and hostile influences under which the canal project was initiated, and under which the first, and many subsequent stages of the work were executed. It might fairly be said that every step of the way was a battle against the obstinacy of nature and the opposition of man. It had been a long, costly and arduous campaign. The victory had come at last, and to-day was undoubtedly a day of triumph. In the long struggle some had passed away. They missed in this auspicious hour the stalwart personality, the giant energy, the unbounded enthusiasm of Daniel Adamson.

After saying the fame of Mr. Leader Williams was assured, and that he would take a place among the great engineers in history, Dr. Pankhurst went on to say :—

Nor could he forget the noble attitude of the 40,000 shareholders who had stood by the canal with an infinite patience and an unquenchable hope. In their constantly growing enthusiasm, they were the great powers of success. In the directorate we had an illustrious body of public servants ; there was a mass of power rarely found in any public body, territorial influence, wide and various business powers, public spirit, responsible relation to public life. There was one thing which ought to give unbounded public confidence in the Ship Canal. No public work of equal magnitude or difficulty was ever subjected to such fierce, constant, persistent and relentless criticism. Out of the fire of that criticism it had emerged, stamped, he said confidently, as sterling gold. The great work lay before us in its completed splendour. He felt that he was expressing the emotions of all when he said their hearts swelled with a good hope for the future of this great district. How could the life of this place be other than larger and loftier when it was now brought into direct contact with all the ends of the earth ?

He concluded :—

On the first day of the new year the great procession of ships ascending the canal and nearing the city, the wide outbursts of applauding enthusiasm from assembled thousands, echoed and re-echoed, as it would be, by millions of this populous district, would dedicate spontaneously and with universal enthusiasm to public uses, to human uses, this noble, this imperial, this heroic work of human effort and endeavour. The Manchester Ship Canal marked a new epoch in the life of the community and a new hope for our country's future. "Floreat flumen navigerum Mancuniense !"

The Chairman, Mr. Addison—

rejoiced it was his duty to pronounce the work complete, and this enabled him to be the first person to wish prosperity to the Port and Harbour of Manchester.

With the 1st January, 1894, commenced a new era in the history of the Ship Canal.   The unwavering public spirit and dogged determination of Manchester had carried to a successful issue a unique and magnificent engineering work, which at first was jeered at, then sneered at, and existed to be cheered at.   On such an occasion promoters and opponents could well afford to be generous, and shake hands; indeed, the former could admit that passing through the refiner's fire had improved the scheme.   All agreed that there was no parallel to the loyal and ungrudging support accorded to the enterprise.

For weeks prior to the 1st January, arrangements of all kinds were in progress, and though this was the business, and not the formal opening day, enthusiasm rose to a high pitch, and there was a general determination to make the procession of ships representative, and to give it a hearty reception.

When the Town Hall struck midnight, hundreds of people were trooping about Albert Square to let in the New Year and with the intention of commemorating the opening of the Ship Canal.   Before that time, trains had departed for Liverpool full of people who hoped to snatch a little rest, and then secure a passage up the canal.   Those who wished to sail up the whole length had to leave very early in order to catch the procession leaving Latchford Locks at ten o'clock.

Close on 10,000 visitors left the Liverpool landing stage on passenger boats to view the canal.   Long before it was light, crowds were rushing about trying to find the boats by which they were booked.   Many people had been up all night, or had left their homes by 3 A.M. in order to catch the special trains from Oldham and elsewhere.   The first boat to leave the landing stage was the *Despatch*, chartered by the Cotton Buying Company to convey their "spinner" friends, and timed to depart at 7.30 A.M.   Starting in the dark, daylight broke upon them as they neared the Cheshire shore.   It was very cold, with a biting air, but a cup of hot coffee soon made the representatives of about 6,000,000 spindles feel at peace with themselves and the world.   The Eastham Locks, and the ships there, were gay with bunting.   On arriving at Ellesmere Port hearty cheers were raised by the people lining the banks, and this demonstration was repeated even more enthusiastically at Runcorn.   This boat, and many others of which it was a type, had a capital view of the whole canal, but in consequence of getting behind the long array of merchant ships in the procession, they did not arrive at Old Trafford till late in the evening.   The *Claughton* was still more unfortunate.   A passenger described her as the most inveterate steeple-chaser he ever met with.   Through bad steering, or not answering the helm, she

The Opening Day, 1st January, 1894.

Ambler.

was constantly trying to jump the banks, but even she got up safely, though the Arts Club on board had to break the monotqny of a tedious journey as best they could.

The day was all that could be wished, and more than could have been expected in mid-winter. The keen air of early morning was succeeded by sunshine, and if this augurs well for a bride, many there were who judged it a good omen for the canal.

Even before it was quite daylight, the South Junction platform at London Road Station was crowded by the Conservative Club party, and others, who were going down to Latchford by the earliest trains. Then followed the train conveying the Manchester and Salford Aldermen and Councillors, who were busy wishing each other "A Happy New Year," and exchanging mutual congratulations about the weather and the canal. The journey to Latchford did not take long, and the train, running up the deviation railway, deposited the passengers at the station. The sight was one not easily to be forgotten. The place was packed with visitors, every house showed some kind of rejoicing, and the ships waiting for the procession were literally a mass of flags and bunting. With much difficulty a way for the directors was found to the *Norseman*, and to the *Snowdrop*, for the members of the City Council.

Mr. Sam Platt's beautiful steam yacht *Norseman*, with the directors on board, was to lead the procession. She was to be followed by the *Snowdrop*, with the Corporation on board, and by the *Crocus*, with ladies and the directors' friends; the *Great Britain*, with the Salford Corporation; the *Skirmisher*, chartered by the Manchester Corporation for their friends; the *Brackley* and *Dagmar*, with the officials of the Ship Canal and Bridgewater Canal. Then came the *Claughton*, chartered by the Manchester Arts Club; the *Despatch*, chartered by the Oldham Cotton Buying Company; the *Wirrall*, chartered by Rylands & Sons, with Mrs. Rylands on board; the *Water Lily*, chartered by the Stretford Local Board; and the *Helvetia*, charterd by the Warrington Corporation. These were followed by vessels chartered by the Manchester Law Society, the Manchester Conservative Club, and Messrs. Peter Walker & Co., of Warrington.

Then came a number of business and pleasure boats, belonging to the Liverpool Tug Company, the Ship Canal Passenger Syndicate, the Salt Union, Ltd., Rd. Evans & Co., Haydock; Lever Bros., Port Sunlight; Brunner Monds, Crosfields, Faircloughs of Warrington, and others.

The coasting steamers from London, Glasgow and Ireland, some of which had been plying to Runcorn, all arranged their sailings so as to be represented at the opening, and for days past larger steamers from abroad had been passing through Eastham Locks and taking positions up the canal so as to fall in when their turn came.

The days were short, and for the convenience of various public bodies, it was arranged that the procession proper should start from Latchford Locks.

In order to accommodate the shareholders, over 50,000 tickets had been issued for dock premises at Old Trafford, chiefly on the Stretford side. So great was the demand for these gratuitous tickets that they commanded a premium, and two men were brought up for selling forged tickets. It was proved a wholesale manufacture had been going on, but the forgery could not be brought home to the prisoners.

Warned by the sad accident that occurred at the opening of the Manchester and Liverpool Railway, and realising the possibility of a helpless crowd of people being pushed into the canal, the authorities had wisely made every provision for the protection of life, and had distributed life-saving appliances all over the docks and line of canal.

At ten o'clock the signal to start was given by a steam whistle from Mr. Platt's yacht, which, painted white and gold, and with its graceful lines, was of itself an ornament to the procession. Immediately a mighty round of cheers was given, but this was quickly drowned by the combined efforts of scores of steam whistles and sirens, ear-piercing in their shrillness; it was a perfect pandemonium when this historic procession began. Slowly the *Norseman* cleared the Lock gates, and she was followed by the *Snowdrop* and other vessels as previously mentioned.

There was enthusiasm and excitement the whole way; every point of advantage on the canal bank had been seized by people, who brought their families, so that even the little ones might be able to say in after years that they had seen the first ships sailing up the Ship Canal.

No sooner was one crowd passed than another came in view, and everywhere were signs of jubilation; handkerchiefs were waved, hats thrown up, bugles sounded, cannon fired, etc. Cheers for Daniel Adamson showed that the pioneer of the canal was not forgotten, and as the *Barry* dredger was passed, the crew joined in the chorus, "For He's a Jolly Good Fellow". For whom the honour was intended was uncertain. On the way it was noticed that sea-gulls were following the boats, and this was considered a good omen.

There had been a little misgiving about the Barton Aqueduct, but when at a quarter to twelve the *Norseman* gave the necessary signal, the immense mass of iron, filled with water, swung gracefully on the pivot, and let the procession pass through.

If the send-off from Latchford had been imposing, the reception of the procession at Mode Wheel, where it arrived a quarter of an hour before time, was hearty and enthusiastic beyond description. Near Barton a band welcomed the procession with "See the Conquering Heroes Come," and directly the *Norseman* was sighted at Mode Wheel, sounds that were weird and deafening issued from the sirens of the various steamers. The blending of them defies description. The ships were passed through the locks in batches. It took ten minutes for the *Norseman, Snowdrop, Crocus* and two other small ships to pass through the big lock, and at 12.40 they were at the top level. Afterwards twelve vessels were all raised at once in the same lock. All the way from Eccles the banks had been crowded, but within the docks was simply a seething mass of humanity, who received the procession with rounds of cheering. A rough estimate gave the number of those here and on the adjoining land as 500,000. Shortly after her arrival, the *Snowdrop*, conveying the Manchester City Council, performed a small function of its own, and passing through the Trafford Bend, visited the portion of the Pomona Docks which is in Manchester. After a few words from the Lord Mayor, the ship returned to the Salford Docks, where those on board witnessed the arrival of the tail-end of the procession. Amongst the visitors of the day were Mr. Jacob Bright, M.P., Mr. Lees Knowles, M.P., Mr. Holland, M.P., and Mr. Pember, Q.C. Another distinguished member of Parliament, hailing from a neighbouring division, ought to have been there, and started by rail for Latchford in the morning, intending to join some friends on the *Water Lily*, but no doubt, fancying he was in Parliament, he fell asleep, and was carried on in the train, and so missed the entire ceremony.

The only ship that unloaded her cargo on the opening day was the *Pioneer*, owned by the Co-operative Wholesale Society. She brought a cargo of sugar from Rouen, and claimed to be the first merchant vessel registered in the port of Manchester.

Messrs. Carrick & Brockbank secured the first consignment of pig iron, and Messrs. Kenyon & Sons of indigo. Messrs. Horrocks, Miller & Co. despatched the first consignment of white calicoes to London.

The *Mersey* steamship was the first to coal at the Partington Tips, and Mr. W. E. Appleton, of James Black & Co., claims to be the first passenger landed at the docks.

Vessels kept arriving all the afternoon, and it was 7.30 before the last of them was safely moored at Manchester.

LINES IN CELEBRATION OF THE OPENING OF THE MANCHESTER SHIP CANAL. BY J. ARCHER MORTON.

Read on Board the Arts Club Steamship, the " Claughton," on New Year's Day, 1894.

To cross no mighty ocean, to seek no foreign shore,
The merchant ships of England their anchors weigh once more.
To-day they sail our Ship Canal, to celebrate the worth
Of British pluck and energy, this triumph of the North.

The "silver streak" they sail to-day was made by British hands,
No foreign workmen did the work, no mercenary bands ;
But British brains devised the scheme and carried out the plan,
Brave Adamson and Walker, who so nobly led the van.

And they who ably fill'd the gap, when Death call'd these away,
Lee, Egerton and Sir Harwood, and all who had a say,
And did their part in this great work, and be they great or small,
Who laid a stone or turn'd a sod, our thanks are due to all.

And if some did depreciate the work when first begun,
They surely will not stint their praise now that the work is done ;
But wish "God-speed" with all their hearts, and work with willing hand,
That trade and commerce may increase and prosper through the land.

And Manchester and Liverpool their merchant fleets combin'd,
New markets may discover, fresh benefits bring mankind.
Long may their commerce prosper, their full sails ne'er be furl'd,
Not till old England's sun be set, the death-knell of the world.

And this our gallant argosy, the good ship *Claughton* nam'd,
Should live in this day's history, should be for ever fam'd.
With our good Arts Club for her freight, "laden with golden grain,"
Workers with pen and pencil, and toilers with busy brain !

And when the story of this day is told in years to come,
And more imposing argosies than ours make here their home,
May we with pleasure call to mind the Arts Club had its share
In this day's work, and, in our hearts be proud that we were there.

The largest steamer to come up the canal was the *Albatross*, 1,450 tons, and the largest sailing vessel to enter it was the *Sophia Wilhelmina*. There was great rivalry as to who should make the first Customs House entry.   It was made for the *Alba-tross* by Messrs. Simpson, Howden & Co., who paid £103 for duty, though the *Ibis*, of Rotterdam, claims to be the first ship to arrive with a foreign cargo.   It speaks well for the arrangements that only two minor accidents occurred during the day. Much sympathy was felt in connection with a sad accident that occurred to the chief engineer of Mr. Platt's yacht, on the Saturday previous to the opening.   Returning to the yacht in the dark, he fell into the Latchford Locks and was drowned.

A Newcastle shipowner, Mr. Renwick, in the *Newcastle Leader* wrote thus :—

What I saw far exceeded my expectations.   The opening ceremony, with its attendant enormous traffic and handling of lock gates, gave me a distinctly favourable impression of the canal's capabilities; and I am convinced that no difficulty will be found in navigating ships right up to Manchester.

Judging by the tone of the Liverpool Press, the opinion in that city was, that the canal must now be considered a necessary evil.   What they could not cure they must learn to endure.   Railways had not caused horses to be as extinct as the dodo, neither would the canal ruin Liverpool.   The *Daily Post* said : " It could not be a rule for ocean-going steamers to bring a full cargo for Manchester, and return with a full cargo to a foreign port.   The majority of steamers arriving from abroad will probably take the bulk of their return cargo from Liverpool, and in a great number of cases will discharge part of import cargo at that port.   Manchester's gain would not be Liverpool's loss.   As to the comparatively few cargoes which will pass Liver-pool, they can signify but little to that port."

They admitted that the taunts of sceptics about construction had been silenced. The question now was, Would it pay?

The *Liverpool Courier* " doubted whether another such splendid example of what can be achieved by local patriotism, zeal, and confidence, can be found anywhere in the civilised world".

However, keeping the flags and cheering out of view, "the canal looks some-what disappointing to the ordinary observer.   In spite of the wonderful engineering feats involved, it is a prosaic enough ditch in appearance, and one is not fascinated by the rate of progress that can with safety be made upon it."

Speaking of "the fifteen million inland ditch," they prophesied : " Most probably the canal will ultimately become an artery of traffic by steam barges upon a scale heretofore unknown"

Certainly no mishaps on the canal were allowed to pass unchronicled.   Hostile newspapers took a deep interest in publishing them with sensational headings such as "Damage to a Steamer; the Masts Smashed"; "Steamer Aground in the Canal"; "Another Landslip at Latchford".

These and many other accidents, likewise of a minor nature, did occur. The *Granada* grounded near Partington, but practically did herself no damage. The *Hazlemere* caught her top mast against Runcorn Bridge, and broke off some feet, through not properly gauging its height. So determined were the owners of the *Hazlemere* to come up the canal, that they refused to go to Garston unless they were paid £100 extra, and given a guarantee they would be discharged in four days.

A landslip occurred at Latchford that at one time threatened trouble to the adjacent locks, and might have undermined a road; but this was dealt with by filling the aperture with cement in bags. None of these accidents were serious, and were merely such as might have been expected in a new system of canal navigation, where everything had to be learned by experience.

When the excitement of the business opening was over, the directors encountered various difficulties. Cargoes had to be landed, housed or disposed of, and new markets created. There was the fact, too, that existing markets were likely to resent the advent of interlopers, and that the railways were hostile.

Tentative preparations had been made for fruit sales, and the first auction took place on the 2nd January. A Liverpool auctioneer told his audience that if the fruit sold well, other cargoes would follow, but if not, the buyers would be treated like naughty boys and given no oranges—they would find no more fruit coming up the canal. Only moderate prices were realised, and then the railways gave much trouble in getting the fruit away; indeed they were so tardy that many buyers lost their markets. Next day Messrs. Farrand, Craze & Goodwin held a public fruit sale in the Shudehill Market, and on the following day the Manchester Fruit Brokers had a sale at the docks.

A large fruit boat, the *Granada*, was just too late for the opening, and came up the next day decorated with festoons of oranges.

In order that ships might be quickly and efficiently unloaded, experienced men were imported from Glasgow, Liverpool, Barrow and other seaports, and these were supplemented by local men. In a few days the latter imagined they had learned sufficient to undertake the work themselves, and along with the unemployed outside the docks they started an agitation to dispense with country labour, asserting that local men should have the preference. The movement grew fast, and about 600 of the malcontents marched to the docks threatening to throw the foreigners into the canal, and asking to interview Mr. Browning, the dock manager. Their spokesman said:—

A number of men had been engaged who belonged to other towns, and he did not think that was fair. The canal was opened for the benefit of Manchester, and men of that town had a right to ask that any work there was should be placed in their hands. Manchester men were perhaps not so experienced as some outsiders, but they soon would be.

Mr. Browning would not discuss the question then, but arranged to meet a deputation at a future date, along with Mr. Marshall Stevens, the manager. At the meeting, Mr. Stevens told the delegates that the Ship Canal Company must have competent labour, and that at all hazards they would protect the experienced men who had come from a distance. They must have men who had done similar work before. Subject to this, they were willing in future to give preference to local labour, but they would not discharge men who knew their duty, and were serving them well. This did not suit the unemployed, and to the number of 1,500 they marched to interview the Mayor of Salford (Alderman Bailey) at the Town Hall. The Mayor told a deputation that the Ship Canal was a private concern and entitled to manage its business in its own way. If Messrs. Burns, of Glasgow, brought their own men to do their stevedoring, he did not see who had a right to prevent them. He had, however, deep sympathy with the unemployed, and would bring the matter before the Ship Canal Directors, and the Salford Unemployed Committee.

Not satisfied with this, about 500 to 600 of them walked in procession and waited on the Lord Mayor of Manchester. Just before they arrived, a deputation of the Trades Council had waited on his lordship, who had promised to try and find municipal work for about 300 of the unemployed.

When the delegates from the processionists were admitted, and had stated their grievances about the outsiders, the Lord Mayor told them he could not interfere with the working of the Ship Canal, but asked whether they were prepared to accept Corporation work if it were provided. On an affirmative answer being given, he promised that the Parks and Cleansing Committees would find work for about 300 Manchester men, and they might start on the Monday following. The deputation expressed themselves satisfied. Nevertheless, in the interim, numbers of men passed into the docks as visitors, and then congregating to the number of nearly 200 round a ship where foreign labour was employed, demanded that all but Manchester men should be discharged. This demand not being complied with, they boarded the ship and stopped all work. Meanwhile, in response to a telephonic message, the chief constable of Salford marched a posse of policemen down, who at once drove away the intruders and restored order. This happened on the 13th January. Up to this time all respectable people had been allowed to pass the barriers and wander about the docks. At the instance of the police, and to prevent further acts of violence, this permission was withdrawn, and only those with permits, or on business, were admitted to the docks.

Much dissatisfaction was caused by the Board's decision, for previously hundreds of people from the city and surrounding towns had been greatly interested in watching the arrival, departure, and unloading of ships. There were thousands of children, and even some grown-up people in Manchester who had never seen a ship before, and they could now feast their eyes to their heart's content. No wonder then that the withdrawal of the privilege caused much disappointment.

Early in January a well-known townsman, Mr. H. T. Gaddum, wrote to the *Manchester Guardian* urging that as the business opening of the canal had been so satisfactory, there was no reason why the shareholders' money should be wasted on a further formal opening ceremony in May, which could not add glory to anybody. This letter gave rise to much correspondence from partisans and critics.

Week by week new developments took place on the canal, but its modest beginning is evidenced by the fact that in the first week only twenty-nine ships, carrying 17,000 tons of merchandise, berthed in the Manchester and Salford Docks. To bring ships all the way up to Manchester and thus secure full tolls was the chief aim of the directors, and to that end it was decided not to charge ships dues on through traffic for the first year. This caused serious complaint from Runcorn, which had not the same privilege. It was argued by the directors that Runcorn had an established business, whilst every inducement was needed to create a new port. In the end, the matter was settled by some minor concessions being made to Runcorn.

It soon became evident that all new traffic would have to be desperately fought for, and that the Rebate System adopted by shipping rings could be wielded in a very arbitrary and cruel manner. An explanation on this point may be useful. A ring decides what rates must be paid to certain ports. In order to prevent competition, each member of it proceeds to bind his customer for a term of years so that he may not take his business elsewhere. He says to him :—

On condition you give me all your trade, I will return you, six months after the termination of each year, say, 10 per cent. of what you have paid for freightage during that year. But if you send a package by any other line, every penny of such rebate will be forfeited.

Now as this compact is in writing, and often yields to the merchant hundreds, and sometimes thousands a year, it is a serious thing for him to break with the ship-owner. If he determines to do so, and be independent, he loses all the arrears of rebate lying in the shipowner's hands, and he knows that the ring will relentlessly fight any new line that may take up his business, and ruin it if they can. How is he afterwards to ship his goods? Thus a merchant is tied more completely to his shipper than a publican to his brewer.

The following is an instance of the difficulties the Ship Canal had to encounter, and of the power wielded by the Conference system. With the opening of the canal, it was the intention to start a new line from Manchester to the West Coast of Africa. The *Albatross* and *Cygnet* and other ships of the General Steam Navigation Company were among the first to come up the canal, and they did the journey from Africa much more quickly than the Liverpool line. They were welcomed by the merchants, both because rates had previously been exorbitant and they carried 30 per cent. cheaper, and because of the convenience. One large mineral-water manufacturer was delighted that he could send lorries loaded with his bottles to the docks, and have them safely housed on the ship, instead of sending them by rail to Liverpool, with all its costs and risks. At first the new line got fair outward cargoes, but then the ring swooped down. They reduced the rates about one-third, and told the merchants they intended to drive away the interlopers. They would carry at less than the new line, come what might. They offered to take 10 per cent. rebate off all freightage. In the end the new line was driven away, and the shippers had to make their peace with the conference liners. In another case a merchant told me his plight: the agent of the ring had urged him to accept the deed offering the rebate, which had been signed for a term of years by many other firms. It was worth £1,200 to £1,500 a year to him. If he signed he would enjoy all the conveniences the old firm could offer, regular sailings, etc. If not he ran the risk of the new firm's irregular sailings, and perhaps of their being run off altogether, and then where would he be? He would have to go cap in hand and beg back again. He had signed with the full knowledge that at the end of the term, if the new firm were run off, very possibly there would be a reversion to the former rates. When it was known the African steamers were going back to Liverpool for outward cargoes, this was quickly proclaimed as the "First defection".

Again, directly a line was started from Manchester to Smyrna and Constanti-nople, the five shipping firms constituting the Levant Ring advertised and issued the following circular—

In the event of steamers being laid on from Manchester to these ports, the five large firms undertake to carry Manchester goods at 5s. per ton from Liverpool and Hull to the above-mentioned ports.

In other words, they would take say one-fifth or one-sixth of their ordinary rates to run off a new-comer, and when this was done the ring would revert to old prices. Only a combination of merchants can break down a ring of this kind. The native merchants of Bombay by joint action broke down a powerful English ring, and other merchants can do the same if they will only combine and be loyal to one another and the Ship Canal. The instances I have quoted show that the shipping rings of England exercise a power more arbitrary, more hurtful to trade, more harassing and more inimical to the interests of England, than even the worst combines of America. The rings cast their net over the shipping merchant so dexterously that he is like a fly in the web of a spider, once in he cannot get out. He feels the thrall and is somewhat ashamed of being so helpless, but in time he tries to excuse himself by the fact that he is only like his fellows in the trade ; and the glittering 10 per cent. which he pockets makes him forget that he is helping to kill the goose that lays the golden eggs.

Only self-sacrifice, unity of action, and sincerity of purpose can break down the shipping rings, which have at their head some of the most wealthy and most powerful shipowners in the land. If they are not checked there will be a Nemesis. German manufacturers and shipping companies, aided by their Government, will in time, and to a certain extent, oust us from the markets of the world, even with our own colonies. Slowly and insidiously they are working to achieve this end. Their well-trained commercial travellers, who speak the language of the people with whom they trade, go to any amount of trouble to meet the wants of customers, and even if they have to sell English-made goods they can send them more cheaply, *via* the free port of Hamburg, than the manufacturer can ship through English ports, and by British ships. The Ship Canal offers an opportunity for Manchester to lead the van, and end the bondage to shipping rings. Their continuance is a menace to the trade of the district.

On the 9th January the fruit saleroom provided by the Corporation in the Camp-field Market, Deansgate, was opened by the Lord Mayor, Sir Anthony Marshall. Speaking on behalf of the fruit trade, Mr. William A. Nicholls enumerated the difficulties they had had to contend with in trying to establish a market. No shipowner would send his ships to Manchester unless he could get more profit than at Liverpool, though it ought not to cost more in sea freight to one port than the other.   Buyers were willing to give a trifle more at Manchester than at Liverpool. They could afford to pay 4d. per case more for oranges, because they saved 8d. in the carriage.   Onions delivered at Oldham *via* Liverpool cost 9s. 9d. per ton, whilst from the Manchester Docks the cost would be 4s. 6d.   Apples 13s. as compared with 7s. by canal; oranges 14s. 8d. as compared with 7s., and so on.   Oldham would save an average of 7s. 2½d. per ton, Nottingham, 5s. 6¾d., and Birmingham, 2s. 4½d. Fruit could be landed at smaller cost and in better condition.

Mr. Craze, another fruit merchant, said cheap inward cargoes depended much on getting plenty of outward cargo, and he asked that Manchester merchants should try and supply it.

Great engineering works often destroy all trace of the pre-existent state, but it is pleasing to think that the old course of the Ship Canal has been faithfully portrayed.   It will be quite possible in years to come to see it as it was, for Mr. Ward, a zealous photographer of Manchester, and others, almost lived on the spot, anxious to photograph any old landmarks before they were removed, also any incidents that occurred and any curiosities that were unearthed.   Mr. Ward alone possesses nearly 2,000 photographs which will some day be of great historical value.

Perhaps the most curious of the early consignments to Manchester was an Egyptian gentleman, by name Nali Gournali, said to be 4,000 years old.   His remains in the form of a mummy, were consigned to Mr. James Magnus of this city. A Liverpool paper thought :—

Business people in Manchester would have preferred to see a ship coming up with cotton, as none of this staple article of commerce had yet arrived by water at their port.

Those who win can afford to laugh.

By the middle of January large steamers laden with cotton began to arrive. The first were the *Finsbury* and *Glen Isle*, from Galveston, and they were quickly followed by the *Venango* and *Ohio*, from New Orleans.   The *Finsbury*, of 1,900 tons register, with 4,170 bales of cotton, was the pioneer cotton steamer; she came up from Eastham in eight hours in charge of an experienced pilot, and escorted by two tugs; the captain said he could have dispensed with tugs in the daylight if he had known what the canal was like.   Great interest was taken in this, by far the largest vessel that had entered the port, and she seemed quite a Goliath among the other ships.   Hundreds of visitors sought the privilege of going over her.   Messrs. Simpson, Howden & Co., her brokers, paid the Canal Company a cheque for £309 15s., being the toll on the cotton and the labour bill for discharging.   The *Glen Isle* ought to have arrived first, as she started ten days earlier, but she was caught in a hurricane, got badly damaged, and had her second mate washed overboard. She accomplished the canal passage quite safely. The succeeding vessel, the *Venango*, belonging to the Neptune Steam Navigation, was not so fortunate. Either through bad steering, or bad pilotage, she stove in some plates at Latchford, and met with other accidents. This was quickly reported in all the papers, and unfavourable critics at once sought to prove that the canal was not safe because an odd ship had suffered slightly. On the other hand, the owners of the *Venango* gave it as their opinion that the navigation was safe, and that a similar mishap might have occurred at any other port.

Sir William Forwood, addressing the Liverpool Chamber of Commerce, adverted to the power of Manchester when she was united and in earnest. Speaking of the efforts of Manchester to make the canal and so secure trade, he said :—

Up to the present time Liverpool had been very apathetic. They had laboured in season and out of season, but they had not got the support from the people of Liverpool that they ought to have had. Manchester worked together as one man, and had a local patriotism which Liverpool had not hitherto possessed. It was because Manchester worked together in that way that they had been able to construct their Ship Canal. No other power but the united people of Manchester would ever have induced Parliament to grant the necessary authority.

Compliments were showered on Manchester for her enterprise and patriotism, but perhaps the most remarkable came from the Lord Mayor of Liverpool, who made a bold effort to stimulate his fellow-townsmen by exalting Manchester before their eyes.

He finds an example of what he desires to see in Liverpool in the behaviour of Manchester, especially in the formation of the Ship Canal, by herculean exertions and sacrifices both vast and minute.

Commenting on this the *Liverpool Daily Post* said :—

We certainly need qualities which both in the case of the canal and in other instances the inhabitants and leading men of Manchester have shown. The canal was begun by Mr. Adamson and other private adventurers, men of great capacity and tremendous faith. It was substantially subscribed to by high and low, the fight was made out of private resources and the promotion was not official. But when financial exigencies arose, the Corporation of Manchester did not hesitate to pledge its credit for the further prosecution of the design. Compare the great exhibition at Manchester with that of Liverpool. The first was a model of management, that at Liverpool was slipshod, haphazard and from hand to mouth. In Manchester it succeeded because the first men of the city never ceased to give the enterprise their personal care and attention.

The commercial and banking interests were so aroused that, at the meeting of the Consolidated Bank in London, the Chairman, referring to the opening of the Manchester Ship Canal, said that it was expected it would lead to a large increase of business to the Bank in the city of Manchester. The Board had therefore elected Sir John Harwood to a seat on the directorate.

On the other hand, it was only to be expected that there would be some grumbling about the canal. Dissatisfied shareholders wrote to the papers apparently under the impression that the main intention of the canal was to reduce freights between Liverpool and Manchester by the Bridgewater Canal, forgetting that the object was to bring the goods direct to Manchester, without breaking bulk at Liverpool. Others expected small deliveries to and from the Continent to be fetched, forwarded, and delivered by ships just as they would be by a railway parcel office.

It was satisfactory, therefore, when the agent of a large company trading with Belgium gave his experience of forwarding goods in bulk, which showed a saving of 12s. 7d. per ton over sending them *via* Goole; he also complimented the canal on quick delivery. Coasting steamers to Glasgow and London, soon began to traverse the canal in five, to five and a half hours, and the London steamers carried for 16s. 8d. what used to cost 25s. Still the difficulty of getting outward cargoes remained, and a committee of merchants and shippers was formed, under the presidency of Mr. William Dorrington, with the view of influencing traffic, and of inducing the Chamber of Commerce, as representing the commercial interests of the city, to take action. When the canal had been opened three weeks, Messrs. Dobell, of Liverpool, offered to send the *Curzon*, 1,860 tons, from Iquique to Manchester, laden with nitrate of soda, provided the Ship Canal Company would run all risks of navigation. The latter, however, declined, and then the Liverpool Press was aflame with the assertion that the Canal Company asked others to take risks they dare not encounter themselves. If you have confidence in it, why not take the responsibility? was their cry.

The great draught, and the steering qualities of the *Curzon*, were the true reasons why the Ship Canal directors declined to accept any responsibility.

Even at the end of January the Ship Canal docks and sheds were only partially ready.   February witnessed the arrival of the Neptune Company's ship *Ohio* with cotton.   Her captain said she had come up with scarcely a scratch, but hostile papers proclaimed "Another Disaster on the Ship Canal".   The fact was, she had just scraped one of the locks, but no damage was done.   The question of "damage" or "no damage" caused an amusing paper warfare.   "A. N." wrote that he had seen the *Ohio*, and there were no traces of injury.   On the other hand, "J. P.," from Liverpool, declared that he had made a special journey to Manchester, and found the figure-head with one hand frayed, the waist damaged, and the lower part of the body *entirely* gone!   To this "A. N." replied, the fraying was simply caused by a tow-rope, and the lower part of the body could not have been knocked off, because Old Neptune, the figure-head, was simply a bust, and *never* had any petticoats.

It is only just to say that the persistent depreciation of the canal was much reprehended by many Liverpool gentlemen, and one in his capacity of Chairman at a meeting in Liverpool, said :—

He did not think that the Manchester Ship Canal would do Liverpool any harm. They owed Manchester a great debt for having made the canal, because it had wakened up the Liverpool people.   He did not believe that the Mersey Bar would have been dredged to this day if the Ship Canal had not been made.   They need not be afraid of Liverpool suffering from the undertaking.

About this time Messrs. D. & C. McIver, of Liverpool, offered to build a fleet to run from Manchester to India, if the merchants of Manchester would show their faith by contributing a considerable share of the necessary capital.   In the middle of February the prospectus was issued of a Manchester and Liverpool Steamship Company, with a capital of £150,000, for the purpose of providing new steamers for a Kurrachee and Bombay line.   There was not an adequate response to the appeal, and nothing further was done.   Mr. Sam Ogden, President of the Chamber of Commerce, in his annual address complimented Manchester on its canal, and said he thought the incidents connected with the canal itself would give them a measure of prosperity which the city and neighbourhood had been strangers to for some time past.   It would put them in direct connection with 150 towns, of which one hundred each had a population of over 100,000.

Even after the canal had been opened for a month the greatest scepticism existed as to its capability for continued use.   Mr. Austen Taylor, Chairman of the Liverpool Shipwrights' Association, comparing Liverpool with Manchester, said that a voyage up the canal meant the possible delay of a week, that it was impossible to guarantee an outward cargo from Manchester, and that steamers would double their port expenses by calling to fill up at Liverpool. Perhaps the most bitter critic of the canal was the *Liverpool Journal of Commerce.* Under the title of " Physical and Engineering Features of the Ship Canal," that paper published a series of five articles, containing thirty-seven reasons why the canal could not succeed. In the writer's opinion it was doomed both in its financial and engineering details; the banks would give way, and the slopes would tumble in, the bottom would silt up, navigation was next to impossible; the width was insufficient for vessels to pass each other, also the headway of the bridges was too low, the rock cutting would cave in, the locks were of faulty construction, floods would ruin the canal, and it would be a hotbed of pestilence, etc., etc. Looking back, it is interesting to see how these predictions have been falsified, and how the Canal has overcome its many difficulties.

Early in February the Canal Company was obliged to invoke the help of the Railway Commission, and ask for an injunction against the Lancashire and Yorkshire Railway Company and the Cheshire Lines Committee, in respect of their alleged refusal to carry traffic delivered to them at the Cornbrook siding. During construction the contractor had formed a junction between the Cheshire Lines sidings at Cornbrook and the dock railways. Over this (the other railways not having made their own junctions) all traffic passed for the first month, and was then handed on to the Lancashire and Yorkshire, and London and North-Western Railways. All at once the Lancashire and Yorkshire refused to accept, and the Cheshire Lines refused to deliver by that route. Mr. Balfour Browne, counsel for the Canal Company, attributed this to the desire of the Lancashire and Yorkshire to force ships to Liverpool, in order to get full terminal charges. He said that the law compelled railways to take and forward goods for distribution on other company's lines, and he asked that the Cheshire Lines should be required to take goods for distribution. He also claimed damages for past delays.

Mr. Littler, Q.C., in reply, said the matter was of small importance. The Lancashire and Yorkshire Railway Company had found the process inconvenient, and the present action was entirely and simply a piece of pig-headed obstinacy on the part of the manager of the Ship Canal. On the President granting an interim injunction, and allowing costs, Mr. Littler said it was quite possible the Court would not be troubled with the matter again. It was not a matter of much importance whether his company sent goods 11 miles for 2s. 6d. or 40 miles for 12s. 6d. Subsequently the Canal Company settled the dispute to their own satisfaction.

During February a lively correspondence was carried on as the outcome of an assertion that "big ships were not likely to use the canal"—one reason given being that they could not get under the bridges.   On this the manager gave the heights of fixed bridges above water-level to be :—

	Feet.	Inches.
Irlam Railway Viaduct	75	3
Partington Railway Viaduct	75	4
Warrington High Level Bridge	75	3
Latchford Railway Viaduct	77	$1\frac{1}{4}$
Latchford High Level Bridge	76	$9\frac{3}{4}$
Acton Grange Railway Viaduct	77	$2\frac{1}{2}$

Great fears were entertained lest a terrific gale that occurred on the 10th February should wreck some iron sheds in the course of construction, or damage shipping. Fortunately, the only damage done was the blowing down of the whole length of dock paling in Trafford Road.

The formal opening of the Mode Wheel Dry Dock took place on the 17th February.   The dock is 450 feet long by 65 feet wide, and is separated from the canal by a pair of heavy iron gates.   When required, the dock can be emptied in forty minutes by a pair of Tangye's 36 in. centrifugal pumps.   The opening ceremony was performed by Mrs. George Renwick, who with a pair of gold scissors cut a tape stretched across the entrance to the dock, and admitted the ferry boat *America*, and the steamer *Martin*.

Mr. George Renwick, the Chairman, said :—

The time for prophecy as to the Ship Canal had gone by, and so had the query, "Will ships come to Manchester?" Ships *had* come and *would* continue to come.   He was amused that a gentleman at the Liverpool Chamber of Commerce had ventured to remark that it would be useful for 600 to 900 ton barges, but large ships could not come to Manchester. They had not built a 450 feet dock with 21 feet of water for 600 ton barges.   As one who knew something of shipping, he told them that the canal and docks were made not for 600 to 900 ton vessels, but of 6,000 to 9,000 tons.   Depend upon it the men who had spent £15,000,000 would not be daunted in overcoming a few difficulties.   The shipowner had done his part, the men who now held the key to the success or non-success of the canal were the *merchants of Manchester and district*.   Unless they determined to support the canal, ships would not come.   It was said of Barry that shipowners would not send their boats there. The collieries near Barry said their coals should be shipped there, and the shipowners *were forced* to come.   Hence the docks paid 10 per cent. from the start.

If the merchants would combine to support the port of Manchester there would be no scarcity of shipowners to carry the goods. Railways could not compete with water carriage. Coal was carried to Bombay at the rate of 1s. per ton for 1,000 miles. When his company asked for £120,000 capital, and Manchester only subscribed £1,500, the hard-headed Tyne-siders were not daunted ; they knew their chance of success and found the rest of the capital. They had a noble waterway, they had splendid docks, they had a huge population and great wealth, and he believed a large, profitable and direct trade would be built up with all the corners of the earth.

On the last day of February, in the Chancery Court, Messrs. James Fairclough & Co., of Warrington, sought for an injunction to prevent the Canal Company interfering with the supply of Mersey water that turned their mills, and which they used for navigation. Mr. Justice Kekewich decided that there had been an inter-ference, and granted an interim injunction, giving time, however, to remove the cause of complaint, and directing that the work to be done should be determined by arbi-tration.

When the directors met the shareholders in February, the chair was taken by Mr. J. K. Bythell, both Lord Egerton and Sir John Harwood being absent through ill-health. After congratulations on the opening of the canal, the Chairman said the engineers assured him that they had not exceeded their estimates. He gave details of the main lines that had commenced running, and expressed his opinion that as regarded traffic a very good beginning had been made. He went on to say that the Board of Agriculture had consented to the construction of lairages and abattoirs at Mode Wheel, and Messrs. Nelson, of London, were erecting large warehouses there for frozen meat. Messrs. Richard Evans & Co. had bought a large piece of land at Acton Grange for coal-tips, and there was every prospect of several large oil companies coming to Old Trafford.

He then dealt with the obstacles they had to encounter from the so-called "Conferences" established by steamship owners, which seemed to bind shippers hand and foot. The Conferences made it difficult for merchants and traders (even when so disposed) to help the Canal Company to the full extent. He instanced the Irish, the Mediterranean and the West African traders as being fast bound by these Con-ferences, and he hoped shippers would not be led away by the offers of the Liverpool lines, for at the best the advantage gained would only be a temporary one. The future of the canal was in the hands of the Manchester merchants and shippers.

To cheapen and facilitate traffic they had done what Liverpool had not done, namely, placed railways round all their docks, but they were a little disappointed the railway companies had not completed their connections in time for the opening day ;

the railways seemed to be holding back, but the ships that had come up should satisfy them that the canal was navigable, and he trusted they would now hurry up. He reminded them that the Chairman of the Chamber of Commerce in London had said that—

If Manchester merchants would only offer sufficiently remunerative freights, he would undertake to promise on behalf of British shipowners that they would only be too glad to send their vessels into Salford Docks.

The Chairman ended his speech by reading a telegram from Sir Henry Ponsonby as follows:—

You may announce that the Queen intends to perform the ceremony of opening the canal if she is able to do so.

A resolution to reduce the directors' remuneration by £900 was negatived almost unanimously by the meeting.

Immediately the canal was opened, the overseers of different townships *en route* met in order to secure unity of rating, and they fixed the assessment at £1,500 per mile. Against this the Canal Company threatened to appeal on the ground that the canal was making no profit, and that they had already reimbursed the different authorities any loss in rates through property being empty. A compromise was effected, and the rate fixed at £500 per mile.

After the March Assizes at Manchester were over, Mr. Justice Day and Mr. Justice Charles chose a novel way of reaching Liverpool. Instead of travelling by railway, they went by a special steamer, accompanied by the High Sheriff, the Lord Mayor, the Lady Mayoress and others, thus getting a good view of the new waterway, with which they expressed themselves much pleased.

Labour troubles soon became a source of much anxiety. No sooner was a difficulty arranged with the Bridgewater boatmen than the dock labourers, through their union in Liverpool, threatened to strike. They objected to piecework, which they said in some cases only yielded them 4d. per hour, and they requested an interview with the directors. The manager, Mr. Marshall Stevens, wrote in reply, that the men in Manchester had made no complaint, and seemed quite satisfied; that a good man earned 10d. per hour by piecework against 6d. per hour by day labour, and he therefore saw no reason for granting an interview. This produced a letter from Mr. Sexton, secretary of the dock labourers, denying that the men worked under a system of co-operative piecework, earning all they could and dividing what they got amongst themselves. He gave instances where, through having to wait for work, small wages had been earned. Following this came an offer from the Salford labourers to meet Mr. Stevens without the Liverpool secretary. Eventually the difficulty was arranged, but not before the unemployed in a body had created a disturbance on the docks and in consequence had been forbidden access thereto. The real home of the dispute was in Liverpool rather than in Manchester.

During March there were rumours of differences on the Ship Canal Consultative Committee of the Corporation, and hints of the possible resignation of Sir John Harwood. At a special meeting of the Council on the 4th April, it was announced that debentures and scrip certificates amounting to £1,493,500, out of the £2,000,000 loan, had been issued. Questions were asked as to when a report on the engineering and financial positions was to be issued. The Lord Mayor stated it was in progress, whereupon Sir John Harwood said it was quite time it was presented, and if he had had his way, it would have been presented long ago. He also made allusions to changes that were being made in the report, which created an uneasy feeling in the Council.

Some doubt having been expressed as to the desirability of inviting Her Majesty if there was any possibility of a financial breakdown, the Finance Committee of the Ship Canal published a report, signed by Sir Edward Jenkinson, its Chairman, showing they had money in hand sufficient for the wants of the year, and at the end would have a balance of £381,515, independent of their large landed estate.

A fortunate line was that commenced by Messrs. Lamport & Holt, of Liverpool. It was started in March, 1894, and is still running successfully (1906) with a largely increased fleet. Their ships load in Manchester for South American ports, thence they take cargo to New York, and there fill up again for Manchester.

About the middle of April the Commander of the Royal Yacht, Admiral Fullerton, inspected the canal with a view to arrangements being made for Her Majesty's visit, and to a report being presented to Earl Spencer as to the escort necessary.

Later on, by command of Her Majesty, the canal was visited by General Carey and Dr. Thompson, of the Local Government Board, who made a report on its sanitary condition.

During the same month the Corporations of Manchester and Salford, with the Ship Canal directors, were arranging for the reception of the Queen, and their efforts received the enthusiastic support of the inhabitants. The working people themselves asked for a general holiday on the occasion.

At the meeting of the Neptune Steamship Company, the Chairman commented on the fact that, whereas the great bulk of the shipping interest at Cardiff, Hull, Newcastle and other ports, belonged to the inhabitants of those towns, Manchester people had not yet gone into shipping as an investment; and he pointed out that it would be necessary for them to do so if the canal was to become a great success. He and his captains had no fear as to the navigation of the canal. They were prepared to provide ships if the Manchester people would support them. Captain Pinkney, one of the managers, said there need be no more accidents on the canal than the average met with elsewhere.

During Easter week the canal was crowded with visitors, who took advantage of the pleasure boats on the canal, the great attraction being the Barton Swing Aqueduct, admitted to be one of the most interesting features of the waterway. The aqueduct forms a tank built of $\frac{3}{8}$-inch plates, and the whole structure can be swung when full of water. The shore ends of the canal have iron gates, made watertight by means of a cast-iron and wood joint. When these are shut the ends of the swinging trough are also closed by watertight doors. The whole is worked by hydraulic power. Never before had a canal been so carried over a navigable river. No wonder then that people flocked to see a canal cut in two when full of water, and a portion suspended in mid-air to allow ships to pass. Still more amazing was it to see the detached pieces joined together and traffic resumed.

The Manchester Ship Canal Bill of 1894, having passed a second reading in the House of Lords, was referred to a Special Committee, with Lord Poltimore as Chairman. The principal object of the Bill was to place the Ship Canal outside the working of the Railway and Canal Traffic Act of 1888, which puts all tolls and charges under the jurisdiction of the Railway Commissioners. Against the Bill were seventeen petitions. The Liverpool Chamber of Commerce lodged a strong one, alleging there was no valid reason for the exemption, and claimed that Manchester ought not to be considered simply as a port, but as possessing a canal just like the Bridgewater Canal. It was urged that as a canal port, having command of railways and canals, Manchester might use its power to the detriment of other ports. With a view to meet the strong opposition to exemption of the Ship Canal from the Act of 1888, the directors withdrew the clause, and inserted a new one to the effect that, "The provision of section 24 of the Railway and Canal Traffic Act, 1888, shall not extend or apply to merchandise traffic conveyed on, or along, the harbour or port of Manchester".

W. H. Hunter, Assistant Engineer During Construction of the
Canal, Afterwards Chief Engineer, 1895 *Seq.*

Lafayette, Ltd.   *To face page* 206.

It was hoped by this concession to keep the Ship Canal out of the hands of the Railway Commission, failing which the directors feared they might be compelled to reduce the minimum freight of goods, and to fix a special tonnage toll for timber, as had been the case with the Bridgewater Canal. This would have meant a serious diminution of income.

Mr. Balfour Browne urged before Lord Poltimore's Committee that the Railway and Canal Traffic Act of 1888 was never intended to apply to a Ship Canal used by sea-going ships.

It is not applicable to us. You have not insisted upon a schedule for the Clyde or the Tyne or the Tees; therefore, don't insist on a schedule for us.

Mr. Balfour Browne went on to say that what was desired was to revise the toll clauses in various old Canal Acts. As yet no one could tell if the Ship Canal schedule was, or was not, sufficient, but it was on the power given by Parliament to levy specified tolls that the capital was raised, and it would be unfair to damage the security. After hearing evidence on both sides, the Committee decided unanimously in favour of the Ship Canal, on the condition that the exemption should only apply to the actual canal, and to no other property of the company. The Bill was then passed.

On the question of the construction of the Eastham jetty, the Dock Board withdrew their opposition on the understanding that the Conservator approved of the extension. As regarded the part exemption from full harbour dues at Liverpool in the case of part cargoes, counsel argued this was only fair, and said Manchester was entitled to the same treatment by the Mersey Dock Board as any other independent port. The Committee took this view, and counsel were instructed to draw up clauses accordingly.

The Ship Canal Company, having got the main canal exempted by their Act from the Railway and Canal Commission, that body dealt with the Bridgewater Canal only. The Commission decided on a scale of tolls, which Mr. Collier said meant a loss of near £5,000 in the yearly income of the Bridgewater Canal.

In the month of May Salford asked the Court of Referees of the House of Commons to let them appear against the Lancashire and Yorkshire Bill. During the discussion Mr. Bidder, Q.C., said:—

We all know there is a sort of combination amongst the railway companies to boycott the canal.

A singular admission from an old opponent!

At this time the city was alive with preparations for the Royal visit. At the May meeting of the City Council the programme was agreed to. A sum of £10,000 was set aside for decorations, and a Reception Committee appointed, *viz.*, the Lord Mayor, and Aldermen King, Sir John Harwood, Leech and Mark. Salford followed suit, and voted £2,000 for reception expenses. Badges were to be provided for the Councils of Manchester and Salford.

Spaces were allotted in Manchester and Salford for children, and it was hoped there would be a repetition of the hearty reception given to the Queen in 1851.

Trafford Road wharf was reserved exclusively for shareholders and their friends. Salford provided a grand stand for the Corporation and other public bodies and their friends in Ordsall Park, and arranged for a display of fireworks at night.

A proposal was made to hold a demonstration of the trades and friendly and temperance societies, but some hitch occurred with the Chief Constable and the Watch Committee, and the project was abandoned.

At five minutes past eleven on Monday, 21st May, 1894, the Queen left Windsor for Manchester, accompanied by Prince and Princess Henry of Battenberg, their children, Lady Southampton and the Ladies-in-Waiting. In the Royal Saloons were the Princess Leiningen, and also Lord Carrington, Sir H. Ponsonby, Sir J. McNeill, and other gentlemen of the Royal Suite. Lord Stalbridge, Chairman of the London and North-Western Directors, with his colleagues, Mr. T. Ismay and Mr. A. Fletcher, travelled in the train. At Wolverhampton it was transferred from the Great Western to the London and North-Western line, and though timed to reach the London Road Station at 4.30 P.M., it steamed in three minutes before.

A tremendous crowd had gathered behind the barriers at the station. Inside the enclosure to receive Her Majesty, there were present amongst others, Earl Spencer, Lord Sefton, the High Sheriff (Mr. J. Radcliffe), the Manchester Reception Committee, with the Town Clerk. Also the Recorder (Mr. J. F. Leese), Mr. J. W. Maclure, M.P., Major-General Hall and his staff. The Mounted Police and Artillery Volunteers lined the roadway.

The Queen was received by Lord Stalbridge, after which Earl Spencer presented the Lord Mayor to the Queen. The Royal Party then entered their carriages and proceeded *via* Piccadilly, Market Street and Cross Street to the Town Hall.

Rounds of hearty cheers from masses of packed humanity greeted Her Majesty. Every inch of ground, every window, and even the warehouse roofs were occupied to welcome the aged and revered sovereign. It was a brilliant and most impressive

J. H. Balfour Browne, K.C., Counsel for the Manchester
Ship Canal.

J. White, Dumfries.   *To face page* 208.

FORMAL OPENING OF THE MANCHESTER SHIP CANAL BY HER LATE MAJESTY QUEEN VICTORIA IN THE *ENCHANTRESS*, 21ST MAY, 1894.
Valentine, Dundee.
To face page 208.

sight. On a stand in front of the Infirmary were the nurses dressed in light blue costumes, which were most effective, also the nurses from the other City Hospitals in their distinctive and simple, yet charming uniforms. Behind them, on the top of the Infirmary, the young medical students displayed their exuberant loyalty. In Piccadilly, and indeed all along the route, the merchants and shopkeepers had vied with each other in making their premises as attractive as possible, with flags, banners and flowers, many of the latter festooned into words of greeting.

When the procession debouched into Albert Square, it was received with vociferous cheering, and the Queen was evidently much pleased with her reception.

In front of the main entrance to the Town Hall a platform had been erected providing seats for guests and for the Council. When the Lord Mayor and the Reception Committee had taken their places, the Queen's carriage drew up abreast of the platform, and close to the Albert Monument. This had been redecorated, and seemed to appeal strongly to Her Majesty's feelings. The Lord Mayor then presented the beautifully bound address which had been adopted by the Corporation earlier on in the day, and which Her Majesty graciously acknowledged in a few kindly words. Earl Spencer then handed the Lord Mayor her reply in which she trusted " Manchester may never cease to be prosperous and thrive, and that all classes and subjects may share in the benefits attending on the success of this most important enterprise".

The Lady Mayoress, the Recorder and the Town Clerk were then presented, and the former handed Her Majesty a beautiful bouquet. The procession was re-formed and proceeded *via* Peter Street and Oxford Street to the Municipal School of Art, All Saints, which was handsomely decorated for the occasion. In front of it were assembled on a dais Dr. Ward, Principal of the College, Sir Charles Hallé, Professors Dixon, Lamb, Leech, Boyd Dawkins, Wilkins, and many other gentlemen connected with Owens College. Hitherto the weather had kept fairly fine, but on reaching the School of Art, shortly after five, there was a sharp shower. Earl Spencer, as Vice-Chancellor, presented Dr. Ward to Her Majesty, and he offered to her, on behalf of the College, a loyal and dutiful address, to which the Queen handed Dr. Ward a gracious reply.

Though many of the Royal Party put up their umbrellas, the Queen declined to have the carriage closed, and nobly braved the weather rather than disappoint the thousands who had thronged to welcome her.

All along Stretford New Road the greatest enthusiasm was displayed. From one end to the other were Venetian poles, and decorated stands occupied every available space; the whole thoroughfare was packed with people who cheered heartily. Opposite the Blind Asylum, the carriages slackened for a minute or two whilst a child from the Deaf and Dumb School presented a bouquet to Her Majesty, which she most graciously acknowledged.

By the time the procession reached Trafford Road the rain had ceased, and once there was a gleam of sunshine. The docks were crowded with shipping displaying a profusion of bunting, and very gay did the scene appear as the carriages turned into the Ship Canal Road alongside the Docks on the Stretford side. A short way down they drew up at the pavilion with boudoir attached. In front of it lay the *Enchantress*, and moored to the Salford side were the gunboats *Speedy* and *Seagull*, which formed the Royal escort. Whilst the people were on the tip-toe of expectation, an amusing incident occurred. Her Majesty's Indian attendant had either been left behind or had missed his way at the London Road Station. As no other conveyance could be found he was sent forward in a London and North-Western Railway closed van, labelled "Parcels collected and conveyed to all parts of the United Kingdom," which drove up just before the Royal Party arrived. Much merriment was created when the door opened and from this plebeian conveyance stepped forth the Indian attendant of the Queen, arrayed in a gorgeous costume of scarlet and gold. This singular "parcel" at once made his way to the Royal Pavilion.

The Queen lost no time in the pavilion, and quickly crossing the carpeted gangway that led to the *Enchantress*, took her seat at the stern under a white canopy. Here the Honorable Mrs. Mitford, sister of the Chairman, presented a bouquet, and Lord Egerton read an address from the directors of the Ship Canal Company, to which Her Majesty handed a reply. The directors, the manager and the secretary were in turn presented, after which the Queen summoned the Lord Mayor of Manchester, Councillor Anthony Marshall, who was then and there knighted with a sword borrowed from Sir John McNeill. The Queen expressed to him her delight with the handsome decorations and with her reception by the people. The Mayor of Salford, Alderman Bailey, was similarly knighted. The engineer, Mr. Leader Williams, was then presented, and was asked to explain the canal to Her Majesty, which he did as the *Enchantress* slowly made her way to Mode Wheel Lock. Here the Queen pressed a button with an electric communication to the gates, which opened under hydraulic pressure, and the graceful *Norseman* steamed out, Her Majesty saying, "I have now great pleasure in declaring the Manchester Ship Canal open". Immediately a salute of twenty-one guns was fired. During the ceremony the *Skirmisher*, with the chief nobility and gentry of the neighbourhood on board, was moored alongside.

The Queen then disembarked at Trafford wharf, and amidst prolonged cheering drove to Salford. Trafford Road was lined with volunteers, and adjoining were stands from which 17,000 children shouted a hearty welcome. Further on was a grand-stand on which were the members of the Salford Council and representatives of various public bodies. Stretching across the road on either side of the stand were pretty floral arches with inscriptions of welcome, and close by was an arch formed of fire escapes and fire appliances. The houses were profusely decorated, and the en-thusiasm was immense. As soon as the borough was entered, the Salford carriages led the way, and when the grand stand was reached, the Mayor of Salford presented, in a silver casket, a loyal address, which was graciously acknowledged. The Mayoress then handed to her Majesty a beautiful basket of red roses, which were accepted in the kindest possible way. The Royal procession passed hence along Regent Road, Egerton Street, Deansgate, St. Ann's Square and Victoria Street to Exchange Station. The train ought to have left at 7.30, but it was nearly eight o'clock before it departed. Before leaving, the Lord Mayor, Sir Anthony Marshall, presented Her Majesty with a handsome copy of the *Manchester Historical Record*, and she assured him that she had been highly pleased with the splendid reception that had been given her.

In the evening a state banquet was given by the Lord Mayor of Manchester. In responding for the guests the Lord Mayor of Liverpool said :—

They in Liverpool had no mean jealousy of Manchester. They wished the people of Manchester every success in their great enterprise. He ventured to say that this municipality, and the people of this great centre, had presented an object-lesson to all great municipalities of a determination and a dogged persistency in overcoming difficulties. Such an example of civic patriotism had not been before exhibited.

At Salford also a banquet was given by the Mayor. At night there were illu-minations in Manchester and Salford, and a display of fireworks at the docks. Sir William Bailey also gave a dinner on the succeeding night to the crews of the *En-chantress*, the *Norseman*, and the gunboats.

Considering the vast influx of visitors to Manchester (40,000 from Oldham alone), the accidents were very few. An officer was thrown from his horse close to the Royal carriage. In Deansgate a mother and child were killed by a coping-stone off a building falling upon them.   When the Queen was informed, she sent £20 to the family of the deceased.   The writer of this book and Alderman Hopkinson nearly met with an accident.   When riding in the procession one of the horses kicked and got its foot entangled in the front of the carriage, which had to be abandoned.   Fortunately, the worst that happened was that the occupants had to walk from Platford's Hotel, Stretford Road, to the docks, dodging amongst the crowd as best they could, and arriving only just in time for the ceremony.

No one can doubt that Manchester is intensely loyal, and Queen Victoria was welcomed as a sovereign whom the people delighted to honour.   Thirty-seven years had passed since her last visit, when she described the Manchester folks as "a very intelligent, but painfully unhealthy-looking population".   Great changes had taken place, and a new generation had grown up who were anxious to see the grand old monarch, whom, till then, they had only known by repute.   At great personal sacrifice, she had braved exposure in uncertain weather to please her people, and perhaps a motto in Market Street best described the popular feeling, "God Bless the Silver Yure".

The Queen's visit came, too, at a time when the directors needed a sunbeam; something, in fact, to cheer them.   Things had not been going happily, and a kindly and encouraging impulse was much wanted.

On board the *Skirmisher*, as a visitor, was Mr. Joseph Lawrence, at one time the very able and energetic chief of the Ship Canal staff, and he supplied the *City News* with an account of the opening, and of his views generally of the position of the company's affairs.   To his mind, the arrangements were lacking in completeness, and sufficient attention had not been paid to distinguished visitors.   He thought the Ship Canal directors, instead of "all this pomp and show and beating of tom-toms, should give place to the active, business-like prosecution of means for securing traffic to the water-way".   He noticed much shrugging of shoulders when men talked of the business prospects of the canal.   Then he fell athwart Sir Wm. Bailey for waving greetings from the *Enchantress*.   "He was for the time being the guest of his sovereign, and etiquette required his temporary effacement."   He proceeded to rate the directors for economic pedantry in not organising subsidiary companies and following the policy of railway companies in launching out into outside businesses.   They were carrying this practice of leaving everything to private enterprise to the verge of fatuity.   If not reversed, it must ultimately spell "ruin".   If the directors had resource, pluck and enterprise, they would be able to put an end to boycotting, and bring the shipping interests of this country to their knees in three months; and he instanced his linotype business to show how he had broken down a monopoly. After chiding the directors for not providing warehouses, for declining to be carriers, and for asking too much for their land, he prophesied success if the enterprise were administered with firmness and enlightenment.

On Tuesday, 22nd May, I received a telegram, dated 21st May, which had been lying at the Ship Canal office since the previous day, informing me that in connection with Her Majesty's visit, she proposed to confer on me the honour of knighthood, and soon afterwards, on joining the SS. *Skirmisher*, which was taking a distinguished party along the canal to Liverpool, I found that my co-director, Mr. Joseph Leigh, M.P. for Stockport, had received a similar message. Among the passengers were Earl Spencer, the Earl of Morley, and Lords Montagu, Bury and Algernon Percy, together with many other visitors who wished to see the new waterway. Everybody seemed to highly appreciate the opportunity of inspecting the canal from end to end.

On the 18th July I was summoned to Windsor, along with a number of other gentlemen, to receive the honour of knighthood. Fortunately, I hit upon the special train which also took down Lords Rosebery, Breadalbane and Tweedmouth. We were met at Windsor by the Queen's carriages, each with four horses and a postillion, and quickly arrived at the Castle. Here our names were called over by a quaint old gentleman in a coat tipped with red, like a bank porter's, an odd-looking waistcoat, and a pair of the most countrified tartan trousers. Certainly none of these garments seemed to have been made for the wearer. Four of our party were late, and we had to wait for the laggards. By mistake they had got into a slow train, and they were half an hour late. The old gentleman turned out to be Lord Playfair, who with Lord Breadalbane and Sir John Cowell were the officials in attendance, and wore the Windsor uniform. Lords Rosebery and Tweedmouth received us, and after a view of the beautiful scenery from the drawing-room we were ushered to luncheon, which consisted of a cold collation served on gold and silver plate. To my mind the fine portly waiters in their gorgeous uniforms were far more imposing than the guests. Afterwards we all went into the green and white room, where we were coached with respect to the ceremony, and told when to bow, and how to kneel and retire. It was specially impressed upon us that we were to offer the back of our right hand on which the Queen would place hers to be kissed. Lord Playfair again called over his list, asking us by what name we wished to be knighted, and gathering some particulars of our history. I had some very interesting companions, *viz.*, Messrs.

Thomas Roe, of Derby, and Robert Hunter, of the Post Office, both small, neat men ; Seymour Haden, the artist; Dr. Russell, the Lord Provost of Edinburgh; Jerom Murch, a very nice old man, over eighty years of age; Isaac Pitman, of shorthand renown, also very old and infirm, who had to be assisted by a companion; Wemyss Reid, the well-known author; Mr. Prinsep, from the Law Courts in India; Thomas Robinson, M.P. for Gloucester, and Moses Philip Manfield, M.P. for Nottingham, who I noticed preferred to assume his second Christian name.   There were in addition, Dr. Grainger Stewart, and Messrs. T. Thornley, Joseph Leigh, M.P. for Stockport; R. T. Reid, Solicitor-General; John Hutton, well known for his civic work in London, and Richard Tangye, of Birmingham.   I sat next to the latter at luncheon, and found that he and two generations before him had been life-long teetotallers.   He said that when the Duke of York visited their works, he was informed that intoxicating drink of any kind had never been permitted on their premises.

A Cabinet Council was being held at Windsor simultaneously with our visit. When it was over, the Queen, accompanied by the Duke of York and the officials and military in attendance, went into a small room off the same corridor.   In turn each of us went forward, made a salaam, dropped on the left knee, received a tap on both shoulders from a very light sword handed to the Queen by an officer standing by, then kissed the Queen's right hand and retired backwards, making a bow as he did so.   As Lord Playfair presented each one, he gave a short reason why the honour was conferred.   Only one recipient was in the room at a time, but as the door was open, his successor could see the man before him being knighted.   The Queen was exceedingly pleasant, and conferred the knighthood in a very distinct voice, and I shall never forget her encouraging smile.   My Christian name puzzled her, and she had to confer with Lord Playfair before she could repeat it correctly. Simple morning dress was all that was required.   The one thing that struck me more than anything else was the extensive and beautiful views from the Castle windows.

Much disappointment was felt that the engineer of the Ship Canal, Mr. E. Leader Williams, did not receive recognition at the opening of the canal.   This was rectified in the middle of June, when the well-deserved honour of knighthood was conferred upon him.

Some time after the Queen's visit, it leaked out that she was not taken to Irlam, as orginally intended, in consequence of an unsatisfactory report on the sanitary state of the canal.   Lord Egerton made this the ground of a strong letter of remonstrance to the Mersey and Irwell Joint Committee, and he suggested they should try and remedy the evil without delay.

Nearly forty years before the advent of the Ship Canal, Manchester was recognised as a port, and had bonding warehouses in Salford, under the supervision of the Custom House. In 1882 these were handed over to the Excise, because there was no business being carried on. History repeats itself, and the premises were now restored to their original purpose, so that goods were again passed at Manchester.

The authorities and traders of Warrington were a thorn in the flesh to the Canal Company during the whole year. First they complained that the river at Warrington was nothing less than an open sewer, and that people could scarcely live in their houses. Then they asserted that the delay in deepening the Mersey caused great loss, and they threatened to claim the £50 per day penalty for non-completion in time. The Queen, too, had not pleased them when she refused to visit Warrington, and the oversight in not inviting the authorities to the Manchester celebration made matters worse.

One of the compensations promised to Warrington was a new lock at Walton, and this was opened at the end of September, the Mayor and Corporation of Warrington and several Ship Canal directors being present. The party afterwards visited the dredgers at work in the channel, and it must have been patent to every one that it was next to impossible to carry out the clauses of the 1885 Act of Parliament, if the interpretation put upon them by Warrington could be sustained, *viz.*, to maintain for ever a dredged channel, having always a depth of 8 feet of water throughout, from Latchford Weir to the works of the Monks Hall Company. Sisyphus never had so hard a task, for directly the full depth of 8 feet was dredged, the next tide filled up the vacuum. Warrington demanded the last pound of flesh, and only a compromise could in the end settle the matter.

In the month of June Messrs. Fairclough & Sons, of Warrington, brought on for trial their action against the Canal Company, who, they alleged, had unlawfully diverted water from the river Mersey, and thereby impeded the plaintiff's business. The defence of the Canal Company was that they had acted only in pursuance of their statutory rights, and had not exceeded them. The further hearing of the action was adjourned, as a settlement seemed imminent. After much negotiation the action was settled in the month of November by a "consent" order, providing that the company should pay Messrs. Fairclough an agreed sum of £7,500 for compensation and damages, and should submit to an injunction restraining the company from abstracting more water from the river Mersey than they were authorised to do by the Act of 1885, but subject to the right (if any) of the Corporation of Warrington to agree with the Canal Company, to vary the already existing statutory provisions in regard to abstraction.

The Warrington Corporation, on being referred to, resolved to confer with the manufacturers and traders of the borough. The result was that the Corporation declined to give their sanction to the settlement between the Ship Canal Company and Messrs. Fairclough & Co., and this led to further litigation.

Subsequently, Messrs. Fairclough moved the court for a writ of sequestration against the Canal Company, on the ground that they had committed a breach of the injunction. Mr. Justice Kekewich held that the breach was proved and made an order. The case, however, was brought before the Court of Appeal, consisting of the late Lord Chief Justice (Lord Russell of Killowen), Lord Justice Lindley, and Lord Justice A. L. Smith, who reversed the decision of Mr. Justice Kekewich and discharged his order.

Hitherto Ireland had been a sealed letter to the Ship Canal as regarded new lines. On the 24th June the S.S. *Quiraing* arrived from Waterford, as the pioneer ship of a new company run by butter and provision merchants in Hanging Ditch, and there seemed every prospect of a successful trade being established. She had a full cargo, and came up from Eastham in four and three-quarter hours.

A new development was the starting of a line of steamers to run weekly between Penzance and Manchester with cargoes of potatoes.

In starting a huge pioneer concern like the Ship Canal, where everything had to be learned by experience, with no similar undertaking to copy from, and where the staff were practically new to the business, there must necessarily be a certain amount of confusion and many mistakes. The warehousing accommodation was insufficient, the railway lines on the docks were incomplete, and there was only one junction to connect them with the general railway system of the country. The distribution and collection of goods had to be organised, a system of canvassing for trade had to be matured, and through rates had to be arranged in face of the unwillingness of the railway companies to co-operate. A few experienced men accustomed to handling cargo had been secured, but the quality of the bulk of the labour available was unsatisfactory. Added to this, the railway and Liverpool dock and town interests were bitterly hostile to the canal, and were only too glad of every opportunity to publish and sneer at any mishaps and shortcomings that might occur.

One of the complaints made by Liverpool was that barges were not permitted to use the Ship Canal. Mr. Marshall Stevens, the manager, publicly announced that this statement was incorrect, and that barges could be taken up on the same terms as ships.

Many of the shippers and merchants of Manchester, instead of supporting the undertaking, threw a wet blanket on it, and even friends of the canal, anxious to help, did positive harm. Some people expected small quantities of goods to be carried cheaply, expeditiously and punctually, and when they were politely told that the Canal Company were not carriers, but simply provided a waterway, they complained.

Of course six or twelve months' delay in opening would have obviated many difficulties, but the remedy would have been worse than the disease. For financial and other reasons, it was absolutely necessary to open the canal even before the equipment was complete.

Only those who have been through a similar ordeal can realise how depressing, nay, almost heart-breaking it is, to be earnestly doing one's best, and then not only to meet with crosses and disappointments, but to have to endure the sneers of some people and the mock sympathy of others.

This was the position of the Ship Canal directors during the whole of 1894. They were, however, slowly emerging from a chaos of misfortunes, accidents and unfulfilled hopes, and it was scarcely to be wondered at that men, who, in their own businesses only knew success, should be tempted to throw up the sponge in disgust. The *Liverpool Courier* wrote in May that :—

Mr. A. L. Jones, of Liverpool, did not attach much importance to the canal being dreadfully malodorous, but those shippers who have tried Manchester have lost heavily, and are dropping off, and it is out of the question to expect a profit. Those who predicted that the canal must prove a failure, so far as attracting vessels of large size was concerned, never ventured to forecast a future one-half so gloomy as that now spoken of as the result of actual observation. They did not anticipate a speedy collapse of the banks, and the consequent enormous dredging and repairing expenditure which is now declared to be inevitable.

The *Liverpool Mercury* said :—

Misfortunes of one kind or another appear to have dogged the steps of the Ship Canal directors. They are committed to an enormous responsibility and have had to suffer a grave disillusionment.

Even the American papers were full of inspired articles intended to damage the canal. A St. Louis paper was afraid, after all the millions that have been spent upon it, that the Ship Canal will never pay expenses.

The *Chicago Herald* said :—

There is a good deal of discontent in Manchester at the slow development of the Ship Canal.   The fourteenth week recorded only sixteen vessels outward, and thirteen inward.

The *Railway Age* reported :—

The cost of dredging alone on the Ship Canal being placed at 1,000,000 dollars, this will bear hard on the investors, whose shares have declined 35 per cent. since the opening of the canal.

Harassed by stinging criticisms, oppressed by the fact that the share capital of the company had decreased in value by £5,000,000, taunted as incapables who ought to be replaced by stronger business men, and suffering from the defection of one of their own number, who seemed disposed to shirk responsibility, the Ship Canal directors were in anything but an enviable position.   But, like Paul of old, "they were cast down but not destroyed".   The shareholders, with splendid magnanimity, never withdrew their confidence, but encouraged them in the efforts they were making to overcome the difficulties which beset them.   Working men never grumbled or flinched though they were losing their hard-earned money.   It was such sympathy, more than anything else, which moved the directors to fresh exertions.

For some time the Manchester Corporation had been anxious to have an independent report on the position and financial prospects of the Ship Canal.   This they instructed Mr. Hill, the Corporation engineer, to prepare, and it was issued on the 27th May.   Originally it was estimated that out of the £5,000,000 lent, there would be left (after completing the structural work) £500,000 for equipment and contingencies.   Mr. Hill stated that the cost of additional works had reduced the sum to £400,000.   This would not be sufficient for the whole of the further requirements. Even the provision of additional extra dredging was placed to capital account. Inasmuch as a further sum of £50,000 was expected from the sale of plant, he proposed that £199,000 should be provided for immediate requirements, and that £250,000 should be laid aside for future wants.

At the Council Meeting on the 6th June, Sir John Harwood moved the adoption of Mr. Hill's report, and made it the occasion of unburdening his mind.   For some weeks past there had been a growing divergence of opinion between Sir John and his Corporation colleagues, who felt that, as they had had a governing majority on the

Executive Committee for a long period, they ought not to exculpate themselves, and throw blame on others for what at the worst were only "errors of judgment".

Sir John Harwood addressing the Council said that in June, 1892, the estimated cost of completion was £2,183,433, showing a deficiency to be met by the Corporation of £1,489,282, leaving, say, half a million of the last loan in hand. The estimates now showed, if interest were paid, that they would not only have nothing in hand, in December, 1895, but that there would be a deficiency of £146,862. He had pointed out to the Consultative Committee that in 1896 a rate of 1s. in the £ would be necessary to meet the deficiency. When at the end of 1890 it became evident the Ship Canal needed financial assistance, Lord Egerton wrote and said Mr. Abernethy estimated that £1,700,000 would be sufficient to complete the work. Mr. Hill, on the quantities then given him by the Ship Canal officials, confirmed the estimate. It was then stated that unless prompt assistance could be rendered, the works would have to stop. Almost a panic was created, and the Corporation got powers to lend up to £3,000,000, or nearly double the sum asked for, and in August, 1891, five members of the Council were placed on the Ship Canal Board. They soon found the previous estimates fallacious, and in October, 1891, they were increased by £863,595, no satisfactory explanation being given. The Corporation directors were dissatisfied because the work was not being done by contract, also with the way in which the stores and materials were bought. As a result, an Executive Committee was appointed on 11th December, 1891, of which he (Sir John) became Chairman. This Committee altered the old system entirely, and introduced contracts for work and tendering for goods, which, in the latter case, effected a saving of 24 per cent. The work was divided into five sections: two were let to Mr. John Jackson for £450,000. The next two were let, under peculiar circumstances, to Mr. C. J. Wills. Nominally, he was the contractor. Really, a schedule of prices showed the cost to be £460,000, and the Corporation found the plant and capital. Any amount saved on that sum was to be divided between Mr. Wills and the Canal Company. The saving came to £150,000, and this was divided accordingly. In the same way the Runcorn section, No. 5, by schedule of prices came to £370,614. The saving was £38,000, a portion of which went to Mr. Jones, the contractor. Sir John Harwood then went on to say that the Corporation directors had been grievously misled by the engineers and others, as regarded the cost of equipment, and that all the estimates put in, including those of January, 1891, and June, 1892, had been insufficient. Further, that his Committee had found statutory obligations at Warrington and elsewhere, costly in character, and of which he had no conception.   The fulfilment of these obligations, certainly not of their creation, would very largely drain the resources, even of the Manchester Corporation.   Sufficient to say, it was very serious.   His connection with the Ship Canal had been a harassing business.

The Corporation had been led on to find £5,000,000, and now he was severely attacked by Lord Egerton for saying more money would be wanted to complete the canal.   He would be no party to advising further loans till a very different arrangement had been made with the Ship Canal Company.   The Corporation had already done all they set out to do when they obtained the Act of 1892, and they then thought half a million of the second loan would not be wanted.   He had striven faithfully and earnestly, in the face of much opposition, to discharge the duties of his office, and the work of the future would require men of a special training.   Under these circumstances he asked to be relieved from further responsibility.

At the adjourned meeting, *Alderman Crosfield* said Mr. Hill's report might be regarded as a post-mortem examination ; that of Sir John Harwood was, however, a clear case of vivisection.   He came on the poor victim with his scalpel, and whether the victim would survive was a matter of doubt.   According to the minority report of Sir John Harwood, there was no hope of salvation, and he spoke as if they were on the down grade and the resources of Manchester were exhausted.   The report presented seemed to come from Sir John alone.   What had the other ten Corporation directors to say for themselves?   Sir John Harwood suggested to one's mind the visit of Macaulay's New Zealander to Manchester to look upon its ruins.   He himself believed the city was in the infancy of its prosperity.   Something told him there was want of unity among the directors, and he would like some information. If divided counsels prevailed, some drastic measure must be taken.   He then went into the question of estimates, and showed a difference of £194,000 between those of Mr. Hill and the Ship Canal directors at the end of 1895.   There was no cause for despair; they need only keep a cool head and a warm heart, and they might hope that the period of darkness would soon pass away, and a bright and prosperous future open out before them.

Councillor Andrews thought when the last £500,000 out of the £2,000,000 was demanded it was time to ask exactly where they were.   Mr. Hill's report was asked for in February.   Why was it delayed till May?   Why was Mr. Hill blamed for producing it?   There seemed to have been a lack of capacity to grasp the situation. The directors must have known they were at the end of their tether financially.

Now, Sir John Harwood told them at the end of 1895 they would be bankrupt. Any way the Corporation must find no more money.

Councillor Southern wished to pay a tribute to the masterly way in which Sir John Harwood had conducted the work of the Executive Committee. No doubt in doing his duty he had made himself unpleasant to individuals. He had made the money go further, and had the work more quickly and better done than before. He explained that though Messrs. Wills and Jones had made large profits, the company had themselves made a large profit on the lowest contract price. He could not understand, however, the depressing attitude of Sir John Harwood. All was going on satisfactorily, and the resources would be equal to the calls made on them. Why then this crisis? Why this alarm? Why these startling disclosures? There had been mistakes made in the estimates, but he did not believe intentionally. In justice to the old directors, let him point out the distracting and perplexing situation they were placed in by the death of the contractor. They suddenly had to become contractors without having had any previous experience, and at the same time had to face appalling financial disasters. In the short interregnum between Mr. Walker's death and the entrance of the Corporation directors, the old Board did their best to encounter the great difficulties they met with, and from which the first loan relieved them. Feeling they were still on a "slippery slope" with nothing to hold to, the directors refused to sanction further expenditure until they knew what the total was going to be. The Traffic Committee's requirements came to nearly a million of money. This was subsequently reduced to £443,000. Again, Sir Edward Jenkinson and Mr. Hill separately revised the estimates for what was absolutely necessary, and came to a very similar conclusion, and both believed their resources would last out till 1896. They were bound to avoid, if possible, a further mortgage of the rates. They had now in Mr. Bythell a most capable man at the helm, and he believed, though there was sufficient cause for anxiety, there was no need to despair. Many regular services to foreign ports had been established, sales of land had been effected, and what remained was a most valuable asset. The Corporation of Manchester could not turn back. Great advantages to the ratepayers had occurred through the canal, and it was not clear there would be any call on the rates; but any way they must face the necessary further sacrifices, and he believed that was the opinion of the citizens of Manchester.

Alderman Leech said he had been pained to listen to Sir John Harwood's speech. The Corporation directors were unanimous as to the report and accompanying explanation, but he felt sure they did not agree with what Sir John Harwood had said, and it would have been fairer if he had consulted his colleagues on several points before speaking so strongly. For some time past they had not had the consideration and confidence they were entitled to, neither had they been consulted in the management of affairs, and without united action they were a rope of sand. Sir John Harwood complained of the weight on his shoulders, yet he would not brook assistance from his colleagues. For three years he had been a prominent member of the Board, and since December, 1891, as Chairman of the Executive, he had been virtual pilot of the concern. He was at once Prime Minister and Chancellor of the Exchequer; he held the keys, so that no money could be spent without his sanction, and his colleagues had never divided against him at any Ship Canal meeting. The engineers had often taken instructions from Sir John alone, and he knew more than his colleagues as to the progress of the works. After taking over the works the reign of the old directors only lasted a year, whilst the Corporation directors, with Sir John Harwood at the helm, had subsequently been directing the ship. Sir John Harwood had no doubt done some splendid work. It was to be regretted that he had recently given way to gloomy fears. It could do no good publicly to point out every weak spot in the Ship Canal, and turn it inside out before its enemies. Proclaiming weakness only provoked a calamity. Like everybody else, the old directors had made some blunders, but could any one believe that, as large shareholders, they would wilfully squander their own money. Miscalculations had been made, but had it not been so with most public works, in the Suez Canal, the Severn Tunnel, and even at the Longdendale Waterworks? Alderman Leech went on to doubt an addition of 1s. 8d. to the rates, but if it were so, who was responsible for refusing the help of Salford and Oldham? It was the auditors who first discovered and called attention to the unsatisfactory way in which goods were being bought when the concern was taken over from Mr. Walker's trustees, and they informed Sir Joseph Lee and Sir John Harwood. If the estimates of October, 1891, and June, 1892, were insufficient, the Corporation directors and engineers were equally to blame with the old directors and their engineers. After explaining the cause of the discrepancies in the estimates, he stated that Mr. Saxon, the solicitor, had often explained the onerous obligations of the Ship Canal, and it was not right to say they were unknown. He was convinced the canal could be made a commercial success, but they must cease washing dirty linen in public, and put shoulder to shoulder to extricate it from its difficulties. He believed the hour would come when Sir John Harwood would wish to blot out the speech of last week as the mistake of his life. It had caused pain to thousands, loss to tens of thousands; it had damaged the credit of Manchester, and benefited nobody. He believed ill-health and anxiety had had very much to do with it.

When the debate was resumed several Corporation directors stated their views. *Councillor Pingstone* said incidentally that the Co-operative Wholesale Society, who had £20,000 in Ship Canal ordinary shares, had, directly or indirectly, received back nearly the whole of that money in saving on carriage.

Alderman Clay thought it was an insult for a director to make an important statement without the consent of his colleagues. No bank Chairman would dream of such a thing. Yet that was what Sir John Harwood had done. He contended it was through that gentleman's instrumentality that the Corporation consented to find money for the Ship Canal. The speaker twitted Sir John with inconsistency in first declaring so much faith, and now indulging in gloomy forebodings. The old directors ought not to be charged with keeping things back, and throughout the negotiations the Corporation had acted under the advice of Mr. Moulton, Q.C. He could only account for Sir John Harwood's conduct by his becoming thoroughly disheartened at the result of the first few months' traffic, and the difficulties they had to encounter. They wanted a clever man with a master mind to act as managing director, and he believed such a man would be found.

The resolution approving Mr. Hill's report was adopted.

At a subsequent meeting Sir John Harwood's resignation as a Corporation director was received. An amendment that it should not be accepted caused an excited debate, and when the vote was taken the numbers were equal. On the Town Clerk's recommendation, the Lord Mayor gave a casting vote against the amendment, feeling sure that Sir John knew his own mind, and had taken the step deliberately. Alderman Lloyd was appointed a Ship Canal director in place of Sir John Harwood.

Naturally the action of the Chairman of the Executive Committee caused a profound sensation. Every one was talking of Sir John Harwood's panic speech, and it was declared that it would cost the shareholders a million sterling.

Mr. Reuben Spencer wrote a strong protest against Sir John Harwood's speech, and reminded him of what he had said formerly ought to be done when the Ship Canal was in trouble. When a man asked Sir John what business the Corporation had to meddle with the Ship Canal, he replied that perhaps they had no legal right, "But if he saw a man who could not swim in 10 feet of water, what was the use of reasoning with him? Pull him out and then talk to him." Mr. Spencer said now was the time to cease vituperation and to act unitedly, and his advice to merchants and manufacturers was—

Forget whatever difficulties and defects appear to you at the moment; pass by the unjust and unsupportable charges about its inability to carry on certain works done by other and long-experienced ports. Through much trouble and labour, through great personal inconvenience, loss of capital and time of the directors, the undertaking has been so far completed as to be able to discharge its functions in a manner which is a surprise to all acquainted with waterways, ports and shipping. Support now is urgently wanted. Don't dwell upon or multiply difficulties, but help the directors to move them by giving them support.

The Liverpool papers revelled in the position. The *Courier* wrote:—

Five months ago the Ship Canal was opened amidst jubilant shouts. It received the advertisement of Royalty. Knights were created, feasting and merrymaking reigned supreme, the argosies of Cottonopolis were to cover the seas. Manchester had taken her position proudly as one of the world's seaports. To-day the narrative given by Sir John Harwood is a record of ignorance, blundering waste and incapacity. By the end of 1895, and perhaps sooner, he says there will not be a penny left.

We should not be surprised if the verdict of posterity upon this enterprise "was magnificent but insane."

The *Daily Post* said:—

The Manchester Ship Canal has been wounded in the "house of its friends". The statement submitted by Sir John Harwood in the City Council implies that the concern is on the verge of hopeless insolvency. This is a mild way of describing the outlook. The Ship Canal has been six months on its trial, and who will not say that it has not been poignantly disappointing.

Herepath's Railway Journal kindly indulged in a series of Ship Canal lamentations:—

The Ship Canal shareholders are human, and it is hard to undeceive them. They believe Sir John Harwood's 1s. 8d. rate is the outcome of a dyspeptic dream. The million or so knocked off share value is a concrete reminder of stern realities. The question is, in fact, whether the canal can be kept open at all. Where is the money to come from? Now, if money is not found, we fail to see what is to hinder the canal from being shut up. The canal people piteously appeal to the merchants to come forward, and say, come and build warehouses. But sentimental chatter of this kind will not suit the business men of Cottonopolis. The cost of dredging alone will absorb 60 to 70 per cent. of the gross revenue. In the face of that, we need not speculate how long the canal could be kept open.

The Board of Directors lost no time in replying to Sir John Harwood's damaging onslaught. They felt it keenly that he should seek to exculpate himself by blaming his colleagues. When the first five Corporation directors were put upon the Board there can be no doubt that they were looked upon with suspicion, and their efforts in the shape of retrenchment and reform did not receive the support it merited. This, however, lasted but a short time, and ever since the Corporation had a majority on the Board matters had gone on pleasantly.

In their report the directors showed that with the canal only partially complete, and not ready for large ships, they had earned £33,710 in five months, and this in spite of being hampered by conferences, insufficient shed accommodation, uncertain labour, no direct railway connections, and no grain warehouses or oil tanks. They then explained what was being done to remedy these and other defects, outlining their proposed plan of working, referring to the negotiations with existing lines of steamers, and to the prospect of opening out new connections all over the world. They gave the receipts of the Suez Canal from the first, and showed how marvellously a big concern had risen from small beginnings, and if that had succeeded why not the ship canal?

At a meeting of shareholders, too, the speech of Sir John Harwood was severely criticised. Dr. Pankhurst telegraphed from London :—

Have read with astonishment and much sorrow the panic speech of Sir John Harwood. United action, with courage, will triumph over every difficulty, and make the canal a commercial and financial success.

Even Mr. Joseph Lawrence, himself a keen critic of the canal and its management, wrote :—

I am willing to criticise the proved faults of the canal management as much as anybody, but I will fight to the death against anybody attempting to ruin the canal. My opinion is, the canal will be finished and be a success in spite of 500 Sir John Harwoods.

In a previous letter Mr. Lawrence had expressed his regret at observations he had made about Sir William Bailey and the opening of the canal.

When all looked gloomy and depressing, a letter appeared in the newspapers from Mr. William Pinkney, of the Neptune Steam Navigation Company, who had brought the two largest cotton ships up the canal. This firm had offered to put on a line to Manchester if the citizens would subscribe the necessary capital, but there had been no adequate response. Mr. Pinkney rated Manchester on having made a canal of which they might justly be proud, and which was quite navigable for large ships, and then allowing it to drift into a barge canal from want of support, as it surely would if they did not stir themselves.   He went on to say :—

No outsiders can prevent success if the public determine upon it, and no outsiders can do much to help it without that public taking the initiative.   After showing so much pluck in making the canal, he could not believe the merchants of Manchester would now let it lie dormant.   Ships would not come unless Manchester exerted herself to dispose of imports and to find back cargo.   Shipowners might be encouraged, but they would not be bullied into coming up.   After giving his personal experience and dealing with the difficulties of creating a port, he gave it as his opinion that Manchester must become a shipowning community like Liverpool if the canal was to succeed.

Colonel Morrison speaking at the June meeting of the Liverpool Chamber of Commerce said :—

Although Manchester was meeting with formidable difficulties in carrying out its spirited project, it would not be safe for them to conclude that these difficulties would be insuperable.   On the contrary, he saw considerable danger looming ahead from the competition of Manchester.

This was the general tone taken by other speakers, also of a lengthy report issued by a special Liverpool Committee on the Ship Canal.   Among other articles the report mentioned cotton.   Out of thirty-three towns, it was found that in twenty-eight of them it was cheaper to import *via* Manchester than *via* Liverpool.   It was 4s. per ton cheaper to bring timber *via* Manchester than *via* Liverpool.   To meet this state of things the report recommended a reduction in dock and town dues, a simpler plan of collecting them, certain exemptions, encouragement and cheapening of overside shipments, and dealing with the abuses and defects of master porterage.   Mr. Cox urged that an attempt should be made to bring down railway rates, and said the Manchester Ship Canal directors were determined to make the thing go.   Other members thought more help was to be expected from Dock Board reductions, whilst another section was in favour of a paid Chairman of the Dock Board.

The debate in the City Council drew the attention of the London shareholders to the divided counsels on the Ship Canal Board, with the result that Lord Rothschild, on their behalf, wrote to Lord Egerton and to the Lord Mayor of Manchester, urging that the affairs of the company should be placed under a responsible head, who should devote the whole of his time to the duties, and receive a salary.   This seemed to find favour with both the directors and the Council, and the name of Mr. J. K. Bythell was suggested.   It was felt on all sides that this step would put an end to divisions on the Board.

Early in July the very important announcement was made that the Ship Canal Company had arrived at an understanding with the railway companies, by which the Canal Company would be able to quote through rates, including railway freight, Ship Canal tolls, and shipping charges, upon all classes of traffic to and from the docks at Manchester, and all stations in the United Kingdom. These rates were sent out broadcast over the district, and enabled traders to see at a glance the charges they would have to pay. The saving as regards cotton will be exemplified by a few instances.

Through rates on cotton from the Manchester Docks :—

	Cost *via* Liverpool.	Cost *via* Manchester.
Bolton	15s. od.	10s. od.
Darwen	17s. od.	12s. 6d.
Hyde	17s. 6d.	11s. od.
Heywood	17s. 8d.	11s. od.
Oldham	17s. 8d.	11s. od.
Rochdale	18s. 6d.	12s. od.
Stockport	15s. 8d.	10s. od.

The Ship Canal Bill after passing the Lords came before a Commons Committee, with Mr. Hoare as Chairman. The North Staffordshire Company having been arranged with, the Bill passed as an unopposed measure.

On 20th July the Board of directors met to consider the question of a salaried Chairman.

Lord Egerton said he felt it was desirable that the affairs of the company should be managed under one responsible head. Whilst thanking his colleagues for the courtesy they had always extended to him, and assuring them of his assistance, should critical circumstances arise, he tendered his resignation as Chairman of the Board, and expressed a hope that Mr. Bythell would be elected to the position. This was agreed to, and £3,000 per annum was voted as salary, which was to be paid out of the £5,000 heretofore voted to the directors. In thanking Lord Egerton for his past services, a very high compliment was paid him for the magnanimous way in which he had assisted the canal in troublous times, and then resigned his office when it was thought desirable there should be a permanent Chairman, who could devote his whole time to the interests of the company. Sir Joseph Lee wrote offering to resign his office in consequence of continued ill-health. His resignation was accepted, and Sir Anthony Marshall took his place as Deputy Chairman.

The last six months of 1894 were largely occupied by efforts to whip up trade for the canal. The promoters were taunted by their opponents with the fact that only 444,000 tons of merchandise, yielding £41,925, had came up the canal during the first seven months, that the merchants of the district were holding aloof, and that shipowners were retiring from the port in disgust. The Press was full of letters making suggestions for increasing trade. All sorts of shipping companies were suggested, and the shareholders, feeling their property at stake, were holding meetings to take counsel as to how trade could be stimulated. The good old friend of the canal, Mr. Jacob Bright, at a meeting of shareholders in the Concert Hall, said by way of encouragement :—

He might say with absolute sincerity that nothing had made him doubt the wisdom of the courageous men who brought this great work into existence. He had no misgivings as to the future of the undertaking. He knew what Lancashire men were. They could overcome great difficulties ; they had overcome great difficulties ; and there was no difficulty so great in the work before them that Lancashire men would not be able to surmount. It was natural there should be some disappointment and some impatience, and there was no great object ever attained easily. There might be men who were faint-hearted about this matter, but the facts as to the necessity for the canal were as true to-day as they were eight or ten years ago. Why then should there be any discouragement? The traffic of the canal was growing and could not but grow. The time would come when they would feel proud they had anything to do with the work, and when the owners of shares would tell their children with pride that they had had their part at an early day in doing that which was for the great benefit of Manchester and Salford.

Mr. J. R. Beard and Mr. Belisha also made stirring speeches, and Mr. W. Briggs, of the Neptune Company, referring to the statement that cotton had ceased to come up the canal, mentioned that his firm had made arrangements to bring several cotton ships before Christmas.

A vigorous effort was determined upon to secure a portion of the Bombay traffic, and a request was also made by the chief houses shipping to China, that the China Conference of Steamship Owners Company should load goods in Manchester instead of barging them to Birkenhead. As an inducement they offered to let the Conference pocket the 4s. per ton saving effected thereby. Here was another example of the rebate system and of the clever way in which the shipping ring bound merchants to them. For instance, the freight of goods and yarn sent to Calcutta was 32s. 6d. and 10 per cent. primage, equal to 35s. 9d. If the shipper sent all his goods by the Conference ships he got a rebate of 5s. 9d. per ton on all he shipped during the year.

This meant a large sum yearly, and it was the fear of losing this that rendered shippers loath to make a change.   A saving of at least 10s. per ton could have been easily effected on Liverpool rates, and the fact that German lines were carrying goods to the East 10s. to 15s. per ton under Liverpool rates was putting a great lever in the hands of foreign competitors, and must in the end seriously affect English trade.   The Conference rates for piece-goods from Liverpool to Hong Kong were 44s. 7d. per ton net, whilst the same Conference carried the goods of Antwerp merchants for 33s. The Conference rate for yarns from Liverpool to Hong Kong was 38s. net, while the German Lloyd boats only charged 22s. 10d. from Hamburg, or little more than half.

Notwithstanding all the efforts of the Eastern merchants, who had also called the Chamber of Commerce to their aid, neither relaxation of the burden nor cheaper shipment from Manchester could be obtained.   The Conference clung to their 45s. rate to China and Japan.   Then Messrs. Brooks & Doxey, of Openshaw, having to ship a large quantity of machinery (sufficient to fill a ship), chartered a vessel of their own in December, the S.S. *Rosary*, to sail from Manchester at little over half the Conference rate.   This alarmed the Conference, and they at once came down from 45s. to 27s. 6d., and to prevent other ships coming up to Manchester offered to bind themselves to that price, if machinists would give them their whole business.   What had been done for machinery could be done for piece-goods, but unfortunately shippers and merchants are a rope of sand.

As regarded the Calcutta sailings an arrangement was made with Sir Charles Cayzer of the Clan Line to run boats from Manchester, and the *Clan Drummond* was advertised to sail on 10th January, 1895, and since then, these boats have run at stated intervals.

The freightage of the valuable cotton goods of Lancashire to Bombay and India has always been a plum for shipowners, and after keen competition they tried to put a stop to cutting rates by forming a Conference and pooling the business. But the Bombay merchants are keen and shrewd traders, and they met combination by combination.   They agreed to work as one man, to throw the old lines over, and to make terms with a new line to do their business for two years at a much reduced rate, and in this they succeeded.   At the advent of the Ship Canal this contract was held by four lines of steamers, trading from Glasgow and Liverpool to Bombay.   Mr. Bythell, being himself an old Bombay merchant, was anxious to secure the traffic for the canal, and he wrote to the secretary of the Bombay merchants, explaining the advantages of the canal, and suggesting that when the next contract was let there should be a clause to the effect that the bale traffic of Manchester must be shipped at Manchester. This was only after the contracting lines had rejected his overtures to load their ships at Manchester. They would not believe it possible for big ships to navigate the canal, although cargoes from India were passing up the Suez Canal, and even through the Ghent Ship Canal which was only 55 feet wide and 21 feet deep. The Manchester Ship Canal, though 120 feet wide and 26 feet deep, was pronounced to be dangerous.

Whilst a reply from the Conference of Shipowners was pending, the Chairman of the Chamber of Commerce, Sir Forbes Adam, at one of the quarterly meetings, said they had been trying to ascertain why the canal was not more used :—

He feared, notwithstanding all the oratory on the subject, no one was prepared to spend a sixpence for the sake of patriotism, and that it was only by getting lower rates that the traffic would come. Even with equal rates a long time must necessarily elapse before existing arrangements worked off and could be changed.

To this Mr. Bythell rejoined at the Victoria Arcade Cotton Meeting :—

The Ship Canal had never asked anybody to give them a penny out of sentiment. They were prepared to show that users of the canal could effect a large saving. Before the canal opened, cotton exporters paid 32s. 6d. freight to Rotterdam and other ports. Cotton could go to-day at 20s. *via* the canal, inclusive of warehouse charges.

Ever since he joined the canal he had seen how merchants were hampered by combinations, and how they were at the mercy of everybody. They would continue to be so till they had backbone to combine themselves. All eastern ports, except Bombay, were handicapped. Calcutta was paying 10s. and China 15s. per ton more than the rates from Manchester would be. The Hindoo merchants of Bombay, by combination, had broken down old rates, and the existing contract expired this year. The Ship Canal had been trying to tender at a price that would secure the shipment of cotton piece-goods at Manchester, and in this they were being assisted by London shippers. He was happy to tell them he had that day received a telegram saying the tender made under their auspices for two years, from 1st January next, to ship all piece-goods from Manchester to Bombay had been accepted at a lower rate of freight than the merchants had been paying, and it compared favourably with the rate to Calcutta from Manchester. This was an earnest of what was to come.

This announcement fell on Liverpool and the shipping rings like a thunderbolt. The latter imagined that by their policy of combination they had the prize within their grasp, and now the Ship Canal, which they had declined to acknowledge, had organised an opposition company and was going to carry the trade to Manchester.

The Liverpool papers were furious. The *Liverpool Daily Post* estimated a loss to Liverpool and Birkenhead of 50,000 to 200,000 tons of the most remunerative traffic, and said that rough goods would follow. It touched shipowners keenly that Messrs. Forwood Brothers & Co., of London, had undertaken to find the shipping. They quite believed that with cheap coals the contract might pay, but they felt sure Indian steamers of the usual size could not navigate the canal. Next day it was reported that Messrs. Raeburn & Verel, of Glasgow, were working with Messrs. Forwood & Co., and had the signing and settlement of the contract in hand. Manchester was to have the first chance of loading cargo, and failing to obtain it, the new line would fill up at Liverpool. Occasionally loading would start at Glasgow and be completed at Manchester. The contract rates were to be 18s. 6d. per ton outwards, against the previous rate of 21s. 6d., with 10 per cent. primage.

The *Daily Post* further suggested that to meet the case, Liverpool should forward imports on the canal by means of dumb barges, and thus carry them 2s. per ton cheaper than on the Bridgewater, or by the railways. This, with reductions in their own charges and a good service, ought to render further defections improbable. Finally, the *Daily Post* said that the contract of Messrs. Forwood & Co. was wholly speculative if not reckless. The next announcement was that Messrs. Sivewright, Bacon & Co. had come into the arrangement. It was stated that their ships were tramps, and would take eight or ten days longer on the voyage than liners.

The *Liverpool Courier* fell athwart the Dock Board, and said it was their mismanagement that had made the Ship Canal a possibility, and that, if the Board were not more helpful and vigilant, this would only be the beginning of trade abstraction. It discovered later that it was a Scotch firm and other steamship owners who were anxious to share in the blessings or assist in brightening the prospects of the Ship Canal, and that outward goods were to be carried at 3s. 6d. per ton less than under the old contract. It further stated that the Ship Canal Company were the contractors employing the shipowners.

For some time the details of the arrangement were very obscure, but the history of the transaction is, that Mr. Bythell asked the Bombay Native Piece-Goods Merchants Association if they would consider a tender to carry all their cotton goods from Manchester, and they consented. Messrs. Forwood Brothers & Co. then undertook to form a company. They associated themselves with Mr. Christopher Furness, of Furness, Withy & Co., and other shipowners, and put in a tender for two years, which was accepted. Not less than three, and not more than six steamers per month were to be despatched from Glasgow, calling at Manchester, and, if they wanted to fill up, also at Liverpool and Birkenhead. Intervals of sailing to be not more than ten, or less than fourteen days. Though all the details were arranged, the contract was not signed at the time. The old figure was 21s. 6d. and the new 18s. 6d. per ton.

As soon as the old conference found they had missed the contract they moved heaven and earth to upset the new arrangement, and it is said they offered to take 2s. to 3s. per ton less than the new contract, but without success.

A deputation of Liverpool traders also waited on the London and North-Western Railway Company to point out that if the new arrangement went through, the railway would not only lose the carriage of goods between Manchester and Liverpool, but would also lose the carriage of coals to the ships at Birkenhead, which would be coaled at Partington. They strongly contended that it would be in the interest of railway companies to make some concessions in rates, and so help to keep the Bombay trade at Liverpool. But the request of the deputation met with a direct negative, the railway company refusing to reduce the rates one halfpenny. The aid of the Marine Insurance people was next invoked, and they put up the insurance from 6s. per cent. to 10s., partly because of the risks of the canal, and partly because the new line would at first employ tramp steamers. But all was of no avail. It is understood that then the Anchor, City, and Clan Liners went to Messrs. Forwood and said to them: "Are you fixed up definitely with the Bombay traders? If not, are you prepared on terms to transfer your position to us?" The Bombay agreement was not signed, and in the end the old lines bought the position of the new company for a very large sum which would scarcely be prudent to mention, and completed the agreement at somewhat increased rates, one condition being that the arrangement as to ships coming up to Manchester should stand. As a result, the Anchor S.S. *Hispania* was advertised to leave Manchester for Bombay on the 5th January, 1895, to be followed ten days after by the S.S. *Clan Fraser*. It is a question whether the Bombay traders were well used, for in consequence of the contract not being signed they had to pay 20s. per ton instead of 18s. 6d., but it was they themselves who delayed signing. The rate from Glasgow was being fixed at 16s. 6d. Independent shipowners at once reduced their rate to 12s. 6d.

At the August meeting of shareholders, Mr. Bythell presided for the first time as Chairman of the company. The report mentioned that the dredging to 26 feet was nearly completed. The Chairman said £64,000 had been spent on dredging, and that the work at Warrington had been both costly and difficult. Six three storey sheds and twenty hydraulic cranes had been erected. The Chairman went on to give a list of the savings to Manchester effected by the canal. Yarns to the Continent had been reduced from 32s. 6d. to 20s., and sugar from 22s. 6d. to 12s. 6d.   He warned shippers, however, that they must not bleed shipowners or they would be driven away, and he gave the following instance.   When a line was started to the Persian Gulf, Liverpool freights went down from 40s. first to 30s. and then to 20s. per ton.   But instead of the Manchester shippers being thankful, they went to the new line and said, "Oh, we can ship by the old route for 20s., you must take less". It was difficult to believe that Manchester people should be so absolutely suicidal as to do that.   After detailing the efforts being made to secure new lines, the Chairman ended by emphasising the fact that the success of the canal depended almost entirely upon the importers and exporters of the district.

The approaching completion of the lairages at Mode Wheel aroused considerable alarm at Birkenhead, where much dissatisfaction was felt at the Dock Board's delay in extending the accommodation.   A Liverpool correspondent wrote: "The inaction of the Dock Board has produced so much irritation that it is quite on the cards that Manchester may, before very long, secure this most valuable trade, and the Birkenhead lairages become a relic of the past".

In September it was announced that the General Steam Navigation Company were withdrawing their line to West Africa.   This, following a statement by the Chairman of the Company that the line was progressing satisfactorily, caused considerable sensation.   On being interviewed by the *Evening News*, the secretary of the company said: "It's no use our going to Manchester.   Manchester people will not give us any support.   They have all gone to sleep, and are simply playing the fool with us.   There is not the slightest inducement for us to go to Manchester, and it is simply a waste of money our continuing to do so."

Another version, emanating from Liverpool, was that a Liverpool Conference had intimated that if the General Steam Navigation Company continued to run from Manchester to West Africa, they would oppose the line from London to Bordeaux, of which the General Navigation people had a monopoly.   Then followed a statement made by Mr. Nelson, the manager of the company, that in future Liverpool would be the objective, but that their ships might still come to Manchester it occasion required.

Subsequently, the African trade section of the Chamber of Commerce interviewed Mr. Nelson, and it was arranged that he should send his ships again to Manchester for one or two trial trips, but their continuance would depend on the support they received.   In the end, however, the rebate system of the Conference lines was too strong, and the line was withdrawn.

In August of this year died Mr. W. J. Saxon, of Messrs. Grundy, Kershaw, Saxon & Samson, solicitors to the Ship Canal Company, who had practically had charge of their business during the long and anxious Parliamentary fight. There can be no doubt that his breakdown and premature death were attributable to over-work, mainly in connection with Ship Canal business. He was a thoroughly able lawyer, a courteous gentleman, and the legal successes of the company were mainly owing to his ability, care and attention. He was only forty-three when he died.

On the 8th September the second floating Pontoon Dock was opened at Mode Wheel. Mr. Renwick, the Chairman of the Pontoon and Dry Docks Company, said :—

Following on the Dry Dock opened last February, they were now about to open a Pontoon Dock. Few people knew the importance of these docks to Manchester. The Pontoon was 260 feet by 63 feet 3 inches square at both ends without rudder, " an ugly, red monster " our Liverpool friends called it, and yet they had navigated it safely along the coast from the Tyne and up the Ship Canal at a cost of £1,500, including insurance. The canal had a big fight before it, and he was quite certain of its success, though it had traducers, and their name was legion. If one of the railways would begin to help the canal, others would follow, and before long they would see the railways fighting to do their best to help the interests of the Ship Canal. Mr. Bell, another director, said there could be nothing more convincing to Manchester than that the hard-headed capitalists of the North had risked their money and entered into the important enterprise of building dry docks, believing the canal would be a success.

At a meeting of shareholders Mr. Hilton Greaves, of Oldham, expressed the opinion that a Cotton Exchange should be opened in Manchester. He said the shipping of cotton to Manchester would increase next year from all sides, as many shippers were already preparing to send it by the canal. This was a good sign, but cotton warehouses must be built, and it was folly to delay. If the Canal Company could not build them, they should sell some land reasonably, and others, connected with the cotton industry, would build. In conclusion Mr. Greaves said he had as much faith as ever in the canal, and never once regretted putting his money in it. It had been a great advantage in the past, and would be in the future.

Several attempts were made to arrange with Liverpool as to the sale and delivery of cotton stored in Manchester, and some progress was made, but it was felt that something more was required. If Manchester was to become a great cotton centre, a local spot market was essential. With that view a meeting of those engaged in the cotton industry of Lancashire was held in the Victoria Arcade on the 31st

October, with *Mr. C. W. Macara* in the chair. The Chairman said it was vital to the importers of cotton, and the exporters of cotton goods, that they should not be behind our foreign competitors as regarded freightage, and that they ought to embrace to the full the opportunity now offered by the Ship Canal. Unless they provided every convenience and accommodation, shipowners would not come up to Manchester.

Mr. W. H. Holland moved :—

That as Manchester is now a port, accessible to the sea-going vessels trading from all the various cotton ports, and is the centre of the cotton, spinning, and weaving industries of the United Kingdom, it is desirable and necessary that steps should be taken forthwith to establish a cotton market at Manchester.

This being passed, it was followed by a resolution moved by *Sir Joseph Leigh*, *M.P.* :—

That it is hereby resolved that a Manchester Cotton Association be formed at once.

Alderman Emmott, of Oldham, in moving a Provisional Committee consisting of many of the leading Lancashire Cotton Spinners, said that consequent on the canal, the rates on raw cotton from Liverpool to Manchester had been reduced 1s. 10d. per ton, and goods between the two towns 2s. The saving on the cotton consumed in Lancashire amounted to £150,000 in the course of a year. Including the reduced railway rates and the savings on cotton and cotton goods, he estimated the cotton traders of the district were advantaged to the extent of £500,000 per annum.

It was encouraging to hear from a large paper maker at Old Trafford that the canal would save his concern alone £2,000 in freightage.

In the beginning of November the new saleroom for fruit, arranged by the Corporation on the Campfield Market Ground, was opened by the Lord Mayor, Sir Anthony Marshall. Previously the sales had been conducted in a building temporarily fitted up. Mr. W. A. Nicholls, Chairman of the North of England Fruit Growers' Association, complimented the Corporation on the commodious and convenient building they had erected, and he felt sure the Ship Canal would make Manchester one of the best and cheapest fruit markets in the kingdom. At first they had great difficulties to contend with, and even the railways conceived it to be their duty to throw obstacles in their way, but now all was changed, and the railways would help them in the distribution of goods. A few instances would show the effect of the Ship Canal on the food consumed in the district; taking Oldham as a typical case—

	Via Manchester.	*Via* Liverpool.	
Onions	4s. 6d.	9s. 1od.	per ton.
Oranges	7s.	14s. 8d.	„

In addition to this the goods would be sold from the steamer's side instead of encountering the damage caused by at least two transhipments and a railway journey. Purchasers in Manchester would have a great saving in carriage, and the fruit would be in a better condition.

Whilst every effort was being made to cheapen freight and improve trade, the attitude of some merchants and shippers was almost heartbreaking. They could not see that cheap production was vital in our competition with foreign producers, whose goods we had to meet in the markets of the world. The following is a quotation from the letter of a shipper :—

I admit shippers have been apathetic because they could see no material benefit from the canal. If Mr. Bythell was to succeed in diverting all traffic to-morrow, it would make little or no difference to us. Our clients might save £500,000 per annum, but we should not. It makes no difference to a shipper financially whether the freight is 80s. or 20s. per ton, so long as we all pay the same. If one shipper could ship at 20s. whilst others paid 80s., he would naturally save 60s., but Conferences have stopped all that. Shippers are human and know their own interests. If they saved 500,000 pence, let alone that amount of pounds, they would not look on the canal with the contemptuous indifference you complain of.

Here was a man advocating Conferences and dear freights, because he would be no worse off than his neighbours!

Contrast this with the view of "A Merchant" who, writing to the *Manchester Guardian*:—

hopes something may be devised to avoid shippers being trapped by the rebate system put in force by shipowners in 1895. Concerted action ought to be taken before the next year's shipments are commenced. I can say that I have good reason for believing that if the rebate system were abolished, I could effect an economy in the freights I pay of £1,200 per annum. And if it were not for the rebate system, which makes it useless for other steamers to attempt to compete with those of Conference lines for Manchester cargo, I do not doubt that I should be able to make most of my shipments by steamers loading in the Ship Canal.

Towards the end of the year the directors were cheered by the arrival of several vessels with large cargoes of cotton, both from America and Egypt. An unfriendly article in *The Times* had done its best to discourage the canal, and made much of the non-receipt of cotton. It was said of this article : "What is new is not true, and what is true is not new ".   One amusing prediction was that the carriage by canal of cotton goods for printing in Glasgow was threatened with destruction by Manchester's new supply of water from Thirlmere, which was suitable for calico printing.

During the last year Sir Joseph Lee, Ex-Deputy Chairman, had been in declining health, and had been obliged to give up active work in connection with the canal.   It was hoped that a visit to a warmer climate would result in his recovery, but it was of no avail, and he passed away on the 18th of December, 1894.   His memory will be held in affectionate remembrance as one who, in his early life, obtained honour in fixing up the French Commercial Treaty.   Later on he threw his whole energies into the Ship Canal enterprise, and was the chief factor in arranging its finances when they seemed in a hopeless condition.   If Daniel Adamson was the most striking personality, certainly Sir Joseph Lee was the financial genius of the canal.   Without his help the concern might have been wrecked.   On three special occasions Sir Joseph Lee did invaluable service to the company :—

1. When he induced London capitalists to underwrite £4,000,000 preference stock on condition that the district took up £3,000,000 ordinary shares.

2. In getting rid of the claims of Mr. Walker, the contractor, and in determining the contract in 1890 without litigation.

3. In inducing the Corporation of Manchester to come to the assistance of the canal.

It is to be regretted that he was not permitted to see the fruition of his labours.

By the end of the year the dock employees had become fairly competent, the quality of the work was better, and was more quickly done.   Indeed records were established in unloading cargoes at the docks, and in loading coals at Partington. When the timber ship *Honiton* came up from Pensacola, the *Liverpool Courier* was responsible for the statement that she saved £110 by using the canal.   People began to observe bright spots on the Ship Canal horizon, and, speaking of its foul water, a ship engineer gave it as his opinion that it was so good for ship boilers when using it that they never had an ounce of scale.

One of the visitors of the year was Mr. Millar, of Albany, Western Australia, who came in his fine 400 ton yacht to see the canal and to ascertain if the forests of Western Australia could not produce hard wood-paving suitable for Manchester streets.

On the last day in February the Mayor of Salford, Sir William Bailey, invited me to dinner to meet the Marquis of Lorne, and afterwards go over the canal with him.   It is not often that Sir William gets cornered, but he certainly was nonplussed when the Marquis asked him the Latin name for Salford.   But this was not the only amusing incident, for, when on the canal, a spiteful gust of wind sent the Mayor's best silk hat flying into the filthiest part of the canal.   As he stood bareheaded, the Marquis humorously observed he was glad the Mayor paid proper respect to his constituents in Salford and did obeisance to them.   We fished out the hat totally ruined, but as there was a spare cap on board, no harm ensued.

Another visitor in May was Lord Rosebery, who travelled the length of the canal to the Barton aqueduct with the evident desire to form his own opinion prior to the Queen's visit.   Afterwards I took him over the Chetham College.   The canal was a surprise to Lord Rosebery, and he was delighted with the quaint old college.

During the year an agitation was commenced by Mr. Peter Spence, of Manchester, who had conceived the idea that the eleven Corporation directors were not the men who ought to be at the helm of the concern, and that instead of being nominated by the Corporation, they ought to be elected by the votes of the peopl This met with support, and was brought forward at the meeting of the associatec shareholders, who resolved to ask the Lord Mayor to call a public meeting to consider the subject.   He, however, declined to do so on the ground that there was no necessity, and that apart from this, the constitution of the Ship Canal Board had been fixed by Parliament after much consideration, and nothing but a new Act could alter the position.   The Chairman of the Shareholders' Association, Mr. Reuben Spencer, had all along been opposed to Mr. Spence's proposition, believing it to be ungracious to attempt to supersede men who were working hard and doing their best for the canal.

This brings me to the end of 1894, and my labours are now nearly at an end. The canal had been finished, so far as being in good working order.[1]   Completed it can never be, for additions and improvements will always have to be made.   The year ended hopefully and without the financial crash which had been prognosticated. The tables in the final chapter, which carry results to 1906, will show how far various hopes and fears have been realised.

[1] See Plan No. 16.

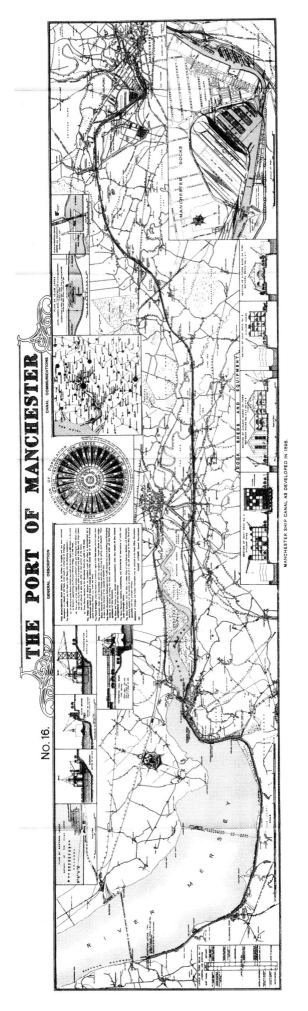

The material originally positioned here is too large for reproduction in this reissue. A PDF can be downloaded from the web address given on page iv of this book, by clicking on 'Resources Available'.

CHAPTER XXV.

DRAMATIS PERSONÆ—RESULTS OF THE SHIP CANAL—
STATISTICS OF SAVINGS TO TRADE—PROGRESS OF THE
PORT—COMPARISON WITH THE SUEZ CANAL.

A thorough development of the canal system of the country would have a beneficial effect, and he proposed last year that Parliament should purchase, improve and extend the canal systems of the three kingdoms.—SIR EDWARD WATKIN, M.P.

A HISTORY of the Ship Canal would not be complete without a sketch of the chief actors in its inception and completion. I purpose, too, in this last chapter to give some figures and facts to demonstrate that the canal has not been a vain show, but that to a very large extent it has fulfilled the expectations of its promoters. It has cheapened both imports and exports; it has enabled our traders better to compete with their rivals; it has brought a host of fresh industries into our midst; it has improved the labour market, increased the value of property and given fresh impulses to business life, and in fact it has practically remade our city. It has not, as was hoped, been able to pay a dividend to the 40,000 shareholders whose determined and heroic enthusiasm has few parallels in the history of the country. Undismayed by defeat and misfortune, they stood by Mr. Adamson in the canal's darkest hours, and their fidelity to the cause and their determination to maintain the credit of Manchester untarnished by failure, will be an example of local patriotism for generations to come. I must again remind my readers that it was not the capitalists or the rich merchants of the district who made the canal; a majority of these feared damage to their vested interests and stood aloof. It was the rank and file of the people, headed by a very few farseeing capitalists who, by their indomitable courage and almost bull dog pertinacity carried the enterprise through. The Cooperative Societies, the Limited Liability Companies and the Trades Unions all gave liberal assistance, as also did thousands of working men, who were clear headed enough to see the advantages of the scheme. Twelve tedious years have passed

(239)

since the opening of the canal, and a dividend is yet in the far distance, but that it will come one day, perhaps sooner than expected, I feel certain.

I have been assured by many traders that indirectly they have had an ample return on their money, and both shopkeepers and working men have been benefited indirectly by increased earning power and cheaper food. In the dock area, where once a few labourers looked after cattle, there are now thousands of men receiving good wages. It is to be hoped that the time is not far off when the whole body of shareholders will receive a reward for the sacrifices they have made.

The canal has never been short of capable men to guide its destiny, and the right men seem to have risen as occasion required. The grand old leader, Daniel Adamson, steered the canal interests when pluck, courage and engineering talent were the main requisites. Sir Joseph Lee followed, and he piloted the concern through its financial troubles. Then came Lord Egerton, whose very name gave confidence and commanded respect. As Chairman of the Executive Committee Sir John Harwood put into the undertaking all his impulsive energy and vast experience in constructive work. Last of all Mr. J. K. Bythell came to the front when a Chairman to take command was deemed necessary. He has been the right man in the right place, and has conducted the affairs of the canal with consummate ability.

Then the leaders also had a splendid staff, chief amongst whom were Sir Leader Williams, Mr. Saxon, solicitor, and at the death of the latter, his senior partner, Mr. James Kershaw; all these gentlemen have devoted themselves heart and soul to the interests of the canal. The first Provisional Committee was formed in 1882. I am surprised to find that only five others of that body, besides myself, are now living.

The names of the members of the Provisional Committee, empowered by the Old Town Hall meeting on 26th September, 1882, to raise a fund and to apply for Parliamentary powers to carry out the scheme, have been already given on pages 81 and 83 (vol. i). I desire, however, to mention others who joined the Provisional Committee at a later date and did excellent work in the cause of the Canal :—

> Henry Boddington, Esq., Manchester.
> Robert B. Goldsworthy, Esq., J.P., Manchester.
> Reuben Spencer, Esq., J.P., Manchester.
> Samuel R. Platt, Esq., J.P., Oldham.
> James E. Platt, Esq., Oldham.
> William Fletcher, Esq., Salford.

No. 15.

BIRD'S EYE VIEW OF THE MANCHESTER DOCKS.

The material originally positioned here is too large for reproduction in this reissue. A PDF can be downloaded from the web address given on page iv of this book, by clicking on 'Resources Available'.

When the Canal Act of 1885 was obtained, photographs of many of the veterans in the fight were secured, and I have reproduced a number of these with a short sketch of their work for the canal. I have also added the photographs of a few of the chief men who came to the front after that date.

Mr. Daniel Adamson must always take the foremost place in the history of the Ship Canal. A man of fine presence, full of an enthusiasm which he had the faculty of imparting to others, of indomitable courage, a shrewd and capable engineer, a vigorous speaker, full of self-confidence, ambitious, and an energetic worker, he was just the man to pilot a difficult and dangerous enterprise. Yet like all other great men he had his weaknesses, of which one would like to speak gently. He never had patience to deal with the details of finance. He was, moreover, uncompromising, did not believe in expediency, and overrated his powers as a financier. Lancashire, however, can never be too grateful for the services rendered by Mr. Adamson, which ought to have some lasting recognition.

Sir Joseph Leigh has been an indefatigable worker from the commencement. Quiet and unassuming in manner he has brought to bear on the business of the company great capacity and skill. It was in recognition of his long services to the canal that the honour of knighthood was conferred on him.

Sir Leader Williams crowned a long and valuable life by designing and carrying out a work that will always be considered one of the wonders of England, and that will hand down his name to posterity. It is easy to be a copyist, but original design, especially when huge difficulties have to be overcome, ought always to command special respect. Burrowing, as it does, through a *terra incognita*, the canal in its formation needed much ingenuity and engineering skill. The Barton Aqueduct and the various swing bridges and locks demanded constructive ability of no ordinary kind. They have been a complete success, and Sir Leader Williams may well be proud of his work.

Sir Bosdin Leech pleaded for a Ship Canal in the public press long before Mr. Adamson's famous meeting, in 1882, and has never severed his connection with the undertaking.

Mr. W. J. Crossley was one of the earliest supporters of the canal and one of the oldest directors. He is also one of the largest shareholders. His matured engineering experience has been most valuable to the Board, and though his many calls have prevented him from giving much time to the canal, he was always available when his counsel and advice were needed. He retired from the Board in 1906.

Mr. Richard Peacock was a singularly quiet, gentlemanly man, whose very name commanded respect. He was Deputy Chairman of the Provisional Committee, and did excellent service in that position. Though essentially a railway man, he saw so clearly the benefits of a Ship Canal that he would never allow private interests to interfere with public duty. Death cut short his career of usefulness early in the history of the canal.

Sir William Bailey, one of the earliest promoters, has rendered yeoman service to the canal. His devotion to the cause has never wavered, but in season and out of season he has been ready to give counsel and financial help ; it is well known he is one of the largest shareholders in the canal. Every one was delighted when his labours received from Her Majesty the recognition of knighthood.

Jacob Bright, M.P. No man ever did more faithful and valuable work for the canal. His was not the ordinary interest of a Parliamentary representative. He threw himself heart and soul into the scheme because he thought it would benefit his city. In the House of Commons, or at public meetings, he was always ready to champion the cause, and his speeches generally carried conviction with them. Manchester owes a deep debt of gratitude to Jacob Bright.

Mr. W. J. Saxon. A model Parliamentary solicitor, and a very amiable and modest man. Few knew the extent of his labours, but those who did, will agree they were invaluable. He was always on the alert, and during the Parliamentary fights, many of his nights were ungrudgingly devoted to the preparation of the Parliamentary work of the succeeding day. He had, too, a very happy way of conciliating opponents. Mr. Saxon was admired and beloved by everybody, and it is to be feared his devotion to canal work hastened his death.

Mr. Marshall Stevens is another veteran still living. A man of great natural ability, he was of immense service to the canal in its early days. He possessed a unique knowledge of railway rates and charges, as well as of shipping usages, and these were very valuable in forming the schedules of canal rates. As a Parliamentary witness Mr. Stevens has few equals. He was made general manager on the formation of the Ship Canal Company.

Sir Joseph Lawrence joined the canal staff at the latter end of 1882, and brought with him a vast amount of valuable practical experience. He had just been fortunate in helping to secure the Hull and Barnsley Railway Act, and he quickly showed his ability as an organiser. I never came across a man who could get through as much work in a given time. The fact that he did not like any restraint on expenditure occasionally caused a little friction, but there is no doubt Sir Joseph Lawrence did very valuable work for the canal.

Mr. John C. Fielden came as a kind of meteor. He roused the middle and working classes in an extraordinary way to support the canal, and when he gave evidence in Parliament his well-arranged statistics commanded respect and attention. Under cross-examination he was equally ready, and when the opposing counsel found they had met their match they generally left him alone. Mr. Fielden was a very gifted man who had risen from the ranks. His end was untimely. He was found drowned in the very canal he had helped to make, but how he got there no one could tell.

Mr. W. H. Hunter has been connected with the Ship Canal from its commencement. As assistant engineer under Sir Leader Williams, he had charge of most important constructional work, and filled a very responsible position. When the canal was finished he became engineer-in-chief and has since done admirable work. No difficulty daunts him and he is full of resource. He is an excellent Parliamentary witness, and has been often complimented on the terse and convincing way in which he gives his evidence. A great compliment was paid when he was selected as one of the European engineers to advise as to the possibilities of completing the Panama Canal.

Mr. James E. Platt has been an unfaltering supporter of the canal and stuck to his post as a director in its darkest hours. When fresh blood became absolutely necessary he graciously stood aside in order that new directors (members of the Consultative Committee) might be added to the Board.

E. H. Pember, Q.C., was the leader of the Ship Canal counsel and did his work magnificently A born orator, his speeches before Parliament were masterpieces of eloquence, and will always be read with pleasure. There can be no doubt his persuasive powers helped materially in at last securing the Bill.

J. H. Balfour Browne, Q.C., fought the Ship Canal battle with an energy and tenacity that will never be forgotten. As a cross-examiner he was perfection, and his able assistance was a material factor in the ultimate Parliamentary success.

Mr. Samuel R. Platt, of Oldham, was a liberal and courageous supporter of the canal from the commencement. He was a man after Mr. Adamson's own heart, and in its darkest hours he had always an abiding faith in its ultimate success. His sage counsel, especially on engineering matters, was always listened to and appreciated by his brother directors. It was a proud moment for him when, as Chairman of the

Works Committee, he led the way in his steam yacht *Norseman*, at the opening of the canal on 1st January, 1894.

Mr. George Hicks has the credit of re-introducing the Ship Canal to public notice after the scheme had lain dormant for many years. In conjunction with Mr. Fulton, he first brought the project before the Manchester Chamber of Commerce, and he it was who induced Mr. Daniel Adamson to captain the undertaking. As a member of the Provisional Committee he has rendered most valuable assistance to the Ship Canal. His work has been unceasing and persistent.

I now pass to the actors who joined in 1887, chiefly in consequence of the Consultative Committee's action.

Lord Egerton joined the Ship Canal Consultative Committee in 1886, and in 1887 was made Chairman of the Ship Canal Company. He took the position unwillingly, but having put on harness he made an excellent leader, sparing no pains to master the details of the work. As Chairman of shareholders' meetings, and in conducting negotiations with the Corporation, he displayed ability and adroitness of no ordinary kind. Much of the success of the canal is due to the confidence placed in him.

Sir Joseph Cocksey Lee was in his time a tower of strength. When it became evident that the first directors had failed to get the capital, Sir Joseph Lee, by his financial skill, succeeded in placing the finances on a sound basis. He showed, too, his administrative capacity in arranging a most complicated dispute with the executors of Mr. Walker, the contractor. Unfortunately, his health broke down, and after a long illness, he died before the canal was developed.

Sir John Harwood was admirably suited to take command when the work on the canal needed push, energy, and experience, to secure its early completion. When the first application was made by the Ship Canal Company for monetary assistance, Sir John could sway the City Council, and without doubt it was his powerful intervention that induced it to assist the canal. But when the estimates were exceeded, and more money had to be found, Sir John became seriously alarmed. He had been used to success all his life, and could not brook disappointment or defeat. Hence he became restive and perhaps a little unjust to his colleagues, especially to the old directors. It was much to be regretted that a man who had played such a useful and important part should retire from the directorate just at the wrong moment, and should fail to have his valuable services acknowledged as they deserved to be.

Sir John Mark was Mayor when Sir Joseph Lee applied to the City Council for aid.   He has never received due honour for helping to make it possible to finish the canal.   Had he assumed a churlish attitude the enterprise might have collapsed. He recognised that the credit of Manchester was at stake, and that failure would be disastrous.   He received the Ship Canal directors most sympathetically, and rendered them valuable aid at a time when it was most needed.

Mr. Charles J. Galloway was a member of the Consultative Committee in 1886, and was appointed a Ship Canal director the following year.   He did good work as Chairman of the Bridgewater Committee, and his advice and financial support could always be depended upon.

Sir Edward Jenkinson had served with distinction in India, and was made K.C.B. in 1888.   He represents the interests of the London shareholders, and has done excellent work for many years as Chairman of the Finance Committee.

Mr. J. K. Bythell joined the Ship Canal Consultative Committee in 1886, and became a director in 1887.   In the first place he was Chairman of the Finance Committee, and was afterwards appointed Chairman of the Board.   His Indian experience has helped to make him a most excellent administrator of dock affairs.   He is fertile in resource, dexterous and adroit in negotiating business, of untiring industry, and he has been a most successful canal manager.

Alderman Southern, though always an ardent supporter of the canal, only took upon himself a responsible position in 1890, when the Corporation came to the assistance of the undertaking.   He was selected as one of their representatives on the Board, and at once became an active factor in pushing on the completion of the work. Subsequently he was made Deputy Chairman.   His services have been most valuable both to the City Council and to the Ship Canal.

Mr. Charles Moseley was originally on the Consultative Committee, and when he became a director he was a tower of strength to his colleagues, and ably assisted Sir Joseph Lee in his financial arrangements.

RESULTS OF THE CANAL.

This History, written as it is twelve years after the completion and opening of the canal, affords the opportunity of judging how far the anticipations in respect to it have been fulfilled.   Broadly they were :—

A. That the canal would cheapen the carriage of material both for food and manufacture.

B. That it would free commerce from the excessive charges levied at Liverpool Bar, and be a powerful lever in removing the Mersey Bar.

C. That it would pay a fair dividend to investors.

D. That it would prevent floods devastating the districts adjoining the Mersey and Irwell, and provide a good navigation to the traders of the city.

E. That it would cause new industries to settle in the district, and both find labour for working men and increase the ratable value. Further, it would enable lairages to be established and so create a new market at Manchester for foreign cattle.

F. That it would be a means of purifying the Mersey and Irwell and be of assistance in draining and reclaiming the moss land adjoining.

G. That the canal would put fresh vigour into the people and trade of Manchester, and benefit every class of the community.

H. That by cheapening the cost of production, it would assist manufacturers and merchants in meeting the competition that exists both in home and in foreign markets.

A. Subsequent to the introduction of Fulton's Ship Canal scheme, the Dock Board reduced their charges—this was in 1880. Again, to checkmate the canal a further considerable reduction in dock dues and rates was made in 1883. The railways followed suit, evidently for the same purpose. This will be seen by the following tables, which show the saving on goods which continue to reach Manchester *via* Liverpool; also on those that come direct *via* the canal. In the latter case, in place of transportation charges across Liverpool and railway carriage, the Ship Canal toll only has to be paid. But a further saving is effected by better appliances and railways to the ship's side, which very much reduce the landing charges. Again, dock and town, and ship dues are less at Manchester than Liverpool. As a rule sea-rates are the same to both ports, and so is the insurance. A few Liverpool liners, such as Lamport and Holt's, are exceptional, and make a small extra charge for coming up the canal; nevertheless, the advantages of Manchester are so great that even these lines are enabled to do a rapidly increasing trade.

TABLE I.

COMPARATIVE COST PER TON OF DELIVERIES IN MANCHESTER *via* LIVERPOOL AND *via* THE SHIP CANAL.

	Cost *Via* Liverpool.		*Via* Ship Canal.		
			Estimated.	Actual Charge.	Saving on Charges paid in 1883.
	1883.	1904.	1883.	1904.	
Cotton	15s. 0d.	12s. 8d.	7s. 0d.	6s. 3d.	8s. 9d. or 58½ per cent.
Grain (Wheat)	10s. 5d.	9s. 5d.	4s. 10d.	3s. 9d.	6s. 8d. or 64 „
Timber (Deals)	10s. 0d.	8s. 11d.	4s. 9d.	3s. 0d.[1]	7s. 0d. or 70 „
Refined Sugar	14s. 8d.	13s. 3d.[2]	6s. 8d.	5s. 0d.	9s. 8d. or 66 „

[1] No porterage included in Manchester or Liverpool.

[2] If in 5 ton lots the cost at Liverpool is further reduced to 12s.

The above table demonstrates a few of the savings effected by the canal on produce that comes directly to Manchester.

Particulars of how bread is cheapened have been previously given on page 115, Vol. I.

B. The relief given to commerce by the Ship Canal is exemplified in the appended table, which shows how competition has compelled the Dock Board and the railways to reduce their charges.

TABLE II.

REDUCTION PER TON MADE IN RAILWAY FREIGHTS AND DOCK CHARGES BETWEEN LIVERPOOL AND MANCHESTER ON GOODS SHIPPED *via* LIVERPOOL, 1883-1904.

			Liverpool Charges.	Railway Freight.	Total.
Cotton	1883		7s. 0d.	8s. 0d. S.S.[1]	15s. 0d.
„	1904		5s. 6d.	7s. 2d. S.S.	12s. 8d.

A saving of 2s. 4d. per ton, or 15½ per cent., on cotton even when not brought by the canal.

			Liverpool Charges.	Railway Freight.	Total.
Grain (Wheat)	1883		3s. 1d.	7s. 4d. C.[2]	10s. 5d.
„ „	1904 . .	Liverpool	2s. 6d.	6s. 11d. C.	9s. 5d.
„ „	1904 . .	Birkenhead	2s. 0d.	—	8s. 11d.

Through Liverpool a saving of 1s. 0d. per ton, or 9½ per cent.
„ Birkenhead „ „ 1s. 6d. „ „ or 14½ „

			Liverpool Charges.	Railway Freight.	Total.
Timber (Deals)	1883		2s. 6d.	7s. 6d. S.S.	10s. 0d.
„	1904		2s. 3d.	6s. 8d. S.S.	8s. 11d.

A saving of 1s. 1d. per ton, or 10¾ per cent.

			Liverpool Charges.	Railway Freight.	Total.
Refined Sugar	1883		6s. 4d.	8s. 4d. S.S.	14s. 8d.
„ „	1904		5s. 4d.	7s. 11d. S.S.	13s. 3d.
		If in 5 ton loads		6s. 8d. S.S.	12s. 0d.

A saving of 1s. 5d. per ton, or 9½ per cent.
„ of 2s. 8d. on 5 ton deliveries, or 18 per cent.

The above table is intended to show that beyond the reduction in charges effected on goods brought directly over the canal, a great saving has been secured to traders who still get their supplies through Liverpool.

[1] S.S. signifies station to station rates.
[2] C includes collection from Liverpool warehouses.

C. That the canal has disappointed the shareholders as regards dividends cannot be denied. There were people sanguine enough to believe that it would at once pay a percentage on the outlay. But they had not sufficiently considered the enormous difficulties that must be encountered by any concern entering into a new business, where fresh connections and new markets have to be established in the face of un-exampled opposition. When it is remembered that the most powerful railways in the kingdom were combined with Liverpool, with the Liverpool Dock Board, and with an army of vested interests to wreck the canal scheme at its inception, or to bear it down by competition, the marvel is that it was not strangled in its infancy. It certainly would have been, had not the Manchester Corporation magnanimously come to the rescue.

Twelve years after the opening we behold the tree wlth its roots firmly fixed in the ground, gaining strength as it grows, and more likely than ever to bear fruit. Sceptics said the canal would not be used. What do we find? A fair share of the cotton of the world, especially Egyptian, coming to Manchester, the timber trade growing enormously, the corn trade firmly established, the present elevator (the largest in England) too small for its traffic, and another in the course of construction. The fruit trade, especially for Spanish fruit, and for bananas, firmly fixed and going to remain. The cattle trade is increasing so fast that it will be necessary to add to the lairage accommodation. The oil trade has developed to an astonishing extent, and is conducted by the most powerful firms in the world. And so on with a number of other trades, the result being that the company has from time to time been compelled to increase the dock and warehouse accommodation.

I fully believe it is in the power of the merchants and shippers of Manchester to make the canal a dividend-paying concern within a few years. The goods shipped are among the most profitable cargo carried, and yet the great majority of them are still exported *via* Liverpool or London, though they could be taken at much less cost from the warehouse doors.

There only needs combination, and ships, other than passenger steamers, would be forced to come up to Manchester, because if shipowners declined to send them, there is sufficient business from Manchester, as a centre, to support lines of steamers to almost any part of the world. A shipowning company, headed by the leading shippers of Manchester, would secure any amount of financial support. I should like to see the rebate system extinguished, but this will be a matter of time. Old shipowners' prejudices will only die out with themselves. Not long ago, when a shipping conference was discussing the propriety of sending ships to Manchester, a Liverpool shipowner threatened to retire from the conference and fight his colleagues rather than consent to his ships coming up the canal.   Return cargo is necessary for the success of the canal.   If the patriotism of shippers will provide it, the canal will soon pay a dividend.

D. Let any one go down the valleys of the Mersey and Irwell and make inquiries as to the effect of the canal, and he will be told that no one could have formed any idea of the change already wrought.   Hundreds of acres in the valleys of both rivers, that used to be flooded, and the crops spoiled or damaged, have now perfect immunity. As an instance, I can remember the Flixton meadows being flooded for days together, and the main road to Flixton Church rendered impassable.   This very land has new houses upon it, and affords one instance out of many, of flooded land being converted into building land.   Landowners and farmers have benefited to the extent of tens of thousands of pounds by the security from inundation now given to their crops and cattle.

E. There can be no greater misfortune than for a town to rely on a single industry. Manchester used to depend chiefly on the cotton trade.   In the wake of the canal have come all kinds of business before unknown to the district.   Many of the old mills in the city, which had been empty for years, have been brought into use, and a large number of new works have been built on the banks of the canal.   The following are some of the industries that have located themselves in Trafford Park, which is adjacent to the docks :—

Acme Lathe and Products Company, Ltd., Engineers.
American Car and Foundry Company, Car Builders.

Bailey, W. H., & Co., Engineers.
Baxendale & Co., Lead Merchants.
Bentley, Isaac, & Co., Ltd. (10,000 square yards), Oil and Colour Merchants.
British Westinghouse Electric and Manufacturing Co., Ltd. (60 acres, 6,000 hands), Machinists.
Brooksbank, E., & Co., Ltd., Oil Refiners.

Colley & Cureton, Engineers.
Cooke, Laidman & Leech, Ltd., Timber Merchants.
Co-operative Wholesale Society, Importers.

Dawson, R. S., Saw Mill.

Fram Fireproof Flooring Company.

General Oil Storage Company, Ltd., Oil Importers.
General Petroleum Company, Ltd., Oil Importers.
Gill, F. E., Builder.
Glover, W. T., & Co., Ltd. (21,760 yards), Electrical Engineers.
Gresham, James, Engineer.
Griggs, Joseph, & Co., Ltd., Timber Merchants.

Hall & Pickles, Iron Merchants.
Higgin, Lloyd & Co., Drysalters.
Higgins, William, & Son, Brickmakers.
Homelight Oil Company, Ltd., Oil Importers.
Hovis Bread-Flour Company, Ltd., Flour Mill.
Howard Asphalt Troughing Company, Ltd., Engineers.
Hulbert, T., & Sons, Grain Merchants.

Illingworth, Ingham & Co., Ltd., Moulding Mill.
Imperial Lumber Company, Ltd., Timber.

Jacob, W. & R., & Co., Ltd., Biscuit Makers.

Key Engineering Company, Ltd.
Kilvert, N., & Sons, Ltd., Lard Refiners.
Kirkpatrick Brothers, Stone Polishers.

Lancashire Dynamo and Motor Company, Machinists.
Leyland, Barlow & Co., Engineers.
Liverpool Storage Company, Ltd.
Liverpool Warehousing Company, Ltd.

McKechnie Brothers, Metal Refiners.
Manchester Banks :—
    The Manchester and County, Ltd.
    The Manchester and Liverpool District, Ltd.
    Williams, Deacon & Co., Ltd.
Manchester Brewery Company.
Manchester Patent Fuel Works, Ltd.
Manchester Ship Canal, Ltd., Warehouse.
Morrell, Mills & Co., Shipwrights.
Morrison, Ingram & Co., Ltd., Sanitary Engineers.

Newsum, H., Sons & Co., Ltd., Timber Merchants.
Nuttall, Edmund, Contractor.

Oakley Steel Foundry Company.
Owens European Bottle Machine Co.

Pickfords, Ltd., Carriers.
Pochin, E. D., Gear Cutter.
Powell & Stabler, Timber Merchants.
Pyke, Joseph, & Son, Provender Mill.

Redpath, Brown & Co., Ltd., Engineers.
Royce, Ltd., Dynamos.
Russell, Thos. E.

Sandars & Co., Maltsters.
Skipwith, Jones & Lomax, Engineers.
Southern, J. W., & Son, Timber Merchants.

Trafford Park Dwellings Company.
Trafford Park Enamelling Company.
Trafford Park Steel Works Company.
Trafford Power and Light Supply, Ltd.

United Electric Car Company, Ltd.

It is a remarkable fact that whilst in 1867 the ratable value of Trafford Park
was £2,000, in 1904 it had increased to £84,000.

New works outside Trafford Park :—

Manchester Dry Docks Company, Ltd., Mode Wheel.
Lancashire Patent Fuel Company, Ltd., Partington.
Cold Air Stores of the Colonial Consignment and Distributing Company, Mode Wheel.
Manchester Corporation Oil Gas Depot, Mode Wheel.
Liverpool Storage Company, Mode Wheel.
Anglo-American Oil Company, Mode Wheel.
Consolidate Petroleum Company, Ltd., Mode Wheel.
Andrew Knowles & Sons, Coal Dealers, Mode Wheel.
Richard Evans & Co., Ltd., Coal Dealers, near Warrington.
Co-operative Wholesale Society's Soap and Candle Works, Irlam.
The China Clay Company, Docks.
The Smelting Corporation, Ltd., Smelting Works, Ellesmere Port,

and many others.

F. At the inception of the canal it was intended to run a culvert parallel with
it, that should either convey the sewage of Manchester to some suitable place outside

the city, to be there dealt with by irrigation, or to carry it to tidal waters. Parliamentary powers were included in the first Bill, but it was subsequently thought better to take the sewage by a more direct route to low lands at Davyhulme.

The crude sewage of Manchester and Salford has been removed out of the waterway, and a serious nuisance greatly modified. The Ship Canal Company's own interest lies in purifying the waterway and they have been persistent and energetic in calling on the Joint Rivers Board to punish all polluters of the water. As a result, the river has most materially improved, and though at times the propellers of steamers and the working of the sluices cause odours to rise in their immediate neighbourhood, the general state of the river has been much better since the advent of the canal. This will still go on. By means of the canal the Manchester Corporation has been enabled to take hundreds of thousands of tons of refuse and to tip it in the ocean. The lowering of the canal bed below Barton has also been of great advantage in draining the moss lands of the district.

G. It is scarcely necessary for me to dwell exhaustively on the benefits which Manchester has derived from the canal. A few instances, chiefly taken from an able pamphlet, written by Mr. A. Woodroofe Fletcher, LL.B., will, however, assist me. The pamphlet was published in 1899, and the improvement has since been going on at an accelerated pace.

Mr. Samuel Ogden, the former Chairman of the Chamber of Commerce and of the Board of Overseers, stated that in the Ancoats district alone, within ten months prior to the opening of the Ship Canal, establishments employing 12,650 workers had been closed, and that there were streets of empty tenements and houses. He had never known so many vacant warehouses. He knew of fifty to let in the centre of the city, which ought to yield a revenue of £14,000 per year. In one township, where, prior to 1879, there had been an increase of £6,000 per year of assessment, the subsequent decade showed a decrease of £900 per year. Another well-known citizen said there were as many empty houses in Manchester and Salford in 1881 as there were full ones in the whole of Stockport.

In Manchester, prior to 1881, the ratable value had increased £60,000 per annum; in the subsequent decade the average decrease was £1,500 per annum.

That the Ship Canal has helped to rehabilitate Manchester and the district is evident from the fact that ten years after its completion there were very few houses or warehouses unoccupied in the city, notwithstanding building operations had been going on extensively, and that 12,000 to 15,000 houses had been built. The consumption of water is a good index of prosperity in a city. This quickly rose from a revenue of £250,000 to £300,000, or an advance of 20 per cent. The development of the Post Office business is still more remarkable. The increase in 1895 was over 10 per cent.; in other words the additional correspondence per annum in Manchester was equal to the yearly sum total of a town like Derby. The increase of the banking business done in Manchester, as shown by the Clearing House returns, is simply phenomenal. In the growth of its business Manchester far exceeds all other local towns.

The improved condition of Manchester as regards prosperity was well illustrated by comparative figures given in an inaugural speech made by the Lord Mayor of Manchester on 9th November, 1906 :—

	1896.	1906.
[1] Area within the city boundaries (acres) . . .	12,911	19,893
[2] Population (estimated)	529,681	637,126
Ratable value	£2,943,000	£4,139,000
Actual yield of a penny rate	£11,449	£15,877
Corporation annual turnover	£6,896,000	£13,587,000
[3] Number of Corporation officers and servants . .	7,488	18,213
Corporation wages	£552,110	£1,442,898

Assets of the Corporation, £30,000,887, and the liabilities, £24,058,550, leaving a surplus of nearly six million pounds.

	1896.	1905.
Ship Canal sea-borne traffic (tons)	1,509,658	3,993,110
Ship Canal receipts for tolls and other charges .	£182,330	£449,436
Post office letters (about)	5,000,000	7,000,000
Manchester bankers' clearing house . . .	£193,573,441	£263,526,003
Post office savings bank accounts . . .	118,750	184,780
Post office savings bank, due to depositors . .	£1,402,000	£2,162,000

The following figures show that in the county borough of Salford the population has increased in the last twenty years nearly 20 per cent., and the ratable value nearly 40 per cent :—

[1] The increase is caused by the addition to the city in 1904 of Withington, Moss-side, Didsbury and Chorlton-cum-Hardy.

[2] In both Manchester and Eccles there have been extensions of area which prevent a fair comparison being made as regards population.

[3] Taking over the tramways is the chief cause of the large increase of employees in Manchester and Salford.

	1886.	1896.	1906.
Area within the borough boundary (acres)	5,171	5,202	5,202
Population (estimated)	196,894	210,707	234,077
Ratable value	£750,679	£835,455	£1,012,193
Actual yield of a penny rate	3,158	3,170	3,645
Assets of the Corporation	1,841,822	2,669,095	5,158,978
Liabilities of the Corporation	1,687,467	2,083,580	3,933,888

If Manchester and Salford have benefited by the Ship Canal, still more so have Eccles and Stretford, the other two neighbours to the Ship Canal Docks.

Stretford stands at the head of the percentage of increase as regards population. Whilst in the last five years Oldham has increased 3·06, and Ashton-under-Lyne 2·14 per cent., Stretford has increased 15 per cent. The same rapid advance is exhibited in the ratable value :—

	Increase 12 years, 1882 to 1894.	Increase 12 years, 1894 to 1906.
Stretford	9·14 per cent.	[1] 106·71 per cent.
Eccles	25·60 ,, ,,	40·93 ,, ,,

Mr. Marshall Stevens, when dealing with the increased traffic of Manchester compared with other ports, showed that as regards imports and exports—

In 1894 she was the 16th port in the kingdom . . . . traffic, £6,918,678
,, 1903 ,, ,, ,, 6th ,, ,, ,, ... ,, 29,576,320
As regards total imports—
In 1894 she was the 17th port in the kingdom . . . . ,, 2,790,129
,, 1903 ,, ,, ,, 4th ,, ,, ,, ... ,, 20,279,255
As regards total exports—
In 1894 she was the 10th port in the kingdom . . . . ,, 4,019,344
,, 1903 ,, ,, ,, 8th ,, ,, ,, ... ,, 8,856,100
He gives the imports—

		Cwts.
Wheat and Indian Corn	1894	132,113
,, ,, ,, ,,	1900	3,864,200
,, ,, ,, ,,	1903	7,033,737
Refined Sugar	1894	373,342
,, ,,	1903	727,782
Cotton	1894	251,778
,,	1903	2,769,145
		Loads.
Timber, Hewn and Sawn	1894	94,655
,, ,, ,, ,,	1903	377,872

[1] Chiefly due to the building on the Trafford Hall estate adjacent to the canal.

At the end of 1904 Manchester was second only to London as an emporium for petroleum. Her imports were 491,462 barrels, against 284,014 at Liverpool and Birkenhead.

A comparison of the Ship Canal with the Suez Canal ought to be an encouragement :—

	Original Capital.	Estimated Cost.	Cost at Opening.	Including Subsequent Expenditure.
The Suez Canal	£8,000,000	£4,588,800	£16,631,953	£19,782,000
The Ship Canal	10,000,000	7,292,972[1]	14,347,891	15,276,570

	Vessels.	Tonnage.	Receipts.
The Suez Canal 1870	486	435,911	£214,972
,, ,, 1880	2,026	4,344,519	1,660,000
,, ,, 1884	3,284	8,319,967	2,599,088
The Ship Canal 1894	4,551	925,659[2]	97,096
,, ,, 1904	5,765	3,917,573[2]	418,043

This includes all sizes of ships, foreign, coast-wise, local and barge traffic.

In 1893, twenty-four years after the opening of the Suez Canal, the dividend paid was 20 per cent., and the £20 share was quoted at £107 8s.

Mr. Marshall Stevens says that whilst in 1894 Manchester only held the twenty-fifth position as regarded steamers from foreign countries, in 1904 she was the tenth port in the kingdom.

The accompanying diagram and statistical tables show the progressive receipts, expenditure and tonnage on the Ship Canal from the commencement.

[1] Figures given in evidence by Mr. Adamson, who said it did not include £1,710,000 for the Bridgewater Canal, the cost of the Act or the purchase of warehouse property. Mr. Walker's contract was £5,750,000, exclusive of land, etc.

[2] Foreign ships only, 1894— 720,425.
,, ,, 1904—1,802,408.

MANCHESTER SHIP CANAL STATISTICS.

Statement as to Receipts and Expenditure for the Ship Canal Department as from the Opening of the Canal.

	Receipts.	£	£		Expenditure.	£	£
1894—1st half	. .	43,803		1st half	. . .	36,285	
2nd half	. .	53,293		2nd half	. . .	42,595	
Total for year .	. .		97,096	Total for year	. . .		[1]78,880
1895—1st half	. .	63,037		1st half	. . .	56,009	
2nd half	. .	74,437		2nd half	. . .	59,320	
Total for year .	. .		137,474	Total for year	. . .		[2]115,329
1896—1st half	. .	81,214		1st half	. . .	88,643	
2nd half	. .	101,116		2nd half	. . .	93,623	
Total for year .	. .		182,330	Total for year	. . .		182,266
1897—1st half	. .	97,330		1st half	. . .	98,224	
2nd half	. .	107,334		2nd half	. . .	88,326	
Total for year .	. .		204,664	Total for year	. . .		186,550
1898—1st half	. .	105,865		1st half	. . .	85,292	
2nd half	. .	129,531		2nd half	. . .	91,605	
Total for year .	. .		235,396	Total for year	. . .		176,897
1899—1st half	. .	124,183		1st half	. . .	91,357	
2nd half	. .	140,592		2nd half	. . .	99,807	
Total for year .	. .		264,775	Total for year	. . .		191,164
1900—1st half	. .	141,346		1st half	. . .	99,946	
2nd half	. .	149,483		2nd half	. . .	107,134	
Total for year .	. .		290,829	Total for year	. . .		207,080
1901—1st half	. .	146,508		1st half	. . .	104,475	
2nd half	. .	163,008		2nd half	. . .	102,980	
Total for year .	. .		309,516	Total for year	. . .		207,455
1902—1st half	. .	166,603		1st half	. . .	106,132	
2nd half	. .	191,888		2nd half	. . .	111,404	
Total for year .	. .		358,491	Total for year	. . .		217,536
1903—1st half	. .	189,422		1st half	. . .	113,446	
2nd half	. .	207,603		2nd half	. . .	117,403	
Total for year .	. .		397,025	Total for year	. . .		230,849
1904—1st half	. .	198,548		1st half	. . .	118,355	
2nd half	. .	219,495		2nd half	. . .	121,940	
Total for year .	. .		418,043	Total for year	. . .		240,295
1905—1st half	. .	208,745		1st half	. . .	120,450	
2nd half	. .	240,691		2nd half	. . .	126,296	
Total for year .	. .		449,436	Total for year	. . .		246,746

[1] Dredging and Maintenance charged to Capital.
[2] Dredging charged to Capital.

MANCHESTER SHIP CANAL.

Statistical Diagrams, 1894 to 1905 (December 31st) relating to Ship Canal Department.

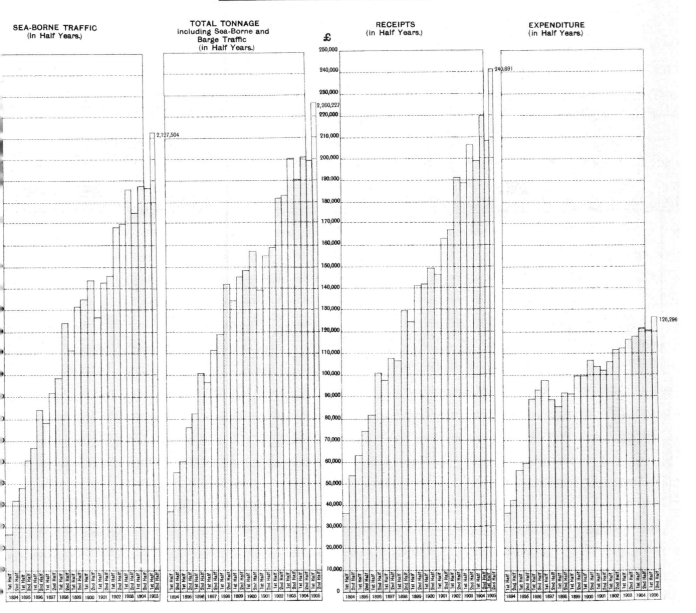

The material originally positioned here is too large for reproduction in this reissue. A PDF can be downloaded from the web address given on page iv of this book, by clicking on 'Resources Available'.

In 1900 a computation was made of the saving to Lancashire effected by the Ship Canal, and it was stated to be £700,000. In 1904 it was estimated at over a million sterling, and this saving is likely to go on increasing.

H. The fact that cheap material and cheap food are necessary to success, scarcely needs demonstration. Competition is now so severe that producers will gravitate to the most convenient and advantageous centres. Works require cheap land, cheap and abundant labour, good water and low local rates. But they want most of all cheap and convenient carriage for raw produce and manufactured goods. During the process of conversion, the same article has often to pay many freightages. Take, for instance, cotton. The raw material travels from Liverpool to Oldham to be spun. As yarn it is sent to Manchester to be sold, then to Burnley to be manufactured, back to Manchester in the shape of cloth, it then goes to Bolton (and back) to be bleached or dyed, and finally, when sold in its finished state, it is forwarded, say to London, etc., to be shipped. Thus it has to pay seven different carriages. If, as is often the case in England, carriage is 50 to 70 per cent. dearer than in America, Belgium or Germany, how can our manufacturers compete with their rivals in those countries, especially when they have also to fight against cheap labour?

Therefore it is most important to be near a waterway in order to produce cheaply. The Westinghouse Iron Works, the flour mills of Messrs. Sutcliffe and Messrs. Baxendell, Messrs. Hall & Pickles, Iron Stores, the oil refineries and numberless other works are proofs that the Ship Canal is fulfilling its original object and affording cheap water carriage. Let me give an instance directly to the point. The Manchester Corporation has constructed large sewerage works at Davyhulme. They used 10,000 to 15,000 tons of cement a year. There was keen competition between German, Belgian and English cement manufacturers. The best qualities of foreign cement would suit the purpose, and they could be delivered at 2s. 6d. per ton lower than any English cement delivered by railway. But fortunately the sewage farm abutted on the canal, and south coast cement could be delivered by water at a price that kept the business at home, and saved Manchester nearly £2,000 per annum. Add to this the saving on gravel, coal, bricks, cinders and other heavy materials, and the total gain through having a waterway came to at least £2,500 per annum.

Dear carriage and charges, coupled with the drawbacks of an insular position, drove Manchester to make the canal. She had to wage a fierce battle with vested interests, and she may congratulate herself that she has succeeded in overcoming difficulties sufficient to daunt the stoutest hearts.   May she continue to prosper and ever remember "that God-helps those who help themselves"!

Having now reviewed the anticipations of friends of the canal and shown how amply they have been realised, I will refer for a moment to the dismal and oft-repeated prophecies of opponents who characterised the canal as a filthy ditch, which never could be navigated except by small coasting vessels that would take at least two days to come up to Manchester.   Over and over again Liverpool shipowners jeered at the idea of Manchester ever becoming a port, and declared they would never dare to send their ships up the canal.   A single fact will show how far they were mistaken, and how prejudice warped their opinions.   In October, 1904, the *Suffolk*, belonging to the Federal Steamship Navigation Company, 500 feet over all, with a beam of 58 feet, came up the canal in nine sailing hours.   Her cargo capacity was 10,000 tons dead weight, and her registered tonnage was 7,313 tons gross, and 4,680 tons net.   This monster of the deep brought a heavy cargo from Australia, and with two propellers came up the canal without a scratch.   She had her funnels altered at Eastham, and tied up at Latchford when it was dark.   Her voyage has made it clear that big ships can easily come up the canal in a day.   A large new dock, 900 yards long, has just been completed, and the depth of the entire canal will shortly be increased by 2 feet.

That Liverpool was unduly alarmed by the advent of the canal is proved by the fact that it has done her no harm.   Ever since it started she has outstripped all other ports in the increase of trade.   By removing the bar and reducing her port charges, she has stimulated and encouraged business.   This is shown by the following figures published by the Mersey Docks and Harbour Board :—

	Total Tonnage.	Amount of Rates.
1895	10,777,146	£1,056,741
1899	12,534,116	1,093,066
1904	15,626,241	1,206,535

The annual report for 1905 of Vice-Admiral Sir George Nares, Acting Conservator of the Mersey, gives the *Customs* tonnage of the Port of Liverpool :—

Year.	Inwards Tonnage.	Outwards Tonnage.
1895 . . .	9,061,648	9,406,180
1905 . . .	11,787,445	12,578,074
	2,725,797	3,171,894
Increase in ten years . .	30 per cent.	33 per cent.

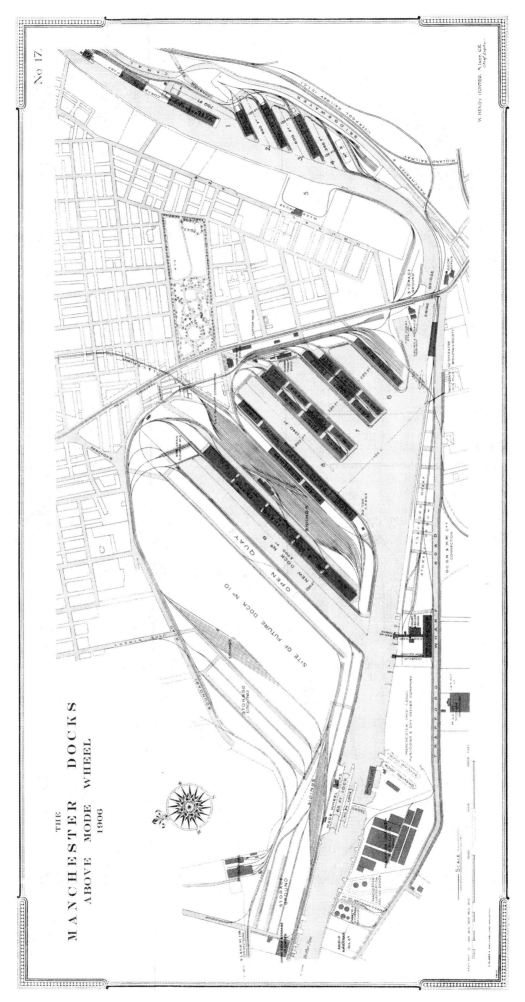

The material originally positioned here is too large for reproduction in this reissue. A PDF can be downloaded from the web address given on page iv of this book, by clicking on 'Resources Available'.

The same report gives the tonnage of shipping to and from Manchester, foreign and coastwise :—

Year.				Inwards Tonnage.	Outwards Tonnage.
1900	.	.	.	1,235,505	1,267,138
1905	.	.	.	1,731,607	1,830,788
				496,102	563,650
Increase in five years	.	.		40 per cent.	44½ per cent.

In the Port of Manchester are included Runcorn and Ellesmere Port, and if these were added to the Manchester returns the increase would be still more astonishing—the latter port has almost doubled its business in the last six years.

By offering increased facilities and cheapening shipping charges there can be no question that trade has been induced to come to the ports on the Mersey that in all probability would otherwise have gone elsewhere.

The figures clearly show that the ruin that was prophesied would overtake Liverpool in case the Ship Canal were made was a groundless fear.

To Liverpool the canal has apparently been a blessing in disguise. It is to be hoped that in future Manchester and Liverpool will cultivate a friendly and generous spirit, and join hands in making the Mersey the cheapest and most convenient avenue for trade in the world. There is plenty of room for both.

I have now finished my task. My aim has been to give a full and complete account of the inception and completion of the Manchester Ship Canal, and I must leave it to some other historian to chronicle its further progress.

APPENDIX I.

MANCHESTER TIDAL NAVIGATION COMPANY.

CIRCULAR ISSUED AFTER THE FIRST MEETING OF THE PROMOTERS IN 1882.

IT is notorious that, under existing conditions, the Mersey and Irwell Navigation is practically useless to its present proprietors.

They are unable to maintain a good depth of water, owing to the enormous quantities of rubbish swept into its bed in times of flood.

The only effective waterway between Manchester and Liverpool is thus subordinated so as to deprive the public of the advantages which this mode of transit would afford.

As a consequence, public attention has been directed to the improvement of the navigation of the Mersey and Irwell, so as to connect Liverpool with Manchester by a tidal waterway.

The problem to be solved by such an undertaking would appear to be the cost of transit in its relation to the commercial and manufacturing industries of this district, and the claim which such industries have upon the recognition of Parliament.

To Lancashire and the surrounding districts this question of the cost of transit is of the first importance, and it is becoming more evident every day that they can no longer afford to neglect it.

To lessen the cost of transit is to increase the power of competition. Reduce transit charges, and you extend your operations to markets previously closed against you.

To turn out goods at a low figure from the looms of Lancashire is of little account unless you can also carry them at a low figure to the place where they are wanted.

In consequence of competition the freight of a ton of goods from Manchester to Calcutta has recently been as low as 19s. 3d., of which 12s. 6d. is charged before the goods leave Liverpool.

The carriage of raw materials used in the English iron industries costs nearly double what it does in France, Germany, Belgium or Austria.

The directors of Messrs. Chas. Cammell & Co., of Sheffield, announce that to avoid excessive railway charges, they have found it necessary to remove their works from Dronfield to the coast of Cumberland, hard by the sea. Surely in this movement there is a moral for Manchester.

Water carriage can be effected at one-third the cost possible by rail. It is well known that goods can be conveyed a thousand miles upon the ocean for the same cost as along one hundred miles of railway.

With the advantages afforded by the Clyde, Glasgow bids fair to eclipse Manchester. Food and raw produce are landed there, and manufactures shipped at a nominal cost compared with the charges between Manchester and the sea-board. These advantages have been secured to Glasgow by the energy of her citizens in bringing up the sea by the deepening of the Clyde.

By converting the Mersey-Irwell into a Lancashire Clyde, Manchester has it in her power to secure advantages as great, nay, even greater than Glasgow.

The shipment and discharge of goods in Manchester would be effected at a nominal cost, as compared with the freight between Manchester and Liverpool, and the heavy port dues of the latter city.

The dues collected by the Mersey Docks and Harbour Board exceed one million sterling per annum.

The annual cost of inland carriage between Liverpool and the Manchester district is between £2,000,000 and £3,000,000. At least £1,000,000 sterling would be saved by the Tidal Navigation in inland carriage alone. To this large saving must be added a very large proportion of the £1,000,000 levied yearly as dues by Liverpool on all the shipping, as well as on the goods they carry.

Merchants would be enabled to ship their goods under the direct supervision of their own staffs, and thereby save the commissions now paid to forwarding agents. In terminal charges (which form a heavy item in cost of transit) a further saving would be effected if goods could be taken direct to the ship in the lorries which collect the packages from the warehouses.

The present railway system involves no less than seven separate handlings of the packages between warehouse and vessel. It is not to be wondered at, therefore, that as a consequence the bales and cases of Manchester goods sometimes suffer serious injury before they are shipped.

CORN.—At the present time there are hardly any corn mills in and around Manchester, notwithstanding that nearly a million and a half tons of corn are annually imported into Liverpool, the greater portion of which, after being ground at that port, finds its way to this city for distribution to the numerous towns of South-East Lancashire and the West Riding of Yorkshire.

SUGAR.—As for sugar refining, with the exception of a small one in Salford, there is not a sugar refinery in the Manchester district. The reason is not far to seek. It is cheaper to grind corn and refine sugar at the port of discharge, and thereby save the greater expense involved in transmitting the bulkier raw material. When corn and sugar can be brought here in the vessels themselves, corn grinding and sugar refining will at once become important industries in Manchester.

COTTON.—Ships being able to discharge their cargoes in Manchester would probably lead to the cotton market being held on the Manchester Royal Exchange, which, looking to the relation of Manchester to the cotton districts, would seem to be the most natural and appropriate arrangement. Liverpool terms have induced Oldham spinners to try buying cotton in Bremen and Havre. Purchases made in those places and imported *via* Hull, after paying heavy railway charges from Hull to Oldham, have effected a saving of one farthing per pound in the price asked for the same quality of cotton in Liverpool.

COAL.—Until the cost of carriage to the sea is greatly reduced, the coal-owners cannot look forward to any great expansion of the foreign trade, or of the demand for coal for steam vessels. The pits from which the main supply of Lancashire coal to the Liverpool market is obtained are quite close to the Irwell, and a canal system proceeds direct from the levels of the coal workings to the river bank.

FLOODS.—The spacious channel required for the Tidal Navigation would provide such an outlet for the rainfall as would entirely prevent the possibility of floods in the lowlands between Manchester and Warrington.

If the statistics of the ports of Liverpool, Glasgow and the Tyne be examined, it will be found that from 1861 to 1881, whilst the shipping of the natural port of Liverpool has only increased 59 per cent., that of the Clyde has increased 103 per cent., and of the Tyne 85 per cent.

A radius of forty miles from Manchester includes a greater population than any similar area in the United Kingdom, the population of North-East and South-East Lancashire, Mid and East Cheshire, Derbyshire, North Staffordshire, and the West Riding of Yorkshire being in round numbers 5,500,000—Manchester, therefore, as a convenient food-distributing centre has no equal in this country.

The advantages of Garston over Liverpool are as nothing compared with the advantages which would result if coasting steamers from the numerous ports on the west and south coasts of England and Scotland could bring their produce to so important a food centre as Manchester. When steamers can take in cargo at the wharves in Manchester, many vessels which now load in London would find it infinitely to their advantage to load at the port from which the goods are originally despatched.

It is proposed to straighten, widen and deepen a portion of the rivers Mersey and Irwell so as to provide a direct tidal channel of sufficient capacity to allow vessels of the largest draught which now navigate the Suez Canel to reach Manchester.

It is intended to use the existing river-bed where it is tolerably straight, and to cut off the bends so as to reduce the length of its course between Trafford Bridge, Salford and Runcorn by about nine miles.

There is no city in the world, not situated on the sea-board or on a navigable river, which has a shipping trade at all to be compared with that of Manchester.

Of the 14,000,000 tons of shipping leaving and entering Liverpool, at least 5,000,000 would use the proposed navigation, and goods would find their way to Manchester either directly in the ships or by means of local steamers.

Allowing 15 per cent. for working expenses, which is a large estimate, the proportion in the case of the Suez Canal being only 6 per cent., then there would be a net revenue of £765,000. That sum would pay 10 per cent. if the undertaking cost £7,000,000, and 5 per cent. if it cost £14,000,000.

OPINIONS OF THE PRESS.

The Manchester Guardian says: " The scheme for the construction of the Ship Canal to Manchester, which has recently been the subject of discussion in our correspondence columns, is not so much in opposition to the spirit of the age as may at first sight appear. With the development of the railway system a tendency to regard inland navigation as a thing of the past set in; but, latterly, in many parts of the world, a reaction in favour of the more ancient means of transit has been observable."

The Manchester Courier says: " Let us hope that now the matter has been once more fairly mooted it will be thoroughly investigated. There cannot be any doubt that if the Tidal Navigation can be introduced to Manchester, it will be the means of materially promoting the commercial and manufacturing interests of the city and the surrounding district. The scheme is certain to be pronounced impracticable by some persons, but it needs not be abandoned on that account. The making of the Suez Canal was strongly discouraged in this country, but it was made for all that."

The London Times says: " The mind has become familiarised with colossal schemes of this sort, that a Manchester Ship Canal is modest and feasible by comparison. The opportunities of making fun out of engineering speculations have been decidedly lessened since the Suez Canal was finished. Whatever the possibilities of this scheme, it may at least receive more respectful consideration than former ones."

The Builder says: " Some of the principal trades in England are being transferred to Glasgow or to Paisley, owing to the superior cheapness in transport, both of raw materials and finished goods afforded by the Clyde. Pig iron has long shown this influence. Heavy iron castings from Sheffield followed. Now the shoe trade is leaving Stafford and Northampton for the valley of the Clyde, and finally the cotton thread trade is leaving Lancashire for Paisley."

The Iron and Coal Trades Review says: " It would certainly be a change in the order of things if we could include Manchester and Leeds among our seaports, and the project appears more feasible as regards the former. It would undoubtedly increase the prosperity of both places materially. We need only look to Glasgow as an example, and note the enormous development of that city since the Clyde has been so far improved as to allow of large ocean-going vessels loading at the city wharves. Every industry has expanded until Glasgow has risen to the position of the second city in the kingdom. To Manchester merchants and manufacturers the saving in railway transit and the heavy charges at Liverpool would be large if they could ship at their own doors, so to speak, and the district round about would derive great benefit. It is a question which deserves the full attention of the traders of the Manchester district."

The Liverpool Mercury says: " It has been suggested by a section of influential persons in Liverpool and Manchester that the Mersey Docks and Harbour Board should take into consideration the policy of adopting the plans of the navigation so ably designed for effecting the great improvements which, it is believed, to a great extent would obviate the block of traffic which now exists along the thoroughfares and railway termini in the vicinity of the

Liverpool docks. It is a well-known axiom that every increased facility of transit augments traffic, and would doubtless eventually add largely to the revenue of the dock estate. This would indeed be the accomplishment of a great national undertaking and worthy of the Board's distinguished antecedents, and in the meantime would give the Board the control, and so enable it to see that this grand work was efficiently accomplished. The interests of Liverpool and Manchester are so bound up together that there ought not to exist any feeling of jealousy, but they should cordially co-operate for their common advancement."

EXAMPLES SET BY OTHER NATIONS.

France : The attention of the French Government is being paid to the consideration of a projected canal to connect Paris with Dunkirk, so as to increase the facilities for traffic in the departments of the Nord, Calais, the Somme and the Oise.

They have also under consideration other projects for widening and deepening the Seine, the improvement of the ports of Rouen and Havre, and the construction of a great ship canal from Bordeaux on the Atlantic sea-board to the Mediterranean. It would extend from Bordeaux, by way of Toulouse, to Narbonne, a distance of 270 miles. It would be of sufficient capacity to accommodate ships-of-war, and at its highest point would be 500 feet above the level of the sea.

The estimated cost of this great work is put down at £56,666,160.

Holland : The great ship canal connecting Amsterdam with the North Sea has proved a great success. It was opened for traffic on the 30th October, 1877. Money has also been voted for the construction of maritime canals between Amsterdam and Rotterdam, and from Amsterdam to Utrecht, and thence to the Merwede River, near Gorcum.

Belgium : A meeting was recently held on the Brussels Bourse to discuss a proposal for a maritime canal from Brussels to the Scheldt. Great improvements of the harbour accommodation at Ostend and Antwerp are being carried out. At Antwerp whole lines of streets which front the Scheldt are being demolished, in order to construct a magnificent extension of the quays. The projects which the Belgians have on hand in connection with the extension of docks, quays, and the improvement of canals and rivers, amply prove that they are fully alive to the necessity of keeping pace with the requirements of commerce.

Germany : A maritime canal for the connection of the Rhine, the Weser and the Elbe is projected. Its construction would provide the Rhenish Westphalian coal and iron districts with a cheap waterway to the north seaports, and it is expected would secure a brilliant future for the Westphalian coal industry. The actual construction of the canal is expected to cost about £5,600,000, independently of £1,800,000 which is to be expended on the Elbe-Spree Canal, making the total estimated expenditure £7,400,000, the greater part of which is to be laid out in the next four years. *As illustrative of the public spirit manifested in these enterprises, it may be mentioned that all the land required is being given free of cost by the municipalities, communes and provinces through which the canals will pass.*

Austria : It is proposed to cut a canal between the Danube and the Oder, thus connecting the Black Sea with the Baltic. A scheme of even greater magnitude, which would afford another line of communication between the Black and Baltic Seas, is the cutting of a canal to effect the juncture of the Dniester and the Vistula, bringing Odessa and Dantzic into direct communication.

America : The United States Government is engaged in deepening Buffalo harbour, improving the channels between the lakes and the tide waters of the Hudson River. The State Engineer of New York is strenuously urging that a foot of water be added to the Erie Canal by raising the banks, so that the transport of freight may be cheapened by being carried in larger cargoes.

The capitalists and merchants of Baltimore have organised a company for the construction of a ship canal to be cut through Maryland and Delaware. By this canal Baltimore will have direct access to the Atlantic.

Canada : Canada has commenced, and will in a few years complete, the finest system of inland navigation in the world, which will make seaports of all our lake cities. British steamships of nearly 2,000 tons will lie at the docks of Chicago and other lake ports unloading their merchandise or receiving their cargoes of grain, provisions, etc. The British will not have to tranship or elevate their grain ; they can continue their voyage through Lake Ontario, the St. Lawrence and the ocean to Europe, only having to pay toll on the Welland Canal and on the river above Montreal.

Greece : The International Corinth Maritime Canal Company has just been formed for the purpose of cutting through the Isthmus of Corinth, which connects the Morean Peninsula with Hellas, the mainland of Greece. This canal will unite the Gulf of Lepanto with the Ægean Sea, and Corinth will be brought within four hours' steam of the Piræus. At present, vessels making for Athens from the west have to proceed right round the Peninsula of the Morea, and have to double Cape Matapan, involving a *détour* of 400 or 500 miles. The depth of cutting for this canal in some parts will be four times that required for the terminal basin of the proposed Tidal Navigation at Throstle Nest.

APPENDIX II.

MANCHESTER SHIP CANAL.

SUBSCRIPTIONS TO THE PARLIAMENTARY FUND OF £100,000.

	£		£
Daniel Adamson, The Towers, Didsbury	1000	Charles Moseley, Victoria Park, Manchester	500
S. W. Clowes, Norbury, Ashbourn ...	1000	Joseph Leigh, Brinnington Hall, Stockport ...	500
John Rylands, Longford Hall, Stretford	1000		
Peter & Frank Spence, Manchester Alum Works	1000	E. C. Potter, 10 Charlotte Street, Manchester	500
G. B. Worthington, 2 Porchester Gate, London	1000	John H. Gartside, Manchester ...	500
George Benton, Clyne House, Stretford	1000	The Workmen of Messrs. Daniel Adamson & Co., Dukinfield	430
The Mayor of Salford (R. Husband) ...	1000	Trustees of the late F. D. Astley, Dukinfield ...	300
Richard Peacock, Gorton Hall, Manchester	1000		
G. & R. Dewhurst, Manchester	1000	Boddington & Leigh, 20 Shudehill, Manchester	300
Wm. Rumney & Co., 53 Portland Street, Manchester	1000	W. J. Crossley, Great Marlborough Street, Manchester	250
Councillor Boddington, Strangeways, Manchester	1000	William Richardson, 4 Edward Street, Oldham	250
S. R. & J. E. Platt, Oldham ...	1000	John Lowcock, Greengate Mills, Salford	250
Edward Mucklow, Bury, Lancashire ...	1000	J. G. Adami, Albion Hotel, Manchester	250
Platt Brothers & Co., Oldham ...	1000	James Kay, Lark Hill, Timperley ...	250
Alderman Davies, Latchford, Warrington ...	500	Charles E. Schwann, 117 Portland Street, Manchester	250
William Butcher, 69 Princess Street, Manchester	500	Bryce Smith, Manchester ...	250
Councillor Goldsworthy, Hulme, Manchester	500	James Boyd, 30 Cannon Street, Manchester	250
Councillor Leech, Back Pool Fold, Manchester	500	Reuben Spencer, Whalley Range, Manchester	250
R. Johnson, Clapham & Morris, Manchester	500	Ellis Lever, Manchester ...	250
		Thomas Bradford, Salford ...	250

(267)

	£		£
Strines Printing Co., 19 George Street, Manchester	250	Robert Kay, Bury, Lancashire ...	100
Horrockses, Miller & Co., 55 Piccadilly, Manchester	250	Councillor Clay, Manchester	100
Deane Stanley, Manchester	250	J. Sawley Brown, Cateaton Street, Manchester	100
Beith, Stevenson & Co., Manchester ...	250	Isaac Neild & Son, 19 Chapel Walks, Manchester	100
Frank Crossley, Manchester	250	Marshall Stevens, Liverpool & Garston	100
John Heywood, Deansgate, Manchester	250	George Woodhouse, St. George's Road, Bolton	100
Colonel Blackburne, M.P., Hale, near Liverpool	200	Malcolm Ross & Sons, Cromford Court, Manchester	100
J. T. Emmerson, Peover, Knutsford ...	200	Tedbar Hopkin, Moreton Street, Strangeways	100
William Robinson, Darlington Lodge, Warrington	200	John Taylor, Newcroft House, Urmston	100
Lord Winmarleigh, Garstang, Lancaster	200	Jacob Behrens, Manchester	100
Dr. Mackie, *Guardian* Office, Warrington	200	R. Scott Collinge, Oldham	100
Henry J. Candlin, 44 Swan Street, Manchester	200	Dods, Ker, & Co., 104 Albert Square, Manchester	100
T. G. Hill & Co., 86 Major Street, Manchester	150	Councillor Southern, Store Street, Manchester	100
William McConnel, Brooklands, Prestwich	100	Councillor Howarth, Manchester ...	100
Eliza Hall, Hope Hollow, Prestwich ...	100	William Sumner, 2 Brazenose Street, Manchester	100
Thomas G. Stark, Ramsbottom ...	100	The Town Clerk, Salford	100
Barlow & Jones, Manchester	100	J. Fairclough & Sons, Mersey Mills, Warrington	100
Thomas Barlow, 25 Major Street, Manchester	100	James Fildes, 37 Brown Street, Manchester	100
J. Charlton Parr, Grappenhall Heys, Warrington	100	H. Verity & Sons, Mosley Street, Manchester	100
W. Wilson & Son, 29 Fountain Street, Manchester	100	Aaron Heartwell, 19 Princess Street, Manchester	100
Edward Collinge, Oldham	100	S. Gratrix, Junior, & Brothers, Manchester	100
James Reilly, St. James Hall, Manchester	100	John Dugdale, Llanfyllin	100
Frederick Flint, Aman Villa, West Street, Scarborough	100	James Lowe, Manchester	100
Lieut.-Colonel Sowler, *Courier* Office, Manchester	100	William A. Turner, Kingston Mills, Pendleton	100
Brockbank, Wilson & Mulliner, Manchester	100	Alderman Walmsley, Higher Broughton	100
		John George Tiller, Manchester ...	100
Samuel Kershaw & Sons, Manchester	100	George Fraser, Son & Co., Manchester	100

	£
J. Woolley, Sons & Co., 69 Market Street, Manchester	100
George Robinson & Co., 109 Princess Street, Manchester	100
The City Property Co., 19 Princess Street, Manchester	100
Peter Whitley, 24 Curzon Street, London	100
J. R. Pickmere, Warrington	100
Higgin, Lloyd & Co., Manchester	100
T. R. Wilkinson, Manchester	100
B. & S. Massey, Openshaw	100
William G. Crum, Knutsford	100
Joseph Lindley, 59 Grosvenor Street	100
Monks, Hall & Co., Limited, Warrington	100
Charles Clegg, Oldham	100
The Mayor of Warrington (J. Crosfield)	100
George Little, Hartford Iron Works, Oldham	100
Adolphus Sington & Co., St. Peter's Square, Manchester	100
Samuel Chatwood, 11 Cross Street	100
William Sharp, 94 Corporation Street	100
M. Kaufmann, 43A Lower Mosley Street, Manchester	100
— Holt, Didsbury, Manchester	100
S. Andrew, Oldham	100
Weston, Grover & Lees, Solicitors, Manchester	100
R. N. Michaelis, 22 Fountain Street, Manchester	100
Jno. Tetlow Lewis, 22 Fountain Street, Manchester	100
C. C. Dunkerley, Manchester	100
J. R. Bridgford & Sons, Manchester	100
De Jersey & Co., Manchester	100
Boddington & Ball, Princess Street, Manchester	100
Hickson, Lloyd & King, Manchester	100
William Brown, Little Bolton, Eccles	100
Christopher Wood, 25 Portland Street, Manchester	100
William Scott, 12 York Street, Manchester	100
Alderman Bailey, Albion Works, Salford	100
Henry Marriott, 72 Mosley Street	100
Edwin Guthrie, 32 Brown Street, Manchester	50
Henry Nevile, 8 John Dalton Street, Manchester	50
Tillotson & Son, Bolton and Eccles	50
John Goodier, Fulshaw Bank, Wilmslow	50
Arthur McDougall, City Corn Mills, Manchester	50
Edward Rowland, 2 Penrhyn Terrace, Old Trafford	50
Edward Lund, 22 St. John Street, Manchester	50
Edward B. Lees, Kelbarrow, Grasmere	50
Stewart, Thomson & Co., Manchester	50
William Beamont, Orford Hall, Warrington	50
J. & H. Patteson, 36 Oxford Street, Manchester	50
S. Oppenheim, 111 Portland Street, Manchester	50
W. W. Kirkman, 8 John Dalton Street, Manchester	50
Isaac Thorp & Sons, 25 Church Street, Manchester	50
James Collins, 76 King Street, Manchester	50
Henry Harrison, Blackburn	50
A. & G. Fox, Manchester	50
John Wainwright, Old Trafford	50
Henry Bleckley, Warrington	50
Edward Walmsley, South Reddish, Stockport	50
George Falkner & Sons, Deansgate	50
D. V. Stewart, Headlands, Higher Broughton	50
P. Goulden & Son, Seedley Ptg. Co.	50
Councillor Potts, Leamington Street	50

	£		£
Sam. Mendel, 43 Hill Street, Berkeley Square, London	50	William Ball, 20 Cooper Street, Manchester	25
W. Sandbach, 227 Oxford Street ...	50	John Norris, Ellesmere Street, Hulme	25
S. T. Woodhouse, Queen's Buildings, John Dalton Street	50	William B. Shorland, 25 Albert Street, Manchester	25
Rylands Brothers, Limited, Warrington	50	James Hailwood, 100 Fairfield Street	25
William Hughes & Sons, Parsonage, Manchester	50	Thomas Wright, 47 Hilton Street ...	25
Isaac Storey & Sons, Little Peter Street	50	Frederick Harrison, 3 Grange Road, Didsbury	20
John Worthington, Old Trafford ...	50	W. A. Darbyshire, Penygroes, N. Wales	20
Alderman Booth, Manchester	50	T. Lister Farrar, 79 Fountain Street, Manchester	20
Leech Brothers & Hoyle, Manchester	50	Edward Keating, 19 Rosamond Place, C.-on-M.	20
E. Aston, 15 Tib Street	50		
George Hicks, Albert Square	50	John H. Harrison, 15 Spring Gardens, Manchester	20
John Gladney ... ... · ...	50		
William Booth, Holly Bank, Cornbrook	50	Allen Richardson, Millgate Lane, Didsbury	20
Hankinson Luke, 14 Brazenose Street	50		
Councillor Mark, St. Ann's Square, Manchester	25	Henry Vollmer, Spear Street, Manchester	20
William Owen, Cairo Street, Warrington	25	William Grantham, Old Trafford ...	20
		John Grantham, Old Trafford	20
J. H. Holmes, Manchester	25	Henry Fisher, 270 Deansgate, Manchester	20
Lord & Holland, Fenny Street, Higher Broughton	25	George H. Courtis, 166 Stretford Road	20
Charles Reynolds, 96 Shakespear Street, C.-on-M.	25	John Wilson, Oakfield, Withington Road	20
E. J. Nicholson, Warrington	25	H. Heap, 99 Portland Street	20
James Stott, Westhoughton, Bolton ...	25	H. Frankenburg, 189 Stock Street, Cheetham	20
Henry Milling, Warrington	25		
Albert Norris, 19 Dickinson Street, Manchester	25	Isaac Marsden, Cable Street, Oldham Road	20
N. P. Sandiford & Son, 14 King Street, Manchester	55	J. Shaw & Sons, Wellington Street ...	20
		Samuel Hooley, Patricroft	20
Groves & Whitnall, Regent Road Brewery	25	Thomas Jackson, 91 City Road, Manchester	20
Charles Entwistle, Irwell House ...	25	Isaac Paulton, Levenshulme	20
John Robinson, 26 Church Street ...	25	J. Shaw & Sons, Wellington Street ...	20
Joseph Donnell, Junior, Staleybridge ...	25	Fletcher & Murphy, Garratt Street ...	20
John Allen, 26 Victoria Street, Manchester	25	James Pollitt, Ashton Road, Openshaw	20
		Thomas Lawson, 72 Embden Street ...	20
James Craven, Stretford	25	John Shaw, 126 Stockport Road ...	20

	£
James Robertshaw, 51 Bury New Road	20
Thomas Goulden	20
Thomas Perch, 27 Bradford Road ...	15
Samuel Crowcroft, 2 Cooper Street ...	15
A. Marshall, 64 Nelson Street	10
Councillor Griffin, Manchester	10
Mrs. Mead, Danebrook, Prestwich ...	10
Jacob Hadfield, Eldon Cottage, Cheadle	10
Councillor Batty, Manchester	10
E. A. Roberts, 64 Princes Street, Edinburgh	10
Lees & Graham, 77 King Street, Manchester	10
Charles Fletcher, 100 Oldham Street, Manchester	10
Rev. J. Cummins Macdonna, Cheadle	10
Ellis Jones, 23 Corporation Street, Manchester	10
C. R. Jennings, 122 Lower Broughton Road, Manchester	10
R. C. Richards, Kirkham	10
Thomas Whitworth, Manchester ...	10
Robert Gibson, Erskine Street, Hulme	10
Matthew Emsley, Pudsey, near Leeds	10
William Oldham, 25 Church Street, Manchester	10
Joseph Oldham, 25 Church Street, Manchester	10
John B. Thorp, 25 Church Street, Manchester	10
Henry Thorp, 25 Church Street, Manchester	10
G. F. Carrington, Guernsey	10
William Henry Bird, 35 Long Millgate, Manchester	10
Elizabeth Wilkinson, 26 Downing Street	10
Jane Widdow, 22 Rumford Street ...	10
Joseph Broome, 28 Downing Street ...	10
Harriett Broome, 28 Downing Street	10
Joseph Broome, Junior, 28 Downing Street	10

	£
Mary Ann Broome, 28 Downing Street	10
J. C. Broome, 28 Downing Street ...	10
Jno. Brearley, 64 Downing Street ...	10
Mrs. E. Geldart, Southport	10
Benjamin Robinson	10
J. H. Higgins, 8 Markland Street ...	10
John F. Sullivan, 41 Howard Street ...	10
Anyon Duxbury, 1 Byrom Street ...	10
A. E. Collings, Town Hall, Salford ...	10
E. Mansfield, 2 Trafalgar Street ...	10
Alderman Burgess, Warrington ...	10
Alderman Harrison, Warrington ...	10
Doctor Gornall, Penketh, near Warrington	10
Warrington Wire Drawers' Association	10
T. D. Benson, 8 York Street, Manchester	10
Richard Brammall, 70 Upper Duke Street	10
Richard Perry, 24 Ambrose Street ...	10
James McMaster, Park Street, Greenheys	10
Matthew Jackson, 218 Stretford Road	10
John Smith, 63 Derby Street	10
William Edwards, 21 Rochdale Road	10
Mrs. Edwards, 21 Rochdale Road ...	10
Edwin Robinson, 186 Stretford Road	10
H. H. Mainwaring, 102 Stretford Road	10
Robinson Dent, 183 Oxford Street ...	10
Alfred Noble, 56 Morning Street ...	10
John Harthill, 1 Sagar Street	10
William Royston & Co., Warrington ...	10
Edward Bolton, Warrington	10
Percy Davies, Solicitor, Warrington ...	10
Johnson & Rawson, Market Street, Manchester	10
E. M. Powell, 33 Chapman Street ...	10
Walter Pratt, 58 Butler Street ...	10
John Crane, Junior, 2 Butler Street ...	10
William Pownall, 133 Stretford Road	10
John L. Ward, 65 Upper Brook Street	10

	£
Elizabeth Ward, 65 Upper Brook Street	10
Hugh Kerr, 65 Brunswick Street ...	10
W. J. Burton, 38 Ardwick Green ...	10
J. C. Renshaw, 81 Rumford Street ...	10
George E. Hammond	10
John Swindells	10
William Dutton, Warrington	10
Edward Neild, Chapel Walks	10
G. Alcock, 95 Rochdale Road	10
W. A. Brown, 47 Oldham Road ...	10
Charles H. Eastwood, 39 Hilton Street ⎫	
Elizabeth Eastwood, 39 Hilton Street ⎪	
M. Leigh Eastwood, 39 Hilton Street ⎬	10
F. W. Eastwood, 39 Hilton Street ⎭	
M. Emsley, Lowtown, Pudsey, Leeds	10
A Friend, Manchester	10
William Colquhoun & Co., 26 York Street, Manchester	10
J. R. Hampson, South Parade ...	10
Marshall, Gibbon & Co., Barton Arcade	10
W. T. Grimley, 70 Oldham Street ...	10
A. Sington, 9 St. James Square ...	10
T. A. Smith, 74 Mosley Street ...	10
S. Brewis & Co., 33 Fountain Street ...	10
J. A. Widdowson, Ordsall Lane ...	10
C. F. Walton, Raglan Hotel	10
Michael Kane, 11 Angel Street ...	10
Humphrey Dyson, 518 Rochdale Road	10

	£
W. Brown, Junior, Eccles	10
Edward Dodd, 11 Upper Birch Street	10
J. H. Newton, 35 Stockport Road ...	10
Henry Stanley, Bury New Road ...	10
Henry Lord, Eccles Old Road ...	10
S. R. Bell, 14 Arlington Street ...	10
James Warburton, Dalham	10
T. B. Carter, Warrington	10
Walter Milner, Warrington	10
Alderman Chandley, Warrington ...	10
T. Grime, Warrington	10
Joseph Carter, Warrington	10
Edwin Richardson, Warrington ...	10
E. Cockshott, 99A Portland Street ...	10
J. H. & B. Hodge, 16 Derby Street ...	10
Joseph Torkington, 293 Stretford Road	10
Thomas Schofield, 99 Chapel Street ...	10
John Hulme, 70 Stockport Road ...	10
Christopher Boddy, 343 Rochdale Road	10
Charles Bennett, 194 Stretford Road ...	10
William Pimlott, 65 Moss Lane West	10
Joseph King, 39 Ash Street	10
Henry Stanley, 67 Bury New Road ...	10
James Fielding, 52 Rochdale Road ...	10
J. Redfern	10
John Blackwell & Co., 17 Palace Street	10
Sums under £10, as per separate list ...	1500
Total ...	£38,026

25th November, 1882.

APPENDIX III.

NOTES ON TABLES PUT IN BY MR. ADAMSON (Vol. i., p. 298).

TABLE No. 1

Shows that the population of Manchester within a radius of ten miles increased 21·13 per cent. between 1871 and 1881, or 57 per cent. more than the average increase of the population of England and Wales. This large additional population will need extended imports of food and increased means of employment.

TABLE No. 2

Shows the imports and exports in 1883 of Liverpool, £206,374,077; of London, £200,369,226; and that the trade of Liverpool exceeded in value all the other ports of Great Britain omitting London.

TABLE No. 6

Comparison of railway rates.

	Distance. Miles. Chains.		Grain. Rate per Ton. s. d.		Rate per Mile. d.	
Liverpool to Manchester ...	31	44	7	8	—	
Collected in Liverpool ...	—	—	1	0	—	
Less cartage in Liverpool...	—	—	6	8	2·53	
Glasgow to Leith ...	49	74	6	0	1·44	
West Hartlepool to Saltburn	32	0	5	10	2·18	
Newcastle to Alnwick ...	39	0	5	6	1·69	
Hull to Bridlington ...	32	39	5	5	2·00	
Cardiff to Abergavenny ...	29	0	5	0	2·06	
London to Buckingham ...	60	60	9	2	1·80	
Bristol to Bridgewater ...	33	31	3	4	1.19	

All station to station rates.

Average rate per ton per mile, 1·76.

The excess of the Liverpool rate over the average rates from other ports = 43·75 per cent.

TABLE No. 7

Shows the railway rate for square timber (station to station rate) to be 7s. 6d. per ton, or 2·85d. per ton per mile between Liverpool and Manchester, whilst for a similar distance from Newcastle, Glasgow, Bristol and Hull the average rate per ton per mile is 1·79d.

The excess of the Liverpool rate over the average of the other rates = 59·21 per cent. In Liverpool an extra charge of 1s. 9d., known as "paid on," is made for cartage in addition to the 7s. 6d., which, as a rule, is not required on timber loaded direct from the ship on to railway trucks.

TABLE No. 8

Shows the average cost of carrying cotton to the chief Lancashire towns to be 3·09d. per ton per mile, whilst if it were carried from London or other seaport towns it would average 1·56d. per ton per mile, an excess of 98 per cent.

TABLE No. 10

Shows the average cost of pig-iron by railway to Manchester from Liverpool and Runcorn to be 2·76d. per ton per mile, or 117 per cent. dearer than the cost of carriage from other ports to inland towns within a radius of 100 miles.

TABLES Nos. 14 AND 15

Show the dues and charges levied at various ports on imports and exports.

	Imports.		Exports.	
	Cotton.	Unrefined Sugar.	Cotton Yarn.	Cloth.
	s. d.	s. d.	s. d.	s. d.
Liverpool	4 5¼	5 4	4 2	3 9
Glasgow	3 6	2 10	3 3	3 3
Hull	2 6	2 4	1 3	2 1
London	5 6	6 3	6 8	5 0

TABLE No. 27

Gives a summary of the estimated entire saving to the cotton trade alone by the Ship Canal, viz. :—

On the imports of raw cotton used by mills within carting distance of Manchester. Carriage... ... ... £183,322
Avoidance of one brokerage and extra charges at Liverpool 165,361

£348,683
In respect to cotton used in Yorkshire 9,359

Total savings inwards £358,042
On exports of cotton manufacturers 98,512

Grand total on cotton and cotton goods £456,554

TABLE No. 28

Gives a comparison of charges per ton for imports *via* Liverpool and *via* Manchester on a few articles other than cotton.

	Liverpool Dock Charges and Railway Freight.				Proposed Maximum Charges to Manchester *via* Ship Canal.			Saving by Canal.		
	Liverpool Dock and Town Dues.	Charges from Liverpool Docks to Railway Station.	Railway Rate to Manchester Station.	Total Charge per Ton.	Canal Toll corresponding with the Railway Rate.	Wharfage corresponding with the Liverpool Dock and Town Dues.	Total Maximum Charges per Ton.	Amount per Ton.	Per Centage.	Amount per 100 Tons.
	s. d.	s. d.	s. d.	s. d.	s. d.	s. d.	s. d.	s. d.		£ s. d.
Bacon and hams	2 2	1 3	10 0	13 5	5 0	1 1	6 1	7 4	54·65	36 13 4
Flour and meal	2 0	1 3	6 8	9 11	3 8	1 0	4 8	5 3	52·94	26 5 0
Fruit, apples	2 8	1 8	8 4	12 8	4 2	1 8	5 10	6 10	53·92	34 3 4
Indian corn	1 0[1]	1 0	6 8	8 8	3 8	0 6	4 2	4 6	51·92	22 10 0
Sugar (refined)	3 0	1 3	8 4	12 7	4 2	1 6	5 8	6 11	55·0	34 11 8
Timber (deals)	1 0	1 9	7 6	10 3	3 9	0 6	4 3	6 0	58·53	30 0 0
Wheat	1 4[1]	1 0	6 8	9 0	3 8	0 8	4 4	4 8	51·85	23 6 8
Eggs	5 0	1 3	11 8	17 11	5 10	2 6	8 4	9 7	53·49	47 18 4

[1] These charges are less than the maximum charges which the Dock Board are entitled to levy.

TABLE No. 29A

Shows the gross increase of tonnage at recently improved ports between 1873 and 1883.

	1873.	1883.	Increase per cent.
Glasgow	3,295,601	5,543,882	68·21
Tyne Ports	9,113,459	13,042,638	43·11
Middlesborough...	1,020,684	2,183,087	113·88

TABLE No. 37A

Gives a comparative statement of arrivals of sea-going vessels at the port of Antwerp since 1873.

	Sailing Ships.		Steamers.		Totals.		Average Tonnage.
	No.	Tonnage.	No.	Tonnage.	No.	Tonnage.	
1873	2,182	650,533	2,615	1,411,703	4,797	2,062,236	430
1879	1,356	620,290	2,892	2,287,721	4,248	2,908,011	685
1883	989	417,860	3,874	3,034,559	4,689	3,857,934	823

TABLE No. 40

Shows comparison with Suez and Amsterdam Canals.

	Mileage.	Total Working Expenses.	Per Mile.
Suez Canal	92	289,254[1]	£3,144
Amsterdam Ship Canal	15⅓	33,330[2]	2,174
Manchester Ship Canal (estimated) ...	35½	112,500	3,169

[1] Includes administration in France and Egypt and sand dredging.

[2] Includes £21,666 for pumping machinery and drainage purposes.

FIRST SHIP CANAL PROSPECTUS ISSUED OCTOBER, 1885.

PRIVATE AND CONFIDENTIAL.

MANCHESTER SHIP CANAL COMPANY.

INCORPORATED BY 48 & 49 VIC., CAP. 188. (ROYAL ASSENT, 6TH AUGUST, 1885.)

AUTHORISED CAPITAL, { SHARE CAPITAL - - - £8,000,000.
{ BORROWING POWERS - - £2,000,000.

Issue of the Company's Share Capital, in 800,000 Shares of £10 each.

Price of Issue : Par, or £10 per Share,
Payable : 10s. per Share on Application,
„ 10s. „ „ Allotment,

and the remainder by instalments, at such periods as may be found necessary, having regard to the time fixed for the completion of the purchase of the Mersey and Irwell and Bridgewater Undertakings, and the progress of the Works; but at intervals of not less than three months, and by calls not exceeding £1 each, and not more than £2 : 10 : 0 in any twelve months.

Interest at the rate of £5 per cent. per annum will be paid on all amounts paid up upon shares which shall from time to time be in advance of the sums actually called up, and payments in full or in advance may be made on allotment, or at any subsequent time, subject to the withdrawal of the option by the Company when they think fit.

The Company being incorporated by Act of Parliament, the liability of a shareholder is absolutely limited to the amount remaining unpaid upon the shares he holds.

DIRECTORS,

The following are by the Act appointed the first Directors of the Company :—

DANIEL ADAMSON, C.E., J.P. (Daniel Adamson & Co., Manufacturing Engineers, Chairman Newton Moor Cotton Spinning Co. and North Lincolnshire Iron Co.), The Towers, Didsbury, near Manchester, *Chairman.*

(276)

COUNCILLOR HENRY BODDINGTON, Jun. (Managing Director Henry Boddington & Co., Limited), Manchester.

JACOB BRIGHT, M.P., Alderley, Cheshire.

WILLIAM FLETCHER (John Fletcher & Sons, Engineers), Eagle Foundry, Salford.

HILTON GREAVES, J.P., Cotton Spinner, Derker Mills, Oldham.

ALDERMAN RICHARD HUSBAND, J.P. (Ex-Mayor of Salford), Seedley, Manchester.

CHARLES PATON HENDERSON, Jun., J.P., Pine Cliff, Torquay.

RICHARD JAMES (Michaelis, James & Co., Manchester, Glasgow, and London), The Grange, Urmston, near Manchester.

JOSEPH LEIGH, J.P. (T. & J. Leigh, Cotton Spinners), Bank Hall, Heaton Mersey, Mayor of Stockport.

JAMES E. PLATT (Director Platt Brothers and Co., Limited, Oldham), Bruntwood, Cheadle.

SAMUEL RADCLIFFE PLATT, J.P. (Chairman Platt Brothers & Co., Limited, and Chairman of the Oldham Chamber of Commerce), Werneth Park, Oldham.

JOHN RYLANDS, J.P. (Governor, Rylands and Sons, Limited, Manchester and London), Longford Hall, Stretford.

The following have consented to join the Board at the first ordinary meeting of Shareholders :—

WILLIAM HENRY HOULDSWORTH, M.P. (Thomas Houldsworth & Co., Cotton Spinners), Norbury Booths Hall, Knutsford.

SIR JOSEPH COCKSEY LEE, J.P., 56 Mosley Street, Manchester.

ALDERMAN W. H. BAILEY (W. H. Bailey and Co., Engineers, Director W. Barningham and Co., Limited), Salford.

JOHN ROGERSON, A.C.E., Croxdale Hall, Durham.

Bankers.

THE NATIONAL PROVINCIAL BANK OF ENGLAND, Limited, Mosley Street, Manchester.

Engineer. | **Consulting Engineer.**

E. LEADER WILLIAMS, M. Inst. C.E. | JAMES ABERNETHY, Past President Inst. C.E., F.R.S.E.

Solicitors.

GRUNDY, KERSHAW, SAXON, & SAMSON, Manchester.

Auditors.

THOMAS WADE, GUTHRIE, & CO., Manchester.

Provisional Manager. | **Secretary.**

MARSHALL STEVENS. | ALFRED H. WHITWORTH.

Offices.

MANCHESTER—70 Market Street. | LONDON—1 Fenchurch Avenue, E.C.

PRELIMINARY PROSPECTUS.

The Company has been incorporated for the following, amongst other, purposes :—

(a) To construct a **Ship Canal** from the **River Mersey at Eastham**, near **Liverpool**, past **Ellesmere Port, Weston Point,** and **Runcorn,** to **Warrington, Salford,** and **Manchester,** available for **the largest class of Ocean Steamers, with Docks at Manchester, Salford, and Warrington, and other Incidental Works.**

(b) To purchase the entire undertakings of the existing **Bridgewater Navigation Company, Limited,** including the **Bridgewater Canals,** the **Runcorn and Weston Canal,** the **Mersey and Irwell Navigation,** the **Runcorn Docks,** the **Duke's Dock in Liverpool,** and all that Company's **Warehouses, Wharves,** Buildings, Lands, Rents, Rights, and Privileges, as a going concern.

Object of the Ship Canal.

The object of the Ship Canal is to afford the cheapest means for the transit of Merchandise and Minerals of all kinds between English and Foreign Ports, and the Manufacturing Towns and Coalfields of Lancashire, Yorkshire, Cheshire, Derbyshire, and Staffordshire, and the adjacent Industrial Districts. The Canal will constitute the nearest Port capable of accommodating steamers of 1,000 tons and upwards to, and will conveniently serve, a district covering an area of 7,500 square miles, and containing a population of over 7,000,000, including nearly all the Lancashire Cotton Trade, the main portions of the Lancashire and South Yorkshire Coalfields, the Cheshire Salt Trade, the Staffordshire Pottery and Iron Trades, and a great proportion of the Lancashire Chemical Trade.

The Canal will enable Ships of the largest—as well as of smaller—tonnage, to trade direct between Manchester, Warrington, and Runcorn, and all Ports across the seas or coastwise, and Merchandise will be conveyed to and from those towns without the cost and delay of transhipment or breaking bulk at intermediate Ports and Railway Stations, and this alone will effect a saving greater than any possible reduction in the Liverpool Dock Dues, or in Railway Charges.

A large Port will be created more than 30 miles nearer to the great manufacturing districts of Lancashire, to the West Riding of Yorkshire, and to Cheshire, Staffordshire, and Derbyshire : and Foreign, and also Irish, Scotch, and other Coastwise Produce will be delivered direct into the greatest consuming district in the world, so that in this, as well as in other ways, the Ship Canal will be a great National benefit.

Manchester is already the centre of the Cotton Trade, and has very large Provision, Corn, Cattle, and Fish Markets. But large quantities of the Provisions, Corn, and Timber consumed in the Manchester District, and actually sold in the Manchester Markets, are stored at the Ports of Liverpool, Hull, Fleetwood, Grimsby, West Hartlepool, and Goole, and the necessity for this cripples the Trade of the district and of the country, by the cost and time expended in transhipment and in the transit to and from those Ports.

There will be tidal gates at the entrance to the Canal, which will be worked as locks at low water, so that large vessels can enter and leave at almost any state of the tide, instead of only during a period of 40 minutes of each tide as at Liverpool. Small vessels will be able to enter and leave at any time.

Vessels will be able to navigate the Canal with safety at a speed of five miles an hour, and it is estimated that the journey from the entrance at Eastham to Manchester will be accomplished in eight hours, which is much less time than is now taken to cart goods from ship to rail in Liverpool, and to carry them thence by rail to Manchester.

The *maximum* dues which could be levied on Ships using the Canal, and which would be included in the sea freight charged by the shipowners, are the same as those now actually charged at Liverpool, and amount to from two to five days' expenses upon the ship, according to the trade in which she is engaged.

The *maximum* Canal Tolls and Dock Rates which could be levied on Goods passing along the Ship Canal and using the Docks, are fixed by the Act *at one-half of the existing rates* charged by Railway or Canal to Manchester, and a' *one-half of the Dock and Town Dues* now charged on goods at Liverpool, and the charges on good: upon the Canal *cannot exceed 50 per cent. of the present cost* from the ship in Liverpool Docks to M nchester.

For example :— Per ton.	Cotton	Woo.	Sugar (Loaves)	Sugar (Raw)	Bacon	Tinned Meats	Wheat	Oranges	Petroleum	Tallow	Timber	
	s. d.	s. d.	s. d.	s. d.	s. d.	s. d.	s. d.	s. d.	s. d.	s. d.	s. d.	
Present cost from the ship in Liverpool Docks to Railway Station at Manchester ... „	14 6	17 3	18 9	12 2	15 10	18 3	9 11	15 10	15 3	13 11	10 3	Present cost.
Cost by the Ship Canal to Dock Quay at Manchester, after paying *maximum* Tolls and Rates „	7 0	7 9	6 8	4 11	6 7	8 0	4 10	6 2	5 11	5 10	4 9	Cost by Canal.
Saving effected by the Ship Canal ... ... „	7 6	9 6	12 1	7 3	9 3	10 3	5 1	9 8	9 4	8 1	5 6	Saving by Canal.
Rates now charged by the Railway Companies for carriage between Liverpool and Manchester, Station to Station ... „	8 0	10 0	11 8	6 8	10 0	10 0	6 8	10 0	10 0	8 4	7 6	Present Rail Rates.
Rates to which the Railway Companies would have to reduce to obtain the traffic *via* Liverpool on even terms with the Ship Canal ... ... „	0 6	0 6	Nil	Nil	0 9	Nil	1 7	0 4	0 8	0 3	2 0	Rail Rates to compete.

Savings to other destinations upon COTTON, per ton :—

Ashton, 7s. 2d.; Bolton, 5s. 1d.; Bury, 5s. 9d.; Clayton, 8s. 4d.; Dewsbury, 6s. 3d.; Dukinfield, 7s. 2d.; Halifax, 7s. 5d.; Heywood, 6s. 8d.; Huddersfield, 7s. 9d.; Leeds, 4s. 5d.; Macclesfield, 6s. 7d.; Mossley, 7s. 7d.; Oldham, 6s. 8d.; Rochdale, 8s. 6d.; Stalybridge, 6s. 10d.; Stockport, 5s. 2d.; Sowerby Bridge, 9s. 6d.; Todmorden, 8s. 11d.; other goods being in proportion.

A saving of even 6s. per ton upon such a valuable article as Raw Cotton is equivalent to a dividend of one per cent. upon the Share Capital of the Spinning Companies of the District, and as one of the two brokerages now charged in Liverpool could be saved by reason of the development of the existing Raw Cotton Market in Manchester, whereby spinners would be enabled to buy their own cotton or to buy through a broker direct from the importing merchant, another one per cent. dividend would be earned by the Spinning Companies, as each brokerage averages 6s. per ton.

As a saving of 50 per cent. in Dock Rates and Railway Charges will be effected on all sea-borne goods destined for Manchester, and a proportionate saving on the same class of goods destined for other towns, according to their proximity to the Canal and to other Ports, every ton of goods brought or sent by the Canal will cause a material saving to the importer or exporter, and the Canal is therefore certain to secure a very large trade.

There will inevitably rise along the banks of the Canal many industries in themselves involving traffic, and with the experience gained in the cases of the Clyde, the Tyne, and other improved waterways, such industries may be expected to be of great magnitude, and the Canal will also doubtless do a considerable local traffic between different points throughout its course.

The Ship Canal will be for all practical purposes a Dock from end to end, 35 miles in length, and additional Side Basins and Lay-byes can be constructed at any point without dock gates and at small expense, which will enable vessels of the largest tonnage to load and discharge alongside works of any description or at railway sidings.

Bridgewater Undertakings.—

The Act provides that the **Bridgewater Navigation Company** shall sell the whole of the Bridgewater Undertakings to the Company for £1,710,000. These undertakings consist of—

(a) "The Bridgewater Canal Undertaking," which includes the Canals known as the Bridgewater Canals and the Runcorn and Weston Canal, also extensive warehouses and wharf accommodation in Manchester, docks and warehouses at Runcorn, and the Duke's Dock and large warehouses in Liverpool; and

(b) "The Mersey and Irwell Undertaking," which includes the Mersey and Irwell Navigation and the Runcorn and Latchford Canal, also extensive warehouses at Manchester and Warrington, and the Old Quay Docks at Runcorn.

The **Mersey and Irwell Navigation** will be absorbed in the construction of the Ship Canals but the **Bridgewater Canals** form a separate Undertaking, and will continue to be a barge navigation between **Manchester** and **Liverpool** *via* **Runcorn,** and will constitute a most valuable feeder and distributor of traffic to and from the **Ship Canal,** by connecting it with the **Pottery** and **Iron Districts** of **Staffordshire** (*via* the **Trent and Mersey Canal**), the **Wigan District** and the **South Lancashire Coalfields** (*via* **Barton**), and with nearly all the towns engaged in the Cotton Trade, and with **Yorkshire** and **Derbyshire** (*via* the **Rochdale** and other **Canals**).

For adjustment of values, £460,000 is treated as the cost to the Company of the Mersey and Irwell Undertaking, and the amount, £1,250,000, remaining as the cost of the Bridgewater Undertaking, will make the purchase a good investment, whether looked at separately or as part of the general undertaking of the Company. The average net income of the Bridgewater Undertakings during the last five years has been upwards of £60,000 per annum. It has been derived from the Bridgewater Canals and those other portions of the Bridgewater Undertakings which will continue in full operation notwithstanding the construction of the Ship Canal. The Bridgewater Canal Undertaking is in good working order. For several years a large sum has been annually spent out of income upon permanent improvements, which are now nearly completed.

The Directors are satisfied that the traffic of the Bridgewater Canals will, under the management of the Ship Canal Company, be capable of considerable development, by the removal of the Bar Tolls which obstruct through traffic, and by the opening of the Canal to general carriers.

The purchase of the **Bridgewater Undertakings** will be completed as soon as the necessary Capital has been subscribed, and the immediate receipt of an income available for the payment of Dividend will be thereby secured.

CAPITAL OUTLAY.

Bridgewater Undertakings.—These Undertakings will cost		£1,710,000
Deduct the cost of the Mersey and Irwell Undertaking, as above ...		460,000
Net amount attributed to the Bridgewater Canal Undertaking ...		£1,250,000
The Ship Canal and Incidental Works.—The estimated cost of these is	£5,328,848	
To which add the cost of the Mersey and Irwell Undertaking	460,000	
And the cost of the Railway Deviations and two Branch Railways	440,892	
		6,229,740
Dock Accommodation.—The estimated cost of the Docks at Manchester, Salford, and Warrington, including Branch Railway at Warrington, is ...		1,403,232
Parliamentary Expenses		150,000
Total Cost		£9,032,972

All the above estimates are based upon the Parliamentary Estimates, and include a sum of £573,528 for contingencies, and if £1,800,000 out of the total of £9,032,972 be obtained under the borrowing powers, the amount to be raised in Share Capital would be £7,232,972, leaving unexhausted Capital Powers of £967,028.

The Directors are advised that the works may be completed within four years from the time of commencement, and that the estimates, which were purposely calculated at high rates to anticipate criticism before Parliament, are sufficient for the construction of the works. The estimates were not questioned during the Parliamentary Contests of 1885.

ESTIMATED EXPENSES OF WORKING AND MAINTENANCE.

Bridgewater Canal Undertaking.—It is unnecessary to take the Working Expenses and Maintenance of this Undertaking separately, as the net revenue is already ascertained.

Ship Canal.—The Suez and Amsterdam Canals afford the best existing data for an estimate, but important differences will exist between them and the Ship Canal. It is estimated that the expenses of working and maintaining the Ship Canal, exclusive of the Docks, but including a part of the General Management expenses, will not exceed £2,000 per mile, which would give a maximum total for the Canal of £70,000

Docks.—For the cost of working and maintaining Docks there are numerous examples, but in most cases the cost of handling the traffic is included in the accounts as working expenses. It is estimated that the expenses of working and maintaining the Manchester, Salford, and Warrington Docks, including the remainder of the General Management expenses, but exclusive of all costs attending the handling or warehousing of traffic, which are paid for separately, will not exceed £400 per acre of water space. This rate is a high one for modern docks, and would give a maximum total for the docks of 34,200

Total Estimate for Ship Canal and Docks £104,200

ESTIMATED TRAFFIC AND REVENUE.

A Traffic of only 3,000,000 tons of General Goods, similar to those named in a former paragraph relating to savings, but including manufactured goods, would pay:—

In Canal Tolls (half the existing Railway Rates) and in Wharfage (half the existing Dock and Town Dues upon Goods) at least	£750,000	
With Ship Dues (less than those charged at Liverpool)...	75,000	
Add average net income of the Bridgewater Canal Undertaking	60,000	
		£885,000
Deduct—Working Expenses, as above	£104,200	
Interest upon £1,800,000 borrowed Capital, at 4 per cent.	72,000	
		176,000
	Net Income.................£709,000	

being sufficient, if ship dues are charged, to pay a dividend of 8 per cent. upon the total share capital of the Company; or, if ship dues are not charged, a dividend of more than 7 per cent. upon the same amount of capital, and in either case to carry £69,000 per annum to a reserve fund.

This estimate is undoubtedly a small one. It does not include any revenue from coastwise traffic or from low-class traffic, such as coal, salt, and iron ore, for which separate accommodation is provided at different points along the line of the Ship Canal, and a very moderate shipment of such traffic would largely increase the net revenue: nor does it include any revenue from porterage and other services, chargeable to merchants under the Act at actual cost plus 10 per cent. as profit to the Company. The total traffic in and out of the port of Liverpool amounts to upwards of 15,000,000 tons of cargo. The amount of traffic included in the above estimate is, in fact, less than the estimated increase during the past four years of the trade of the district which can be served by the Canal, and it is therefore reasonable to expect that before the Canal Works can be completed the increase of the trade of the district will be sufficient to furnish a remunerative traffic to the Ship Canal without removing existing trade from any other port.

The Directors are satisfied that a much larger quantity of traffic will use the Ship Canal, and the traffic expected was described in great detail in the Tables placed before the Parliamentary Committees of 1884 and 1885. They showed a probable traffic in various specified articles, including coastwise, low-class goods, and local trade, at the end of seven years, amounting to 9,649,316 tons, giving a Revenue—

In Canal Tolls and Wharfage of	£1,492,282	
With Dues upon Shipping, less than those charged at Liverpool, of	187,500	
Profit of 10 per cent. upon Porterage, if charged only at an average rate of 6d. per ton ...	24,123	
Average net income of the Bridgewater Canal Undertaking	60,000	
		£1,763,905
Deduct—Working Expenses (say)	£120,000	
Interest as above	72,000	
		192,000

Net Income.................£1,571,905

sufficient to pay a dividend of 18 per cent. upon the total share capital of the Company, and to accumulate a Reserve Fund at the rate of £131,905 per annum.

Waterways have a great advantage over Railways and other modes of transit in respect of working expenses and the cost of maintenance. The Suez Canal revenue returns show that, while the traffic has increased fourfold in the last ten years, the working expenses have remained almost stationary.

The Committee of the House of Commons before whom the first inquiry was held in 1883, acting entirely on their own initiative, inserted the following clause in the preamble of the Bill, and it is said to be without precedent:—

"*And whereas it appeared from the evidence adduced that if the scheme could be carried out with due regard to existing interests, the Manchester Ship Canal would afford valuable facilities to trade, and ought to be sanctioned.*"

The proportion of the Shipping Trade of the country that, according to population, could be allocated to the Ship Canal *as the nearest Port* is at least **21,000,000** tons, and although, for purposes of Revenue, it has been shown that it is not necessary to get more than **3,000,000** tons of cargo, it should be remembered:—

> (*a.*)—That the industrial community of which Manchester is the great centre contributes in value fully two-thirds of the entire exports of the United Kingdom.
>
> (*b.*)—That as regards imports, a district which furnishes so considerable a proportion of the exports made largely from imported raw materials and which is almost wholly dependent for its food supplies upon other parts of the United Kingdom, or the Colonies and Foreign Countries, must of necessity absorb a far greater proportion of the imports and of Coasting traffic than its share *pro rata* averaged according to its proportion of the entire population of the country.
>
> (*c.*)—That the Shipping Trade using the Ports of the United Kingdom amounts to upwards of 3 tons of cargo per head of the population.

The Directors entertain a thorough conviction of the soundness of the undertaking, based upon practical experience of the trades and requirements of the District, and a careful investigation of all the facts and circumstances bearing upon the case, and they recommend the Shares with confidence upon the following, amongst other, grounds:—

> To Investors, because they are satisfied that the undertaking will prove a sound, remunerative, and improving investment.
>
> And to all who are engaged, or in any way interested, in the trade of the district, or in that of the country in general, because the Canal will tend largely to revive and develop trade.

The Revenue of the Bridgewater Canal Undertaking will enable the Directors to pay interest at the rate of £3 per cent. per annum upon all the capital which it will be necessary to call up during the first twelve months.

The Directors will in due course deposit in the Private Bill Office of Parliament a Bill to enable the Company to pay interest out of capital, and it will be for the Shareholders to say whether that Bill be proceeded with. The Directors feel satisfied that if the Shareholders desire it powers can be obtained to pay interest at the rate of 4 per cent. upon the whole of the capital during the remaining period of construction.

Applications for Shares may be made on the enclosed form, accompanied by a deposit of 10s. upon each Share applied for, to **The National Provincial Bank of England, Limited**, at Mosley Street, Manchester, at the Head Offices in London of that Bank, or at any of the various branches throughout the Kingdom ; or to any of the following Banks and their Branches :—

THE BANK OF ENGLAND.

THE MANCHESTER AND LIVERPOOL DISTRICT BANK, LIMITED.	PARR'S BANKING COMPANY, LIMITED.
	THE OLDHAM JOINT STOCK BANK, LIMITED.
THE MANCHESTER AND SALFORD BANK, LIMITED.	THE BANK OF BOLTON, LIMITED.
MESSRS. CUNLIFFES, BROOKS, AND COMPANY.	THE ASHTON, STALYBRIDGE, AND GLOSSOP BANK LIMITED.
THE MANCHESTER AND COUNTY BANK, LIMITED.	
THE UNION BANK OF MANCHESTER, LIMITED.	THE ADELPHI BANK, LIMITED.
THE CONSOLIDATED BANK, LIMITED.	MESSRS. JOHN STUART AND COMPANY.
THE LANCASHIRE AND YORKSHIRE BANK, LIMITED.	MESSRS. JAMES SEWELL AND NEPHEW.
THE MANCHESTER JOINT STOCK BANK, LIMITED.	THE CO-OPERATIVE WHOLESALE SOCIETY, LIMITED.

Where no allotment is made the Deposit will be returned without deduction, and where a partial allotment is made the surplus Deposit will be applied towards the amount payable on allotment.

The Act of Parliament may be inspected at the Offices of the Company's Solicitors, 31 Booth Street, Manchester.

Forms of application may be obtained from the Secretary, at the Offices of the Company in Manchester or London, or from any of the above-mentioned Bankers, or from the Solicitors or other Officers of the Company.

Manchester,
  October, 1885.

Description of Canal Works.—

The **Ship Canal** will commence at Eastham, near Liverpool, on the Cheshire shore of the Mersey, and will terminate in the City of Manchester. It will be about 35 miles in length, will have a minimum depth of 26 feet of water, and will be of ample width for the largest Vessels to pass each other at any point, and may be compared with the Suez and Amsterdam Canals in width and depth, as follows :—

CANALS.				DEPTH—FEET.				BOTTOM WIDTH—FEET.
Suez ...	...	...	...	26	..	...	...	72
Amsterdam	...	...	...	23	...	...	...	89
Manchester	...	...	...	26	...	...	...	120

The level of the Docks at Manchester, which is 60 feet 6 inches above the ordinary level of the tidal portion of the Canal, will be reached by four sets of Locks. The Locks will be of a size sufficient to admit the largest Merchant Steamers afloat. Each set comprises (a) a large Lock, 550 feet by 60 feet; (b) a smaller Lock, 300 feet by 40 feet for ordinary vessels; and (c) one Lock, 100 feet by 20 feet, for small Coasters and Barges : and all capable of being worked together.

Each set of Locks will be worked by hydraulic power, enabling vessels to be passed in 15 minutes. It has been ascertained by careful gaugings that the Rivers Irwell and Mersey (which will be diverted into the upper reaches of the Canal) will supply more than sufficient water for the Locks even in the driest season.

The estimates include large Docks in Manchester, Salford, and Warrington, as sanctioned by the Company's Act, with a water area of 85½ acres, containing more than four miles of Quays. There will also be a mile of Quay space near Manchester on the Ship Canal, in addition to Wharves at many places alongside its course. The Docks will be of the most approved construction, and special provision will be made to secure the rapid loading and discharging of Vessels. Extensive Shed accommodation will be provided at the Docks, and the cost of no less than 50 hydraulic cranes is included in the estimates.

The Manchester Docks will be in close proximity to the Railways of the Midland Railway Company and the Cheshire Lines Committee, and to the Bridgewater Canal, and provision can be made at small expense which will enable goods to be transferred direct from or to the Sheds, or Quays, to or from Trucks on the Railways, or Boats on the Bridgewater Canal: and the whole of the Docks in Manchester and Salford will be within short distances of the Goods Stations of the London and North-Western and the Lancashire and Yorkshire Railway Companies.

At Partington a Basin will be constructed for loading Coal from High Level Tips, with Branch Railways to connect the Ship Canal with the system of the Cheshire Lines Committee (comprising the Midland, the Great Northern, and the Manchester, Sheffield and Lincolnshire Railway Companies) and a Branch Railway will be constructed at the Warrington Dock to connect the Canal with the systems of the London and North-Western and Great Western Railway Companies. Provision will thus be made for a large trade in coal from the Lancashire and South Yorkshire Coalfields.

Barge Lifts and Coal Tips will also be provided at Barton, where the Bridgewater Canal will be carried over the Ship Canal by means of a swing aqueduct, and traffic can be transferred in the barges direct from one Canal to the other, or Coal can be tipped from coal boats direct into the ship. Barton is immediately contiguous to the Worsley and Leigh, and other parts of the South-East

Lancashire Coalfield, and a large supply of coal will therefore be readily available for return cargo or for bunker purposes.

It will be seen from the published map that in addition to the facilities for the collection and distribution of traffic, which as above mentioned will exist at Manchester, Barton, Partington, and Warrington, the Ship Canal will be in direct communication :—

At Ellesmere Port—with the Shropshire Union Canal, and with the London and North-Western and Great Western Railway systems.

At Weston Point—with the River Weaver Navigation.

At Runcorn—with the Bridgewater Canals, and through them with the Trent and Mersey Canal, and *via* the River Mersey with the town of Widnes and the St. Helens Canal, and by means of a Branch Railway (which is part of the Bridgewater Canal undertaking) with the London and North-Western Railway system.

And at Manchester—with the Bridgewater, Rochdale, Ashton, Bolton and Bury, and Peak Forest and other Canals.

In addition to the Docks already existing at Weston Point (the port of the Cheshire Salt District) and Runcorn (the principal port for the Pottery and Iron trades of North Staffordshire), facilities will be provided by widening out the Canal near those Docks, and also near the Lock at Runcorn, to be constructed for the accommodation of the Widnes traffic, which will enable vessels on the inward voyage to discharge materials for use in the Pottery, Chemical and Iron manufactures, and on the outward voyage to load Salt, Coal, and Pottery, Chemical and Iron manufactures at those points.

SECOND SHIP CANAL PROSPECTUS ISSUED JULY, 1886.

MESSRS. N. M. ROTHSCHILD & SONS are authorised to receive Subscriptions for the Issue of 725,000 Shares of £10 each,

IN THE

MANCHESTER SHIP CANAL COMPANY.

Incorporated by 48 and 49 Vic., Cap. 188. (Royal Assent, 6th August, 1885.)

AUTHORISED SHARE CAPITAL £8,000,000 IN 800,000 SHARES OF £10 EACH,
of which 75,000 have been allotted since the formation of the Company.

Price of Issue, Par, or £10 per Share.
Payable, 20s. per Share on Application.
    ,,   20s.   ,,   ,, Allotment.

And the remainder by Instalments, at intervals of not less than Three Months and not exceeding £1 each or £2 10s. 0d. in any Twelve Months.

The Company have entered into a Contract with Messrs. Lucas & Aird for the execution of the whole of the Works within Four Years from their commencement, for the sum of £5,750,000, being £561,137 less than the Parliamentary Estimates.

Interest at the rate of £4 per cent. per annum will, in accordance with the " Manchester Ship Canal Act, 1886," be paid during the period of Four Years required for construction of the Works, upon all amounts called up and paid on the Shares.

The liability of a Shareholder is absolutely limited to the amount remaining unpaid upon the Shares he holds.

DIRECTORS.

DANIEL ADAMSON, C.E., J.P., The Towers, Didsbury, near Manchester—Engineer (*Chairman*).

SIR JOSEPH COCKSEY LEE, J.P., 56 Mosley Street, Manchester—Merchant (*Deputy Chairman*).

ALDERMAN W. H. BAILEY, J.P., Salford— Engineer.

HENRY BODDINGTON, JUN., Manchester— Brewer.

JACOB BRIGHT, M.P., Alderley, Cheshire— Merchant.

WILLIAM FLETCHER, Eagle Foundry, Salford—Engineer.

CHARLES PATON HENDERSON, JUN., J.P., Pine Cliff, Torquay.

WILLIAM HENRY HOULDSWORTH, M.P., Norbury Booths Hall, Knutsford— Cotton Spinner.

ALDERMAN RICHARD HUSBAND, J.P., Seedley, Manchester.

RICHARD JAMES, The Grange, Urmston— Merchant.

JOSEPH LEIGH, J.P., Bank Hall, Heaton Mersey—Cotton Spinner.

JAMES E. PLATT, Bruntwood, Cheadle— Engineer.

SAMUEL RADCLIFFE PLATT, D.L., J.P., Werneth Park, Oldham—Engineer.

JOHN ROGERSON, J.P., A. Inst. C.E., Croxdale Hall, Durham.

JOHN RYLANDS, J.P., Longford Hall, Stretford—Merchant.

Bankers.

THE NATIONAL PROVINCIAL BANK OF ENGLAND, LIMITED, Mosley Street, Manchester.

Engineer.

E. LEADER WILLIAMS, M.Inst.C.E.

Consulting Engineer.

JAMES ABERNETHY, Past-President Inst. C.E., F.R.S.E.

Solicitors.

GRUNDY, KERSHAW, SAXON, & SAMSON, Manchester.

Provisional Manager.

MARSHALL STEVENS.

Secretary.

ALFRED H. WHITWORTH.

PROSPECTUS.

The Company has been incorporated for the following, amongst other, purposes :—

(a) To construct a **Ship Canal** from the River Mersey at Eastham, near **Liverpool,** past Ellesmere Port, Weston Point, and Runcorn, to Warrington, Salford, and Manchester, available for **the largest class of Ocean Steamers, with Docks at Manchester, Salford, and Warrington, and other Incidental Works.**

(b) To purchase the entire undertakings of the existing **Bridgewater Navigation Company, Limited,** including the **Bridgewater Canals,** the **Runcorn and Weston Canal,** the **Mersey and Irwell Navigation,** the **Runcorn Docks,** the **Duke's Dock in Liverpool,** and all that Company's **Warehouses, Wharves,** Buildings, Lands, Rents, Rights and Privileges, as a going concern.

The object of the Canal is to afford the cheapest means for the transit of Merchandise and Minerals of all kinds between English and Foreign Ports, and the Manufacturing Towns and Coalfields of Lancashire, Yorkshire, Cheshire, Derbyshire, and Staffordshire, and the adjacent Industrial Districts. The Canal will constitute the nearest Port capable of accommodating Steamers of 1,000 tons and upwards to, and will conveniently serve, a district covering an area of 7,500 square miles, and containing a population of over 7,000,000, including nearly all the Lancashire Cotton Trade, the main portions of the Lancashire and South Yorkshire Coalfields, the Cheshire Salt Trade, the Staffordshire Pottery and Iron Trades, and a great proportion of the Lancashire Chemical Trade.

The Canal will enable Ships of the largest—as well as of smaller—tonnage, to trade direct between Manchester, Warrington, and Runcorn, and all Ports across the seas or coastwise, and Merchandise will be conveyed to and from those towns without the cost and delay of transhipment or breaking bulk at intermediate Ports and Railway Stations, and this alone will effect a saving greater than any possible reduction in the Liverpool Dock Dues, or in Railway Charges.

There will be tidal gates at the entrance to the Canal, which will be worked as locks at low water, so that large vessels can enter and leave at almost any state of the tide, instead of only during a period of 40 minutes of each tide as at Liverpool. The sill of the largest entrance lock will have ten feet greater depth of water over it than the deepest dock entrance in Liverpool. Small vessels will be able to enter and leave at any time.

Vessels will be able to navigate the Canal with safety at a speed of five miles an hour, and it is estimated that the journey from the entrance at Eastham to Manchester will be accomplished in eight hours, which is much less time than is now taken to cart goods from ship to rail in Liverpool, and to carry them thence by rail to Manchester.

A large Port will be created more than 30 miles nearer to the great manufacturing districts of Lancashire, to the West Riding of Yorkshire, and to Cheshire, Staffordshire, and Derbyshire: and Foreign, as well as Irish, Scotch, and other Coastwise Produce will be delivered direct into the greatest consuming district in the world, so that in this, as in other ways, the Ship Canal will be a great National benefit.

Manchester is already the centre of the Cotton Trade, and has very large Provision, Corn, Cattle and Fish Markets. But large quantities of the Provisions, Corn, and Timber consumed in the Manchester District, and actually sold in the Manchester Markets, are stored at the Ports of Liverpool, Hull, Fleetwood, Grimsby, West Hartlepool, and Goole, and the necessity for this cripples the Trade of the district and of the country, by the cost and time expended in transhipment and in the transit to and from those Ports.

The proportion of the Shipping Trade of the country that, according to population, could be allocated to the Ship Canal *as the nearest Port* is at least 21,000,000 tons per annum, and although, for the Canal to secure a remunerative revenue, it will not be necessary to get more than 3,000,000 tons of cargo, it should be remembered :—

(*a.*)—That the industrial community of which Manchester is the great centre contributes in value fully two-thirds of the entire exports of the United Kingdom.

(*b.*)—That as regards imports, a district which furnishes so considerable a proportion of the exports made largely from imported raw materials, and which is almost wholly dependent for its food supplies upon other parts of the United Kingdom, or the Colonies and Foreign Countries, must of necessity absorb a far greater proportion of the imports and of Coasting traffic than its share *pro rata* averaged according to its proportion of the entire population of the country.

(*c.*)—That the Shipping Trade using the Ports of the United Kingdom amounts to upwards of 3 tons of cargo per head of the population per annum.

The low rates at which the Canal Company will be able to accommodate traffic will make successful competition impracticable, and as a saving of 50 per cent. in Dock Rates and Railway Charges will be effected on all sea-borne goods destined for Manchester, and a proportionate saving on the same class of goods destined for other towns, according to their proximity to the Canal and to other Ports, there will be a material saving to the importer or exporter on every ton of goods brought or sent by the Canal, which is therefore certain to secure a very large trade.

There will inevitably arise along the banks of the Canal many industries in themselves involving traffic, and with the experience gained in the cases of the Clyde, the Tyne, and other improved waterways, such industries may be expected to be of great magnitude, and the Canal will necessarily do a considerable local traffic between different points throughout its course.

The Ship Canal will be for all practical purposes a Dock from end to end, 35 miles in length, and additional Side Basins and Lay-byes can be constructed at any point without dock gates and at small expense, which will enable vessels of the largest tonnage to load and discharge alongside works of any description or at railway sidings.

Bridgewater Undertakings.—

The Act provides that the **Bridgewater Navigation Company** shall sell the whole of the Bridgewater Undertakings to the Company for £1,710,000. These Undertakings consist of—

(a.) " The Bridgewater Canal Undertaking," which includes the Canals known as the Bridgewater Canals and the Runcorn and Weston Canal, also extensive warehouses and wharf accommodation in Manchester, docks and warehouses at Runcorn, and the Duke's Dock and large warehouses in Liverpool; and

(b.) " The Mersey and Irwell Undertaking," which includes the Mersey and Irwell Navigation and the Runcorn and Latchford Canal, also extensive warehouses at Manchester and Warrington, and the old Quay Docks at Runcorn.

The **Mersey and Irwell Navigation** will be absorbed in the construction of the Ship Canal, but the **Bridgewater Canals** form a separate Undertaking, and will continue to be a barge navigation between **Manchester** and **Liverpool** via **Runcorn**, and will constitute a most valuable feeder and distributor of traffic to and from the **Ship Canal**, by connecting it with the **Pottery** and **Iron Districts** of **Staffordshire** (via the **Trent** and **Mersey Canal**), the **Wigan District** and the **South Lancashire Coal Fields** (via **Barton**), and with nearly all the towns engaged in the Cotton Trade, and with **Yorkshire** and **Derbyshire** (via the **Rochdale** and other **Canals**).

For adjustment of values, £460,000 is treated as the cost to the Company of the Mersey and Irwell Undertaking, and the amount, £1,250,000 remaining as the cost of the Bridgewater Undertaking, will make the purchase a good investment, whether looked at separately or as part of the general undertaking of the Company. The average net income of the Bridgewater Undertakings during the last five years has been upwards of £60,000 per annum. It has been derived from the Bridgewater Canals and those other portions of the Bridgewater Undertakings which will continue in full operation notwithstanding the construction of the Ship Canal. The Bridgewater Canal Undertaking is in good working order. For several years a large sum has been annually spent out of income upon permanent improvements, which are now nearly completed.

The Directors are satisfied that the traffic of the Bridgewater Canals will, under the management of the Ship Canal Company, be capable of considerable development, by the removal of the Bar Tolls which obstruct through traffic, and by the opening of the Canal to general carriers.

The purchase of the **Bridgewater Undertakings** will be completed as soon as the necessary Capital has been subscribed, and the immediate receipt of an income will be thereby secured.

Capital Outlay—

Messrs. Lucas & Aird have entered into a contract with the Company to construct the Ship Canal, with the Docks at Manchester, Salford, and Warrington, the Branch Railways, and all the accommodation and incidental works authorised and required by the Company's Act, and to complete the same within four years from their commencement for the sum of £5,750,000

The cost of the Bridgewater Canal Undertaking and the Mersey and Irwell Undertaking is fixed by the Act at 1,710,000

The other Land necessary for the construction of the Works, including the cost of purchase and conveyance, will be according to Estimates carefully prepared and submitted to Parliament 802,936

And the Preliminary Expenses, being the cost of the Act 146,000

£8,408,936

In addition to the authorised Share Capital of £8,000,000, the Company has borrowing powers to the extent of not less than £1,812,000, making the total Authorised Capital £9,812,000, a sum sufficient to enable the Company to complete the construction of the Works, and to pay interest on share and borrowed capital and all expenses incurred during construction.

The Company has power under special Act of Parliament to pay Interest or Dividend out of Capital on the amount from time to time paid up upon the Shares during the time authorised for the completion of the Works, provided the aggregate amount so paid shall not exceed £752,000. This power will enable the Company to pay interest at the rate of £4 per cent. per annum upon all called up capital during the period of four years required for the completion of the Works.

ESTIMATED EXPENSES OF WORKING AND MAINTENANCE.

Ship Canal.—The Suez and Amsterdam Canals afford the best existing data for an estimate, but important differences will exist between them and the Ship Canal, in favour of the latter. Waterways have a great advantage over Railways and other modes of transit, in respect of working expenses, and cost of maintenance, as while the traffic increases, the working expenses remain almost stationary.

It is estimated that the expenses of working and maintaining the Ship Canal, exclusive of the Docks, but including a part of the General Management expenses, will not exceed £2,000 per mile, which would give a maximum total for the Canal of £70,000

Docks.—For the cost of working and maintaining Docks there are numerous examples, but in most cases the cost of handling the traffic is included in the accounts as working expenses. It is estimated that the expenses of working and maintaining the Manchester, Salford, and Warrington Docks, including the remainder of the General Management expenses, but exclusive of all costs attending the handling or warehousing of traffic, which are paid for separately, will not exceed £400 per acre of water space. This rate is a high one for modern docks, and would give a maximum total for the docks of   34,200

Total Estimate for Working and Maintenance of the Ship Canal and Docks £104,200

ESTIMATED TRAFFIC AND REVENUE.

The estimates of Revenue were fully discussed when the Bill was considered before the Select Committees of Parliament, and the following statements are based on the calculations then made :—

A Traffic of only 3,000,000 tons of General Goods would pay :—
In Canal Tolls (half the existing Railway Rates) and in Wharfage (half the existing Dock and Town Dues upon Goods) at least £750,000
With Ship Dues (less than those charged at Liverpool)   75,000
Add average net income of the Bridgewater Canal Undertaking   60,000
£885,000

Deduct—Working Expenses as above £104,200
Interest upon £1,800,000 borrowed Capital, at 4 per cent.   ,..   72,000
176,200

Net Income.....................£708,800

being sufficient, if ship dues are charged, to pay a Dividend of 8 per cent. upon the total Share Capital of the Company; or, if ship dues are not charged, a Dividend of more than 7 per cent. upon the same amount of Capital, and in either case to carry £69,000 per annum to a reserve fund.

This estimate is undoubtedly a small one. It does not include any revenue from coastwise traffic or from low-class traffic, such as coal, salt, and iron ore, for which separate accommodation is provided at different points along the line of the Ship Canal, and a very moderate shipment of such traffic would largely increase the net revenue: nor does it include any revenue from porterage and other services, chargeable to merchants under the Act at actual cost plus 10 per cent. as profit to the Company.

The total traffic in and out of the Port of Liverpool amounts to upwards of 15,000,000 tons of cargo. The amount of traffic included in the above estimate is, in fact, less than the estimated increase during the past five years of the trade of the district which can be served by the Canal, and it is therefore reasonable to expect that before the Canal Works can be completed the increase of the trade of the district will be sufficient to furnish a remunerative traffic to the Ship Canal without removing existing trade from any other port.

The Directors are satisfied that a much larger quantity of traffic will use the Ship Canal, and the traffic expected was described in great detail in the Tables placed before the Parliamentary Committees of 1884 and 1885. They showed a probable traffic in various specified articles, including coastwise, low-class goods, and local trade, at the end of seven years, amounting to 9,649,316 tons, giving a Revenue—

In Canal Tolls and Wharfage of	£1,492,282	
With Dues upon Shipping, less than those charged at Liverpool, of	187,500	
Profit of 10 per cent. upon Porterage, if charged only at an average rate of 6d. per ton ...	24,123	
Average net income of the Bridgewater Canal Undertaking	60,000	
		£1,763,905
Deduct—Working Expenses (say)	£120,000	
Interest as above	72,000	
		192,000
Net Income...................£1,571,905		

sufficient to pay a dividend of 18 per cent. upon the total Share Capital of the Company, and to accumulate a Reserve Fund at the rate of £131,905 per annum.

Applications for Shares may be made on the enclosed form, and must be forwarded, accompanied by a Deposit of £1 per Share to Messrs. N. M. ROTHSCHILD AND SONS, or to any of the Banks named on the other side, from whom Forms of Application may be obtained.

Where no allotment is made the Deposit will be returned without deduction, and where partial allotment is made the surplus Deposit will be applied towards the amount payable on allotment.

The Subscription Lists will be opened on Tuesday, 20th July,
and closed on or before Friday, 23rd July.

LONDON,
    JULY, 1886.

APPLICATIONS FOR SHARES MAY BE MADE ON THE ENCLOSED
FORM TO

MESSRS. N. M. ROTHSCHILD AND SONS, LONDON ;

THE NATIONAL PROVINCIAL BANK OF ENGLAND, LIMITED, AT MANCHESTER ;

Or at any of its Various Branches throughout the Kingdom, or to any of the Following
Banks :—

THE BANK OF ENGLAND.
THE MANCHESTER AND LIVERPOOL DISTRICT BANK, LIMITED.
THE MANCHESTER AND SALFORD BANK, LIMITED.
MESSRS. CUNLIFFES, BROOKS, AND COMPANY.
THE MANCHESTER AND COUNTY BANK, LIMITED.
THE UNION BANK OF MANCHESTER, LIMITED.
THE CONSOLIDATED BANK, LIMITED.
THE LANCASHIRE AND YORKSHIRE BANK, LIMITED.
THE MANCHESTER JOINT STOCK BANK, LIMITED.
PARR'S BANKING COMPANY, LIMITED.
THE OLDHAM JOINT STOCK BANK, LIMITED.
THE BANK OF BOLTON, LIMITED.
THE ASHTON, STALYBRIDGE, AND GLOSSOP BANK, LIMITED.
THE ADELPHI BANK, LIMITED.
MESSRS. JOHN STUART AND COMPANY.
MESSRS. JAMES SEWELL AND NEPHEW.
THE CO-OPERATIVE WHOLESALE SOCIETY, LIMITED.
THE OLD BANK, CHESTER.

APPENDIX VI.

THIRD SHIP CANAL PROSPECTUS, ISSUED JULY, 1887.

MANCHESTER SHIP CANAL COMPANY.

Issue of £4,000,000 Perpetual £5 per cent. Preference Shares of £10 each; but which during the four years required for the construction of the canal will bear Interest at £4 per cent. per annum.

The total Share Capital of the Company is £8,000,000, of which one-half is in Ordinary Shares, and the other half in the Preference Shares now offered. Of the £4,000,000 Ordinary Shares upwards of £3,000,000 have been already subscribed and 20 per cent. paid thereon, a further 20 per cent. will be immediately called, which will make 40 per cent. paid upon the Ordinary Capital. The Contractor has agreed to take £500,000 in paid-up Ordinary Shares in lieu of cash if required by the Company.

MESSRS. BARING BROTHERS & CO., and MESSRS. N. M. ROTHSCHILD & SONS are authorised to receive Subscriptions for the above £4,000,000 Perpetual £5 per cent. Preference Shares of £10 each at par, payable as follows:—

On Application £1 per Share.
Allotment £1 ,,
£2 ,,

The remainder by calls at intervals of not less than three months, and not exceeding £2 per Share.

Subscribers may pay up in full on allotment or on any date fixed for payment of an instalment.

Interest at the rate of 4 per cent. per annum will accrue on the amounts as paid up, and be payable half-yearly in London, on the 1st January and the 1st July. The first payment of interest will be made on the 1st January, 1888.

Faiiure to pay any of the instalments or calls when due will subject previous payments to forfeiture.

Applications for these Preference Shares must be made in the annexed form, and be accompanied by a deposit of £1 per share on the number applied for.

When no allotment is made the deposit will be returned without deduction, and when a partial allotment is made the surplus will be applied towards the payment due on allotment.

Share Certificates will be issued against Letters of Allotment after payment of the £2 per share, and an Agreement has been entered into with the Commissioners of Inland Revenue whereby the Company's Shares are transferable free of stamp duty.

The annexed Memorandum, officially furnished by the Directors, gives particulars of the Undertaking.

A print of the Act of 1887 authorising the issue of the Preference Shares, and of the Minutes of Evidence given before the Parliamentary Committees, can be inspected at the offices of the above-mentioned firms.

Subscription Lists will be opened on Tuesday, the 19th instant, and closed on or before Thursday, the 21st instant.

LONDON, 15*th July*, 1887.

MANCHESTER SHIP CANAL COMPANY.

OFFICIAL MEMORANDUM.

Directors.

The Right Honourable LORD EGERTON OF TATTON (*Chairman*).
Sir JOSEPH C. LEE, Manchester (*Deputy-Chairman*).

W. H. BAILEY, Salford.
HENRY BODDINGTON, Manchester.
JACOB BRIGHT, M.P., Manchester.
J. K. BYTHELL, Manchester.
W. J. CROSSLEY, Manchester.
CHAS. J. GALLOWAY, Manchester.

Sir W. H. HOULDSWORTH, Bart., M.P., Manchester.
JOSEPH LEIGH, Stockport.
CHARLES MOSELEY, Manchester.
JAMES E. PLATT, Oldham.
SAMUEL R. PLATT, D.L., Oldham.
JOHN RYLANDS, Manchester.

Engineer.
E. LEADER WILLIAMS, M.Inst.C.E.

Solicitors.
GRUNDY, KERSHAW, SAXON & SAMSON, Manchester.

Secretary.
A. H. WHITWORTH.

Capital.   The Company was incorporated by Special Act of Parliament, 48 & 49 Victoriæ, cap. 188, with a Share Capital of £8,000,000, and has power to borrow £1,812,000, making the total authorised Capital £9,812,000.

By the Manchester Ship Canal Act, 1886, the Company was authorised to pay interest during construction out of Capital at the rate of £4 per cent. per annum (not exceeding in the aggregate £752,000). This power will enable the Company to pay interest at £4 per cent. per annum upon all Capital paid up during the period of four years required for completion of the works.

By the Company's Act of 1887, authority was given to divide the Share Capital into £4,000,000 of Ordinary Shares and £4,000,000 Preference Shares, bearing a preferential dividend at the fixed rate of £5 per cent. per annum, payable out of the profit of each year in priority to any dividend for that year on the Ordinary Shares.

Of the £4,000,000 Ordinary Shares upwards of £3,000,000 have been already subscribed and 20 per cent. paid up thereon, a further 20 per cent. will be immediately called up, which will make 40 per cent. paid up upon the Ordinary Capital. The Contractor has agreed to take £500,000 in paid-up Ordinary Shares in lieu of cash if required by the Company.

The Company has been formed for the purpose of constructing a Ship Canal 35 miles Works. in length, and with a bottom width of 120 feet, from the River Mersey at Eastham, near Liverpool, past the existing docks at Ellesmere Port, Weston Point, and Runcorn, to Warrington, Salford, and Manchester, available for the largest class of ocean vessels, with docks at Manchester, Salford, and Warrington, and other incidental works ; and to purchase Purchase of the entire undertakings of the existing Bridgewater Navigation Company, Limited, including Bridgewater, the Bridgewater Canals, the Runcorn and Weston Canal, the Mersey and Irwell Navigation, Irwell, and the Runcorn Docks, the Duke's Dock in Liverpool, and all that Company's warehouses, other Canals. wharves, buildings, lands, rents, rights, and privileges, as a going concern.

The price fixed by the Act for the undertakings of the Bridgewater Navigation Company is £1,710,000. The average net income derived from them during the last five years has been nearly £60,000 per annum, and has for the most part been obtained from the Bridgewater Canals and other portions of the undertakings, which will continue in operation notwithstanding the construction of the Manchester Ship Canal. The Directors are satisfied that the traffic will increase, and that the purchase will be profitable to this Company.

A contract has been made with Mr. T. A. Walker, of Westminster, for the construction of the Ship Canal, with docks at Manchester, Salford, and Warrington, the Branch Railways, Contract and and the accommodation and incidental works authorised and required by the Company's cost of Works. Act and the completion of the same within four years from the commencement, for the sum of £5,750,000

The cost of the Bridgewater Canal undertaking and the Mersey and Irwell undertaking is fixed by the Act at . . . 1,710,000

The cost of the other land required for the construction of the Company's works, including the expenses of purchase and conveyance, will be according to estimates carefully prepared and submitted to Parliament 802,936

The preliminary expenses, the preparation of plans and working drawings, and engineering and other expenses to date . . 200,000

£8,462,936

The total authorised Capital being £9,812,000, an ample margin remains available for the provision of Graving Dock accommodation, payment of interest, and all expenses during construction.

Careful estimates of the traffic, and revenue which the Canal will secure as soon as completed, were submitted to Parliament, and showed a revenue, including the income of the Bridgewater undertakings, sufficient to pay a dividend of 7 per cent. on the total Share Capital of the Company.

Estimates for subsequent years were also submitted, showing a much larger revenue.

The above-mentioned estimates having been questioned, the whole project was, in the autumn of 1886, referred to an independent Committee of twenty-three merchants and others acquainted with the trade and commerce of the district, who entered upon the inquiry at the request of the Mayor of Manchester.

This Committee in their Report, dated the 26th of October, 1886, state that not one of them entered upon the inquiry with his mind made up, and that many of them did so with opinions adverse rather than favourable to the commercial prospects of the scheme, and that they had unanimously arrived at the following conclusions :—

1. That the Ship Canal and Works are practicable from an engineering point of view, and would permit vessels of the largest class to be safely navigated to and from Manchester.

2. That the Canal and Works can be completed ready for traffic at a cost within the estimate of £5,750,000, and that the sum of £802,936 set down for the purchase of the necessary land is a safe estimate.

3. That the Canal and Works should be constructed under a contract fixing a maximum sum, in order to prevent any possibility of the estimate being exceeded.

4. That Graving Dock accommodation should be provided, and be ready for use on the opening of the Canal.

5. That the estimate of £104,200 per annum for the expenses of the working and maintenance is ample. There would probably be a material saving on this item during the first few years after completion.

6. That the acquisition of the Bridgewater undertakings for the sum of £1,710,0) fixed by the Act of Parliament would be an advantageous purchase for the Ship Canal Company, and that the present average net income of £60,000 would, af er the completion of the Works, continue to be derived from the Bridgewater Ca al and those other portions of the Bridgewater properties not required for the const ction of the Ship Canal.

7. That the capital powers of the Company, under their Acts of Parliament, amountin to £9,812,000, are sufficient for all the purposes contemplated by their Acts.

8. That the project is a thoroughly sound commercial undertaking, and would speedily become remunerative on the completion of the Works. That a large amount of traffic would be at once secured, and that thereafter the increase in traffic and revenue must be steady and continuous.

The Committee further reported that they had fully investigated the Directors' statements of traffic and revenue, adding—

> After investigating those statements very fully, we have arrived at our conclusions on an independent basis. Including local, coastwise, and heavy traffics, some of which will, we believe, be at once obtained, and also including considerable quantities of traffic which would be new to the district, we are of opinion that during the second year after the Canal and Docks are open for traffic there is a reasonable prospect of securing along the whole length of thirty-five miles of Canal—

4,428,532 tons, yielding a gross income of		£794,173
We deduct 25 per cent. for contingencies, say		198,543
Leaving		£595,630
We add the Bridgewater Canal net revenue		60,000
Making the revenue		£655,630
From which we deduct interest on £1,800,000 debentures	£72,000	
Working expenses (which we consider to be over-estimated)	104,200	
		£176,200
And we thus arrive at a net revenue for the second year of		£479,430

> This sum, which we consider a safe estimate, would be sufficient to pay a 5 per cent. dividend upon the whole Share Capital of the Company (£8,000,000), and to leave a surplus of £79,430.
>
> Our estimate of traffic and revenue for subsequent years is much larger, and points to the undertaking becoming increasingly remunerative under capable administration.

The Directors consider that the Committee's estimate of the traffic of the Canal confirms those submitted to Parliament. They do not see any necessity for the deduction made to meet contingencies, as they feel confident that the full estimated revenue will be realised in the second, if not in the first year's working, which will enable the Company to pay a dividend of 8 per cent. per annum, besides placing a substantial sum to a reserve fund.

DEANSGATE, MANCHESTER, 13th July, 1887.

APPENDIX VII.

COMPENDIUM OF GENERAL INFORMATION CARRIED DOWN TO 1906.

The entrance to the Ship Canal at Eastham is 19 miles from the bar at the mouth of the river Mersey, and the access is from the sea *via* the lower estuary.

The access channel has been excavated to a depth of 20 feet below Old Dock Sill, Liverpool, so as to give access to the Ship Canal at any state of the tide.

The Ship Canal is 35½ miles in length, and the principal Docks are at Manchester, a distance of 50 miles from the sea.

The Ship Canal between the entrance at Eastham and the Docks at Manchester will shortly be excavated throughout to a depth of 28 feet, which depth is maintained by dredging, but in the tidal portion of the Ship Canal between Eastham and Latchford (21 miles) the available depth varies from 26 feet to 33 feet, according to the state of tide.

The bottom width at the full depth is 120 feet, with the following exceptions:—

(*a*) At the curve at the Weaver Outfall the width at the full depth is 180 feet. At the bend at Runcorn, approaching the Runcorn Railway Bridge, the width is 150 feet.

(*b*) For a part of the length between Latchford Locks and Partington Coaling Basin, *i.e.*, from Warburton Bridge to the upper end of Millbank Wharf (about three-quarters of a mile in all), the bottom width is at present only 90 feet, and large vessels are not allowed to pass each other on that portion of the Ship Canal.

(*c*) From Barton Aqueduct to the Manchester Docks the bottom width is 170 feet.

The tidal portion of the Ship Canal from Eastham to Latchford Locks (21 miles) is maintained 14 feet 2 inches above Old Dock Sill, Liverpool (or 9 feet 6 inches above Ordnance Datum, *i.e.*, mean sea level), giving a depth of 26 feet of water.

The three locks at Eastham form three separate entrances, which are open to estuary level whenever the tide rises more than 14 feet 2 inches above Old Dock Sill. When the tide is below this level, access is obtained by means of the locks. The large lock is 600 feet by 80 feet, the intermediate lock is 350 feet by 50 feet, and the small lock is 150 feet by 30 feet. A vessel can be passed through the largest lock in eight minutes or less. Width of Ship Canal at Eastham Locks, 315 feet.

The Lower Sill of the large lock at Eastham is 23 feet below Old Dock Sill, the Upper Sills of all the large locks being 28 feet below normal water level. This will enable the

(300)

depth of water in the Ship Canal to be made 28 feet throughout, and the works for this deepening are in progress.

Eastham Lay-bye—1,450 feet long, with a depth of 26 feet of water alongside, and provided with electrically driven sheer legs, capable of lifting 15 tons to a height of 105 feet above water level at 30 feet 6 inches from line of dolphins.

Mount Manisty, on the north side of the Ship Canal, was formed by the material excavated from the rock cutting during the construction of the Ship Canal, and is available for ballasting vessels.

Ellesmere Port is the terminus of the Shropshire Union Railways and Canal Company whose navigations extend to Chester, and through Shropshire, also by junctions with other canals into North and South Staffordshire, etc. In addition to their docks, the Shropshire Union Company have constructed a long length of wharfage facing the Ship Canal, which has 20 and 24 feet of water alongside.

Near this point are the shipbuilding and repairing yards of the **Manchester Dry Docks Company, Limited,** There is also a Pontoon Floating Dock, 350 feet long and 70 feet wide, capable of lifting vessels up to 2,500 tons weight. The depth of water at the entrance is 16 feet.

The Smelting Corporation, Limited, have erected on the south bank of the Ship Canal at Stanlow (near Ellesmere Port) large smelting works, alongside of which there is a deep water **Lay-bye** and **Wharf** on the Ship Canal available for ocean-going vessels. These works are connected with the London and North-Western and Great Western Joint Railways by means of the Ship Canal Company's Railway which runs from Ellesmere Port Station to the Works.

River Gowy—Passes under the Ship Canal into the river Mersey through two cast-iron syphons, each 12 feet in diameter.

Weston Marsh Lock—229 feet by 42 feet 8 inches. Access from the Ship Canal to the Weaver Navigation—constructed at the cost of the Canal Company, but manned and worked by the Weaver Navigation Trustees at their cost.

Weaver Sluices—Ten sluices, each 30 feet wide with 13 feet lift.

Weston Mersey Lock—600 feet by 45 feet. Access from the river Mersey to the Ship Canal for the free passage of traffic across the Ship Canal between the estuary and the Weston Point Docks, of the River Weaver Navigation Trustees.

Bridgewater Lock—400 feet long by 45 feet wide. Access from the river Mersey to the Ship Canal for the free passage of traffic across the Ship Canal between the estuary and Runcorn Docks, owned by the Ship Canal Company.

Runcorn Lay-bye—This lay-bye abuts on the Ship Canal at Runcorn immediately below Runcorn Docks. It is 1,500 feet in length with a depth of 26 feet of water alongside, and is fully equipped with coal-tip, movable cranes, and other appliances for loading and discharging cargo. Sailing vessels whose lower masts after striking top masts are too high to enable them to pass under the fixed bridges are berthed at this lay-bye, and their cargoes are discharged overside and lightered to Manchester without extra cost to the importer beyond the Ship Canal toll.

Runcorn Docks, belonging to the Manchester Ship Canal Company, cover an area of 70 acres, of which the water space is 15 acres, and there is storage ground in addition covering 77 acres. The Docks include the Tidal Dock, the Alfred, Fenton, Arnold, Francis, and Old Docks, and are chiefly used for pottery materials, timber, roadstone, pig iron, grain, etc.; coal, salt, pitch, hardware and earthenware are the chief exports. The Bridgewater Canal runs inland from Runcorn, where the L. & N.-W. Railway is in direct communication with the Company's Railways. The Docks are well equipped with sidings, coal-tips, salt shoots, hydraulic and floating cranes, and other dock appliances. There is also a small Graving Dock belonging to the Ship Canal Company for the repair of their own, and other small craft by arrangement.

Bridgewater House—Built by the first Duke of Bridgewater and occupied by him as a residence during the construction of the Bridgewater Canal and Docks at Runcorn.

Runcorn Bridge—Carrying the L. & N.-W. Railway over the river Mersey and Ship Canal. Headway, 78 feet above high water spring tides.

Runcorn (Old Quay) Lock—250 feet by 45 feet. Access from the river Mersey to the Ship Canal for barge traffic.

Runcorn (Old Quay) Swing Bridge—Weight of steel work, 650 tons; span, 120 feet.

United Alkali Co.'s Wharves—Each 600 feet long, with a depth of water alongside of 26 feet and 12 feet respectively.

Old Randles Sluices—Two sluices, each 30 feet wide with a lift of 13 feet.

Vyrnwy Subway—An ellipse, 14 feet by 8 feet, conveying the fresh water supply of the City of Liverpool under the Ship Canal.

Moore Lane Swing Bridge—Weight of steel work, 790 tons; span, 120 feet.

Haydock Coal Shipping Wharf—Messrs. Richard Evans & Company, Limited. Acton Grange, near Warrington, 340 feet long, 40 feet wide, with depth of 22 feet of water alongside.

Deviation Railways (Nos. 1 and 2)—Viaduct, weight of steel work, 1,946 tons; span 264 feet. L. & N.-W. and G. W. main line. Headway, 75 feet.

Stag Inn Swing Bridge—Weight of steel work, 790 tons; span, 120 feet.

Warrington Lay-bye, where large vessels can be loaded and discharged; 300 feet long, with a depth of 26 feet of water alongside. The equipment includes a movable hydraulic transporter and a steam crane, each of a capacity of 30 cwt., and one locomotive. Warehouse accommodation has also been provided.

Walton Lock—150 feet by 30 feet. Access from the river Mersey to the Ship Canal for Warrington traffic (19 miles from Eastham).

Warrington Borate Co., Ld., Works—This company's works are connected with the L. &. N-W. Railway at Walton Junction by means of the Ship Canal Railways.

Northwich Road Swing Bridge—Weight of steel work, 1,350 tons; span, 120 feet.

Twenty Steps Lock—85 feet by 19 feet 6 inches. Access from the river Mersey above Warrington.

Latchford High Level Bridge (Cantilever)—Weight of steel work, 783 tons; centre span, 206 feet.

Knutsford Road Swing Bridge—Weight of steel work 1,350 tons; span, 120 feet.

Deviation Railway (No. 3)—Viaduct, weight of steel work, 1,220 tons; span, 250 feet. L. & N.-W. Railway, Warrington and Stockport.

Latchford Locks—Large lock, 600 feet by 65 feet; small lock, 350 feet by 45 feet. Rise, 16 feet 6 inches. Width of Ship Canal at the locks, 290 feet. Latchford Railway Station is about half a mile from Latchford Locks. Electric car service between Warrington and Knutsford Road Swing Bridge at short intervals.

Statham Wharf and Brick Yard (Ship Canal Brick and Tile Company).

Rixton Junction with Mersey and Irwell Navigation; above this point the Ship Canal is a canalised river.

Warburton High Level Bridge (Cantilever)—Weight of steel work, 783 tons; centre span, 206 feet.

Lancashire Patent Fuel Co., Ltd.—Works for the manufacture of Patent Briquettes for both foreign and home markets.

Deviation Railway (No. 4)—Viaduct, weight of steel work, 494 tons; span, 137 feet. Cheshire Lines, Glazebrook and Stockport.

Partington Coaling Basin—Quay space, 20 acres; length, nearly three-quarters of a mile; water space, 6½ acres; and over 19½ miles of railway sidings. There is accommodation for seven tips, six of which are constructed, and in use. They are fitted with hydraulic machinery and all the latest improvements for loading coal; about 160 tons per hour can be loaded from each tip. The equipment includes seven locomotives and two steam cranes, and there is direct railway communication with both the South Yorkshire and Lancashire coalfields. Wharves are also provided for mining, timber, chemicals, etc.

Mersey Weir—Junction of river Mersey with Ship Canal. Two weirs, each 100 feet long.

Deviation Railway (No. 5)—Viaduct, weight of steel work, 550 tons; span, 149 feet. Cheshire lines, Manchester and Liverpool.

Irlam Wharf—300 feet long, 50 feet wide, with 26 feet of water. The Co-operative Wholesale Society's Soap and Candle Works at this point are connected with the Cheshire Lines Railway, at Irlam and Glazebrook by means of the Ship Canal Company's Railways.

Irlam Locks—Large lock, 600 feet by 65 feet; small lock, 350 feet by 45 feet. Rise, 16 feet. Width of Ship Canal at the locks, 370 feet.

Boysnope Wharf and Lay-bye—North side of Ship Canal (6¾ miles from Manchester). Used by the Manchester Corporation in connection with Chat Moss Sewage Farm.

Barton Locks—Same dimensions as Irlam. Rise, 15 feet. Width of Ship Canal at the Locks, 330 feet.

Barton (Road) Swing Bridge—Weight of steel work, 640 tons; span, 90 feet.

Barton Aqueduct (carrying the Bridgewater Canal over the Ship Canal)—235 feet long, 6 feet deep, 18 feet wide; span, 90 feet; weight of swinging span, 1,450 tons.

Irwell Park Wharf, Eccles—1,000 feet long, with 26 feet of water, for the discharge of pit props, timber, etc., consigned by railway to inland towns.

Oil Tanks—The Consolidated Petroleum Company, Limited, have at Eccles five tanks, capacity 22,000 tons (6,160,000 gallons); at Mode Wheel the Anglo-American Oil Company, Limited, have eight tanks, capacity 11,348 tons (3,177,500 gallons); the Liverpool Storage Company, Limited, six tanks, capacity 10,000 tons (2,800,000 gallons); the Manchester Corporation, two tanks, capacity 4,016 tons (1,044,171 gallons); at Weaste the General Petroleum Company, Limited, have four tanks, capacity 8,600 tons (2,460,075 gallons); in Trafford Park, which is alongside the Ship Canal, the General Petroleum Company, Limited, have two tanks, capacity 7,000 tons (2,059,612 gallons); the Homelight Oil Company, Limited, two tanks, capacity, 8,000 tons (2,200,000 gallons); and the General Oil Storage, Company, Limited, one tank, capacity 8,000 tons (2,240,000 gallons).   The total tankage capacity at, or adjacent to, the Docks is now 78,964 tons (22,141,358 gallons).

Oil is conveyed to each of the above depôts through pipes direct from vessel to tanks, whence it can be reloaded into carts, barges, or railway waggons.

Coaling Crane—Messrs. Andrew Knowles & Sons, Limited, have on the north side of the canal a steam crane of 25 tons capacity for loading coal, the crane lifting the waggons, and tipping direct into vessel alongside.*

Coastwise Cattle Wharves—Cattle pens and wharves for the accommodation of the Coastwise Cattle Trade have been provided by the Ship Canal Company at the Manchester Docks, with pens 252 feet long, divided into three sorting yards 84 feet by 15 feet 6 inches, and six pens 42 feet by 10 feet.   Cattle can be transferred (day and night) direct from steamer through the sorting yards and pens into railway trucks, or be driven by road to the adjacent Salford Cattle Market.

Weaste Wharf, connected with the Salford Sewage Works at Mode Wheel, is 150 feet long, 50 feet wide, and has 26 feet depth of water alongside.

The **Union Cold Storage Company, Limited,** have a large refrigerator store, below Mode Wheel Locks, capable of holding 175,000 carcases, and specially constructed for the accommodation of the Australasian frozen meat trade.   A wharf for large steamers, with railway connections has been provided for the berthage of steamers discharging at the store. This company has also a cold air store with a capacity of 80,000 carcases in Miller Street, City.

Manchester Corporation Cold Air Stores—In addition to the storage facilities for frozen meat, etc., referred to in the preceding paragraph, excellent cold storage is provided by the Manchester Corporation in a central position about a mile and a half from the Docks. These cold stores form a valuable adjunct to the extensive meat markets and abattoirs of the Corporation, and are constructed to accommodate 120,000 carcases of sheep.

Manchester Corporation Lairages and Foreign Animals Wharf—The site is twelve acres in extent, with wharfage on the Ship Canal of 800 feet, and a frontage of 850 feet to Trafford Wharf Road.   The Ship Canal at this point is 300 feet wide.   This site is particularly well suited for the purpose, being within 2 miles of the Manchester City Abattoirs and Carcase Market, and having the advantage of direct and convenient approach. These premises enable foreign cattle to be landed and dealt with at Manchester, the centre of the area of consumption.   Rails connecting with the Ship Canal Railways run into the premises, providing facilities for distribution of meat all over the country.

Special meat trains for London and other markets are frequently despatched.

The plant includes equipment for the most expeditious and approved methods of landing and sorting cattle by day or night, machinery for refrigerating, electric lighting, hydraulic lifts, etc.

Extensions have recently been completed which give a total accommodation for 1,850, cattle and 1,500 sheep, and there is land still reserved to provide, when fully utilised, a total accommodation up to 3,000 head of cattle and 3,000 sheep.

Mode Wheel Locks—600 feet by 65 feet and 350 feet by 45 feet. Rise, 13 feet. Width of Ship Canal below the locks, 320 feet. The total rise from the level of a tide rising 14 feet 2 inches above Old Dock Sill at Liverpool to the level of the water at Manchester is 60 feet 6 inches.

Near Mode Wheel Locks are the **Manchester Dry Docks Company's Works,** fitted for shipbuilding or repairing, a graving dock 475 feet long (now being lengthened to 535 feet) by 65 feet wide, with 22 feet of water on the blocks, which are 4 feet high ; another dry dock is in course of construction and will be 425 by 65 feet, and a floating pontoon 260 feet long by 63 feet wide, with 16 feet of water on the blocks, which are 3 feet 9 inches high, capable of lifting vessels weighing up to 2,000 tons.

Grain Elevator—Storage capacity, 40,000 tons (or 1,500,000 bushels), in 268 separate bins. The following operations can be performed simultaneously :—

(*a*) Discharging from vessels in the Docks at the rate of 350 tons per hour.

(*b*) Weighing in the tower at the water's edge.

(*c*) Conveying to the house and distributing into any of the 268 bins.

(*d*) Moving grain about within the house for changing bins or for delivery, and weighing in bulk at the rate of 500 tons per hour.

(*e*) Sacking grain, weighing, and loading sacks into forty railway waggons and ten carts simultaneously.

(*f*) Conveying from the elevator into barges or coasters at the rate of 150 tons per hour if in bulk, or 250 sacks per hour if bagged.

An important feature in the elevator is **Metcalf's Patent Dryer,** which gives most satisfactory results. The dryer is capable of drying 50 tons of grain at each operation, and grain can be moved to or from the dryer from or to any bin in the house. **A powerful pneumatic apparatus** capable of discharging 200 tons per hour from ships into the elevator is provided to supplement the foregoing appliances.

Trafford Wharf—Half a mile long, with 26 feet of water.

Co-operative Wholesale Society, Limited—Wharf and Warehouse.

The **Manchester Dock Estate** covers an area of 406½ acres, including a water space of 120 acres, and quays 6½ miles in length and 286½ acres in extent. The height of the quay walls is about 8 feet above ordinary water level.

The Dimensions of the Manchester Docks are :—

No. 1	...	700 by 120 feet.	No. 6	...	850 by 225 feet.
„ 2	...	600 „ 150 „	„ 7	...	1160 „ 225 „
„ 3	...	600 „ 150 „	„ 8	...	1340 „ 250 „
„ 4	...	560 „ 150 „	„ 9	...	2700 „ 250 „
„ 5	...	980 „ 750 „			

(Partially Constructed.)

The equipment includes 53 hydraulic, 60 steam and 91 electric cranes, with a radius of from 16 to 40 feet, capable of lifting from 1 to 10 tons to a height from rail level of from 13 feet to 59 feet; a 30-ton steam crane; 43 locomotives; 6 floating pontoons of a dead weight capacity of 800 tons each, and all modern appliances for giving vessels quick despatch.

There is also a pontoon sheers capable of dealing with weights up to 250 tons, with a lift of 21 feet.

There is a range of thirteen single floor, one two-floor, six three-floor, five four-floor and twelve five-floor transit sheds, fitted with the most modern appliances, including a cold, transit shed, for the sorting of frozen meat and other produce; also thirteen warehouses, seven storeys each, fitted with 27 friction hoists worked by gas engine; and in Trafford Park the Ship Canal Company have four single-floor warehouses, each 300 feet by 100 feet. The docks, quays, sheds, and warehouses are lighted by electricity, and there are 34 hydraulic and 16 electric capstans on the quays. Bonded accommodation is also provided.

"The Gamewell Fire Alarm" system has been established at the Docks, by means of which the Manchester and Salford Fire Brigades and the Manchester Fire and Salvage Boat *Firefly*, capable of throwing 4,000 gallons of water per minute, can be promptly summoned to any part of the Dock Estate on the outbreak of a fire. The Ship Canal Company also maintain at the Docks and at the grain elevator an efficient Fire Brigade fully equipped.

Trafford Road Swing Bridge—Weight of steel work, 1,800 tons; 211 feet long, 30 feet deep, and 50 feet wide.

DIMENSIONS OF LOCKS.

	Small.	Intermediate.	Large.	Rise.
Eastham	150 by 30 feet.	350 by 50 feet.	600 by 80 feet.	Ft. Ins.
Latchford		350 „ 45 „	600 „ 65 „	16 6
Irlam		350 „ 45 „	600 „ 65 „	16 0
Barton		350 „ 45 „	600 „ 65 „	15 0
Mode Wheel (Manchester)		350 „ 45 „	600 „ 65 „	13 0

BRIDGES.

The height from the normal water level in the Ship Canal to the underside of the girders in the seven fixed bridges over the Ship Canal leaves a clear headway of 75 feet, except the Runcorn Bridge, which is 84 feet 4 inches from the normal water level. As, however, the headway is necessarily a few feet less when high tides or floods occur, to avoid detention masts should clear the bridges at 70 feet above the water level.

RAILWAYS.

The Manchester Dock Railways are 61 miles in extent, and completely intersect the Dock Estate. The total length of railways already completed at the Docks and at many points alongside the Ship Canal is upwards of 128½ miles. The following railways connect with the docks: London and North-Western, Lancashire and Yorkshire, Great Northern, Midland, Great Central and Cheshire Lines.

The Ship Canal Company's Railways alongside the Ship Canal between Manchester and Eastham are connected with the railways of other companies as follows: With the Cheshire Lines at Irlam and Glazebrook; with the London and North-Western at Latchford and Runcorn Docks; with the joint line of the London and North-Western and Great Western Companies at Walton Old Junction (near Warrington) and Ellesmere Port.

Traffic can be conveyed in railway waggons between the various loading and discharging berths at the docks and other places on the Ship Canal, and over the above lines to every railway station in Great Britain.

BARGE CANALS.

The following canals enable direct communication by water to be maintained between the Ship Canal and all the inland navigations of the country:—

Bridgewater.	Huddersfield.	Aire and Calder.
Leeds and Liverpool.	Stockport.	Trent and Mersey.
Bolton and Bury.	Macclesfield.	Weaver Navigation.
Rochdale.	Calder and Hebble.	Shropshire Union.
Ashton.	Peak Forest.	

TRAFFORD PARK ESTATE.

1,183 acres in extent—adjoins the Manchester Docks, and has a frontage to the Ship Canal on the south side of about 3 miles. The estate has easy access to the Ship Canal and Bridgewater Canal and with the Dock Railways, which are connected with the entire railway system of the country.

Capital Powers and Expenditure of the Manchester Ship Canal Company as at 30th June, 1906.

CAPITAL POWERS.

[1] Ordinary Shares, £10 each	£4,000,000
[1] Perpetual 5 Per Cent. Preference Shares, £10 each	4,000,000
[1] Manchester Ship Canal Corporation 3½ Per Cent. Preference Stock ...	988,255
Perpetual 3½ Per Cent. First Mortgage Debentures	1,359,000
4 Per Cent. First Mortgage Debentures (terminable 1914)	453,000
4 Per Cent. Second Mortgage Debentures (terminable 1914) ...	600,000
Debenture Stock under Act of 1904	2,000,000
New Mortgage Debentures (held by the Corporation of Manchester) ...	5,000,000
Mortgage of Surplus Lands	100,000
	£18,500,255

[1] NOTE.—Order of priority.—It is provided by Section 12 of the Manchester Ship Canal (Finance) Act, 1904, that " All profits of the Company after payment of the dividends on Corporation Preference Shares and Corporation Preference Stock shall notwithstanding anything contained in any of the

The expenditure on Capital Account to 30th June, 1906, was as follows:—

Construction of Works (including Plant and Equipment)	£10,730,926
Bridgewater Canals	1,268,089
Land (Purchase and Compensation)	1,484,794
Engineering and Surveying	184,324
Interest on Share and Loan Capital	1,170,734
Parliamentary Expenses	189,483
General Expenses	419,702
Interest on Debentures discharged by the issue of a like amount of Preference Stock to the Corporation of Manchester	988,255
	£16,436,307

recited Acts or other Acts relating to the Company be divisible as follows:—Two-thirds to the holders of the Preference Shares issued in pursuance of the powers of the Acts of 1885 and 1887; One-third to the Ordinary Shareholders. Provided that when the said two-thirds due to the holders of the Preference Shares issued in pursuance of the powers of the Acts of 1885 and 1887 shall in any year amount to two hundred thousand pounds all the remaining profits of that year shall be payable to the Ordinary Shareholders."

APPENDIX VIII.

MANCHESTER SHIP CANAL ACTS.

1. Manchester Ship Canal Act, 1885.
2. Manchester Ship Canal (Payment of Interest out of Capital) Act, 1886.
3. Manchester Ship Canal (Preference Shares) Act, 1887.
4. Manchester Ship Canal (Additional Lands) Act, 1888.
5. Manchester Ship Canal (Alteration of Works) Act, 1888.
6. Manchester Ship Canal (Tidal Openings, etc.) Act, 1890.
7. Manchester Ship Canal (Various Powers) Act, 1890.
8. Manchester Ship Canal (Additional Capital) Act, 1891.
9. Manchester Ship Canal (Further Powers) Act, 1893.
10. Manchester Ship Canal (Additional Capital, etc.), Act, 1893.
11. Manchester Ship Canal (Surplus Lands) Act, 1893.
12. Manchester Ship Canal (Bridgewater Tariff, etc.) Act, 1894.
13. Manchester Ship Canal (Tidal Openings) Act, 1896.
14. Manchester Ship Canal (Acquisition of Land) Act, 1897.
15. Manchester Ship Canal (General) Act, 1900.
16. Manchester Ship Canal (Finance) Act, 1904.
17. Manchester Ship Canal (General) Act, 1904.

APPENDIX IX.

CHRONOLOGICAL ORDER OF EVENTS.

First mention of a Ship Canal from Manchester to the sea ... ... 1710.

Application made to Parliament to cut a Canal *via* Lymm and the Wirrall Peninsula to the Dee Estuary. Application refused ... ... 1825.

Sir John Rennie, the Engineer, reported to a Warrington Committee on alternative proposals for a Ship Canal from Warrington to Liverpool which could be extended to Manchester ... ... 1838.

Report of Engineers to the Mersey and Irwell Navigation Company as to practicability of improving waterway so as to accommodate large vessels ... ... 1840.

Public discussion, which lasted four days, in Manchester Old Exchange with reference to a Ship Canal proposal ... ... 1841.

Letter from Mr. George Hicks to the *Manchester Guardian* suggesting the Rivers Mersey and Irwell should be converted into a Ship Canal ... ... 11th Oct., 1876.

Petition to the Manchester Chamber of Commerce to inquire into the possibility of making the River Mersey navigable ... ... 1877.

Meeting of the Chamber of Commerce when Mr. Fulton, C.E., explained his scheme for an improved waterway from the sea to Manchester ... ... 23rd April, 1877.

Meeting at Mr. Clement Walmsley's office where a resolution favourable to a Ship Canal was adopted ... ... 16th Oct., 1879.

Meeting of the Salford Town Council when a motion "To ask Government to appoint a Special Commission *re* Improvement of the Irwell" was defeated ... ... Feb., 1881.

"Mancuniensis" (Mr. James W. Harvey) published his able pamphlet, *viz.*, "Facts and Figures in Favour of a Tidal Navigation to Manchester" ... ... 30th May, 1882.

Meeting convened by Mr. Daniel Adamson at his house, The Towers, Didsbury, attended by upwards of seventy leading merchants, manufacturers and municipal representatives, when it was decided to form a Provisional Committee ... ... 27th June, 1882.

First inspection of the route of the proposed waterway by the Provisional Committee
17th Aug., 1882.

Provisional Committee and subscribers to the Guarantee Fund decided to adopt Mr. Leader Williams' Lock Scheme and to proceed with the undertaking ... 26th Sept., 1882.

Conference with representatives of the various trades in Manchester and Salford, when the project was enthusiastically endorsed ... ... 3rd and 9th Nov., 1882.

General Meeting of Manchester Chamber of Commerce passed resolution cordially approving of scheme ...   .   13th Nov., 1882.

Great Mass Meeting of working men in the Free Trade Hall. Resolutions in support adopted unanimously ...   .   13th Nov., 1882.

Great Meeting of citizens of Manchester, convened by the Mayor (Alderman Baker), in the Town Hall   ...   .   14th Nov., 1882.

Manchester City Council passed resolution giving the project hearty approval and promising strenuous support   ...   .   30th Nov., 1882.

First Parliamentary Plans and Book of Reference deposited   ...   30th Nov., 1882.

Salford Town Council passed resolution expressing approval   ...   6th Dec., 1882.

Great Meeting of Burgesses of Salford, convened by the Mayor (Alderman Husband), in the Town Hall   ...   .   8th Dec., 1882.

First Bill deposited in Parliament   ...   .   15th Dec., 1882.

Conference of the Mayors and Town Clerks of the boroughs interested in the project of a Ship Canal ...   .   20th Dec., 1882.

Parliamentary Deposit (£229,905) paid   ...   .   15th Jan., 1883.

Alleged Non-Compliance with Standing Orders of Parliament   ...   19th Jan., 1883.

Standing Orders Suspended   ...   .   5th Mar., 1883

[It is worthy of special notice, as illustrating the force and extent of public sympathy, that within a few day 326 petitions were presented by Municipalities, Chambers of Commerce, Limited Companies, etc., including one from the Associated Chambers of Commerce, and one signed by 187,340 inhabitants and ratepayers of Manchester and surrounding districts.]

First Bill passed by Committee of House of Commons   ...   ...   6th July, 1883.

Rejected by Committee of House of Lords   ...   .   9th Aug., 1883.

Provisional Committee and subscribers unanimously decided to renew the application to Parliament in the ensuing Session ...   .   28th Aug., 1883.

Salford Town Council passed resolution expressing regret at failure of first Bill, and congratulating promoters upon decision to make further application   ...   31st Oct., 1883.

Great Public Meeting in the Free Trade Hall in support   ...   ...   31st Oct., 1883.

Similar Meeting in Salford Town Hall ...   .   30th Nov., 1883.

Second Bill deposited ...   .   21st Dec., 1883.

Second Bill passed by Committee of House of Lords ...   .   24th May, 1884.

Great Trades Demonstration at Pomona   ...   .   21st June, 1884.

Second Bill rejected by Committee of House of Commons ...   ...   1st Aug., 1884.

Provisional Committee and the subscribers decided to make a third application to Parliament ...   .   15th Aug., 1884.

Great Public Meeting in support in Free Trade Hall   ...   ...   15th Aug., 1884.

Resolution passed at Town's Meeting, convened by the Mayor of Manchester (Alderman Goldschmidt), consenting to the Corporation becoming Joint Promoters of the Bill, and contributing the proceeds of a 2d. rate   ...   .   6th Oct., 1884.

Similar resolution passed by Salford Town Council ...   .   9th Oct., 1884.

Confirmed by Town's Meeting at Salford convened by Mayor (Alderman Makinson) 28th Oct., 1884.

Similar resolution passed at Town's Meeting at Warrington, convened by Mayor (Alderman Harrison) ... ... 23th Oct., 1884.

Third Bill deposited ... ... 16th Dec., 1884.

Passed by Committee of House of Lords ... ... 7th May, 1885.

And by Committee of House of Commons ... ... 3rd Aug., 1885.

Reported to House of Commons ... ... 3rd Aug., 1885.

Read Third Time ... ... 5th Aug., 1885.

Received Royal Assent ... ... 6th Aug., 1885.

Great Meeting of subscribers in Manchester Town Hall ... ... 19th Aug., 1885.

Great Demonstration at Eccles ... ... 31st Aug., 1885.

Trades Procession to Belle Vue, and Demonstration ... ... 3rd Oct., 1885.

Great Meeting of subscribers and friends in the Free Trade Hall ... 5th Oct., 1885.

Banquet by the Mayor (Alderman Sir John Harwood) and Corporation of Manchester to celebrate passing of Act ... ... 6th Oct., 1885.

Preliminary Prospectus issued ... ... 8th Oct., 1885.

Salford Town Council resolved to subscribe £250,000 of capital (and deposited, in November, a Bill for that purpose) ... ... 15th Oct., 1885.

Town's Meeting in Salford endorsed action of Council ... ... 21st Jan., 1885.

First Meeting of Shareholders ... ... 1st Feb., 1886.

£20,000 deposited in Bank of England on account of purchase of Bridgewater undertaking ... ... 2nd Feb., 1886.

Poll of Salford ratepayers in support of the resolution to subscribe £250,000—For, 16,653; Against, 2,443 ... ... 6th Mar., 1886.

Prospectus issued by Messrs. Rothschild & Sons ... ... 20th July, 1886.

Report of Consultative Committee presented after five weeks' inquiry 26th Nov., 1886.

Contract for the whole of the works let to Mr. T. A. Walker ... 8th June, 1887.

Prospectus offering 400,000 £10 Preference Shares, issued by Messrs. Baring Bros. and Messrs. N. M. Rothschild & Sons ... ... 15th July, 1887.

Bridgewater Canals Undertaking purchased for £1,710,000 ... 3rd Aug., 1887.

Board of Trade Certificate, that two-thirds of the share capital had been issued and accepted, signed ... ... 4th Aug., 1887.

Announcement made that sufficient of the ordinary capital had been subscribed to enable the Directors to issue £4,000,000 of Preference Shares already guaranteed on condition that £3,000,000 of the Ordinary Shares were first taken up; thus providing £7,312,360 of capital and fulfilling the requirements of the Act one day within the limit of the time prescribed ... 6th Aug., 1887.

First Sod cut by the Chairman of the Ship Canal Company (Lord Egerton of Tatton), at Eastham ... ... 11th Nov., 1887.

Death of the Contractor (Mr. T. A. Walker) ... ... 25th Nov., 1889.

Contract with the Exors. of the late Mr. T. A. Walker determined, and works and plant taken over by the Company ... ... 24th Nov., 1890.

Eleventh Half-yearly Meeting of Shareholders held at Concert Hall. Favourable attitude of City Council announced. Report and proceedings unanimously approved by the Shareholders ... ... 3rd Feb., 1891.

Special Meeting of General Purposes Committee of the Manchester City Council to receive "a report from the Mayor (Alderman Mark) upon a subject of great public importance," *viz.*, financial aid to the Company, when the appointment of a Special Committee was unanimously recommended ... ... 3rd Feb., 1891.

This course approved by the Council, and Committee appointed ... 4th Feb., 1891.

First Meeting of Committee ... ... 6th Feb., 1891.

Report of Special Committee of Manchester Corporation presented to Council at Special Meeting, recommending that the pecuniary assistance requisite to complete the undertaking should be rendered, and unanimously adopted amid applause ... ... 9th March, 1891.

Manchester Corporation and Manchester Ship Canal Bills, providing for loan of £3,000,000 by the Corporation, deposited in Parliament ... ... 17th April, 1891.

Water first admitted into the Manchester Ship Canal from the estuary at Ellesmere Port 19th June, 1891.

Water admitted into Eastham Locks ... ... 2nd July, 1891.

Traffic to Ellesmere Port *via* Ship Canal commenced ... ... 16th July, 1891.

Manchester Corporation Bill and Manchester Ship Canal Bill received the Royal Assent 28th July, 1891.

First Issue of Manchester Corporation Stock (£1,500,000) made through the Bank of England ... ... 30th July, 1891.

Five Corporation Directors and Engineer appointed by Manchester City Council, in pursuance of Act of 1891 ... ... 5th Aug., 1891.

Ship Canal opened to Weston Marsh Lock for the accommodation of the River Weaver traffic ... ... 28th Sept., 1891.

[Runcorn Section, 3½ miles, then proceeded with under the superintendence of Mr. E. D. Jones, the Company's Agent.]

Revised Estimates (1st Sept., 1891) submitted to Special Committee of Manchester Corporation; the Mayor (Alderman Leech) in the Chair ... ... 26th Nov., 1891.

Executive Committee, consisting of four Corporation and three Shareholders' Directors, appointed by the Board, with "full power to carry out all works, etc." ... 11th Dec., 1891.

Confirmed by the Shareholders "until the opening of the Canal for traffic to Manchester" ... ... 29th Feb., 1892.

Second Issue of Manchester Corporation Stock (£1,500,000) through the Bank of England ... ... 2nd March, 1892.

Contracts 1 and 2—from Runcorn Old Quay to Thelwell, 8 miles—were let to Mr. John Jackson; and Contracts 3 and 4—from Millbank to Manchester, 10 miles—to Mr. C. J. Wills, in ... ... April and May, 1892.

Saltport established   ...   .   22nd July, 1892.

Further Revised Estimate (1st June, 1892) prepared by the Executive Committee of Directors   ...   .   26th July, 1892.

Special Committee of Manchester Corporation approved Report of Corporation Directors upon the Executive Committee, and recommended the Council to take the measures necessary to enable them to complete the undertaking ...   .   11th Aug., 1892.

Manchester City Council decided to apply to Parliament for powers to advance a further sum, not exceeding £2,000,000, for the purpose of completing the Canal on stipulated terms, and invited other towns to co-operate   ...   .   28th Oct., 1892.

Oldham Town Council considered in Committee a proposal to advance £250,000 to the Manchester Ship Canal Company, and approved, with four dissentients ...   28th Oct., 1892.

Salford Town Council approved (with one dissentient) recommendation of General Purposes Committee to advance £1,000,000 to Ship Canal Company   ...   28th Oct., 1892.

Oldham Town Council rejected (21 to 13) proposal to advance £250,000 to the Ship Canal Company...   ...   .   9th Nov., 1892.

Manchester City Council's resolution to apply to Parliament for powers to advance £2,000,000 confirmed by meeting of citizens   ...   .   16th Nov., 1892.

Town's Meeting convened by Mayor of Salford (Alderman Keevney). Resolution in favour of advance of £1,000,000 to the Ship Canal Company defeated; poll demanded by the Mayor   ...   .   16th Nov., 1892.

Resolution in favour of proposal that Oldham should advance £250,000 to the Ship Canal Company carried, with five dissentients, at Town's Meeting convened by Mayor of Oldham (Alderman Noton); 600 present ...   .   21st Nov., 1892.

Resolution to promote Bill unanimously adopted by Oldham Council ...   .
7th Dec., 1892.

Manchester Corporation and Manchester Ship Canal Bills, providing for further loan of £2,000,000, deposited in Parliament   ...   .   17th Dec., 1892.

Salford Corporation Bill to authorise loan of £1,000,000 to Ship Canal Company deposited in Parliament   ...   .   17th Dec., 1892.

Oldham Corporation Bill to authorise loan of £250,000 to the Ship Canal Company deposited in Parliament   ...   .   17th Dec., 1892.

Salford Poll declared—13,385 in favour of advance of £1,000,000 to the Ship Canal Company, 3,032 against   ...   .   20th Dec., 1892.

Deviation Railway No. 5, at Irlam (Cheshire Lines), opened for goods traffic ...   ...
9th Jan., 1893.

Deviation Railways Nos. 1, 2, 3, at Acton Grange and Latchford (L. & N.-W. and G.W.R.) opened for goods traffic   ...   .   27th Feb., 1893.

Deviation Railway No. 4, at Partington (Cheshire Lines), also opened for goods traffic
27th Feb., 1893.

Salford Corporation and Oldham Corporation Bills to authorise loans to the Ship Canal Company of £1,000,000 and £250,000 respectively, withdrawn...   ...   16th March, 1893.

Deviation Railway No. 5, at Irlam, opened for passenger traffic   ... 27th Mar., 1893.

Manchester Corporation and Manchester Ship Canal Bills, authorising £2,000,000 additional loan capital, received Royal Assent   ... ... 12th May, 1893.

Deviation Railway No. 4, at Partington, opened for passenger traffic 29th May, 1893.

Eleven Directors appointed by Manchester City Council, in pursuance of the Acts of 1893 ... . 7th June, 1893.

Water admitted to Runcorn Section ... ... 9th June, 1893.

Third Issue of Manchester Corporation Stock (£1,500,000) through the Bank of England ... ... 27th June, 1893.

Water admitted into Canal between Runcorn Docks and Old Quay ... 8th July, 1893.

Deviation Railways Nos. 1, 2, 3, at Acton Grange and Latchford (L. & N.-W. and G.W.R.) opened for passenger traffic, thus releasing the last piece of land required to be cut through ... . 9th July, 1893.

Water admitted to the Canal between Runcorn and Latchford ... 17th Nov., 1893.

Canal filled from end to end, 10.30 P.M. ... ... 25th Nov., 1893.

First journey of Directors by water over whole length of Canal ... 7th Dec., 1893.

Visit of representatives of London and Provincial Press ... ... 16th Dec., 1893.

Warrant issued by Lords Commissioners of Her Majesty's Treasury, constituting Manchester a Harbour and Port for Customs purposes (dated 18th Dec.), presented to Lord Mayor by Mr. D. P. Williams, the appointed Collector for the Port ... 22nd Dec., 1893.

Certificate signed by J. E. W. Addison, Esq., Q.C., M.P., Chairman of the Salford Hundred Quarter Sessions, that the Canal and Docks as authorised and defined by the Special Acts are completed and fit for the reception of vessels ... ... 30th Dec., 1893.

Canal opened for traffic to Manchester, when 71 vessels entered Docks 1st Jan., 1894.

Formal opening by Her Late Majesty Queen Victoria ... ... 21st May, 1894.

Bill authorising the purchase of the Manchester race-course and the construction of further dock accommodation thereon received Royal Assent ... ... 25th June, 1900.

Bill for the readjustment of the Company's finances and for raising £2,000,000 additional capital received Royal Assent ... ... 22nd July, 1904.

Bill for increasing the depth of the Ship Canal from 26 feet to 28 feet and other works received Royal Assent ... ... 15th Aug., 1904.

Dock No. 9, half a mile long, with five four-floor ferro-concrete transit sheds constructed on the site of the old Manchester race-course opened by their Majesties King Edward VII. and Queen Alexandra ... ... 13th July, 1905.

APPENDIX X.

THE original intention was to publish the whole of this interesting correspondence, but it proved to be so voluminous that only a *résumé* has been given. The sources from which the information has been gleaned are given in order that readers may refer to them if they desire to do so. No citizen of Liverpool ever offered a more vigorous opposition to the Ship Canal than did Sir William B. Forwood. In the Liverpool Council and Chamber of Commerce, also as a witness before several Parliamentary Committees, he never lost an opportunity of pouring contempt upon the company. Sir William also did his utmost to stimulate the commercial interests of Liverpool to oppose the Ship Canal Bill, and he often chided the merchants for their supineness. By his speeches and by articles in the Press he also tried to stop the necessary flow of capital. This will be exemplified in the following correspondence, which is typical of the newspaper warfare current at the time.

I.—Extracts from Letter of Sir William B. Forwood to the *Liverpool Courier*, 29th May, 1884.

He wrote that Manchester expected to save £100,000 per annum on cotton consumed within fifteen miles of Manchester; 4s. per ton, or £80,000, on manufactures shipped through Liverpool; also to make a saving on imported produce consumed in the above area. She also expected to have a traffic of 3,000,000 tons through the canal.

Sir William stated this was impossible. There was only a total of 2,400,000 tons of traffic in a twelve-mile area round Manchester, and there were already four railways and a canal to do this traffic. It was a fallacy to suppose that ships would not charge a much higher sea freight considering all the dangers, delays and troubles of the canal, and that the advantages to Manchester would be utterly contemptible and insignificant. Liverpool had no objection to Manchester making the canal—"We believe it will be a huge financial fiasco; at the best Manchester could only hope to attract small craft up the canal".

Sir William went on to say that the engineering evidence proved that the proposed canal would render the Liverpool Docks useless except for small vessels, through causing a silting-up of the channel and bar, and that the effect of training walls would invariably be accretion and destruction of tidal capacity. "Does any sane man suppose that the railway companies will stand idly by and allow the canal to take their traffic? A reduction of 3s. per ton in the railway rates of carriage would still leave them a paying trade, but would effectually starve out the canal and reduce it to bankruptcy; and this is the policy the railway would undoubtedly adopt." Further, "At the present moment Manchester goods are conveyed to Calcutta and Bombay, *via* Liverpool, at a lower charge than they can be carried from Manchester to London, or Southampton ".

II.—EXTRACTS FROM THE SPEECH OF SIR WILLIAM B. FORWOOD TO THE LIVERPOOL CITY COUNCIL, JUNE, 1884.

He did not wish in any sense to exaggerate the injury which might be done to the port of Liverpool if the scheme were carried out, but he had no hesitation in saying, after having listened to the long inquiry (lasting fifty days), and to the evidence of all the most eminent engineers of the day, that if the scheme were carried out—nay, only partially carried out, or he would go further and say if only commenced—at the estuary, it would be fatal not only to the prosperity of the port of Liverpool, but to the whole of this part of the United Kingdom. They had no intention of proving that they were afraid Manchester might take any trade away from Liverpool. They had no such fear, but he thought they demonstrated emphatically that the commercial advantages to Manchester were of a very visionary character. He would like to remind the Council that if even £500,000 were borrowed and spent in the purchase of stone, and that stone was deposited in the upper estuary of the Mersey for training walls, the navigation of the port would be destroyed for ever.

III.—EXTRACTS FROM REPLY BY COUNCILLOR LEECH IN A LETTER TO THE *MANCHESTER EXAMINER AND TIMES*, 9TH JULY, 1884.

He charged Sir William Forwood with trying unnecessarily to alarm the people of Liverpool. If there never would be adequate traffic, if no large foreign ships would come up the canal, and if there was no commercial jealousy, why was he so anxious to prevent the canal being made? After allowing the bar to go from bad to worse, and allowing it to become a danger to navigation; after abstracting 1,200 acres from the estuary and narrowing the river, why was Liverpool now objecting to a small abstraction which the practical engineers of the Clyde, the Tyne, and the Tees (who had successfully done similar work) said would do no harm? Supposing the worst that could happen, and damage was done, would not Manchester suffer? Was it likely she would risk ten millions of money? Liverpool had taxed shipping for the improvement of the river, and had not applied the money for the purpose, but with Manchester as a willing helper there was no doubt the bar and other hindrances to navigation would be removed.

If, as Sir William said, Manchester ought to be satisfied with a ten feet canal, foreign vessels must discharge their cargo into barges either in the docks or in the river, and Mr. Hornby, Chairman of the Dock Trust, had stated in evidence that even goods transhipped in mid-river were liable to Liverpool Dock dues. Before the advent of the Ship Canal, Sir William had advocated carting cotton all the way from Liverpool into Lancashire in order to break down dear railway rates. He then said it could be done at 3d. per ton per mile ; now, when the promoters of the canal had estimated the cost as 5d. per ton per mile, he stated it could not be done under 8d. per ton. A reduction by the railways of 3s. per ton in order to crush the canal would have no terror to manufacturers and traders. It did not matter to them if they saved money on their freights, or got a dividend on their shares. Besides, railway shareholders would tire of sacrificing their dividends by a suicidal policy.

IV.—Extracts from a Letter by "Ship Canal" to the *Manchester Examiner and Times*.

The writer taxed Sir William Forwood and his friends with having declared that money would never be forthcoming for the canal, and then when he wanted to stimulate the citizens of Liverpool in turning round and saying "they must not delude themselves into believing that the capital will not be raised. It will be raised," he said, "What a commentary this is upon all the Liverpool people have been saying these past two years"!

V.—Extracts from a Letter to the *Manchester Guardian* dated 16th June, 1884.

Sir William wrote that the opponents of the canal had been charged with changing their views. Instead of doing injury to Liverpool, he believed it would be an advantage to that city if only it could be constructed without injury to the estuary. "Destroy its régime, and you close the sea entrances to Liverpool, and this I venture to say would be a national calamity. Destroy the bar of the Mersey and you deal a death-blow to the prosperity of the manufacturing industries of England."

"We in Liverpool say we do not object to the Ship Canal, but we do most strongly and earnestly protest against the present scheme, because it will destroy our estuary. Carry the deep-water entrance to Garston, or some other point beyond the estuary, and our opposition disappears."

VI.—Extracts from Letter by "Ship Canal" to the *Manchester Guardian*, 20th June, 1884.

The writer joined issue with Sir William Forwood on the statement that in Liverpool they were actuated by no feeling of rivalry. He quoted a variety of Parliamentary evidence and said: "Indeed I think the promoters of the canal have very reasonable grounds for complaint at the tactics of Sir William Forwood and his friends. He has had his say before three Parliamentary Committees, and has been worsted in the encounter. Parliament has in short said his fears are groundless, and yet he continues to repeat his jeremiads of impending disaster to Liverpool, unsupported by an atom of additional proof other than what has already been advanced by the highest skilled witnesses which money could obtain."

Referring to Sir William's statement that the destruction of the bar meant the destruction of England's industries, he went on to say: "Is it not a fact that every pilot in Liverpool, as well as the captains of the largest liners, have been hoping and wishing for years to see something done to move the bar away? And it is not going too far to say that every one of these pilots, if they were free to express their honest opinions—being now in the service of the Dock Board—would gladly give evidence in favour of the Ship Canal scheme, believing, as they do, that the increased scouring power and velocity of current in the proposed improved channel will immensely benefit the bar."

VII.—EXTRACTS FROM THE LETTER OF SIR WILLIAM B. FORWOOD TO THE *MANCHESTER GUARDIAN*, 24TH JUNE, 1884.

Sir William described the serious injuries to the estuary which he conceived would be the result of the Ship Canal scheme, and reviewed the evidence of engineers at the Parliamentary inquiry as follows: "You may improve the bar by increasing the scour, but it will never be improved or be kept open by dredging. It is situated out at sea where the sea is so heavy that for nine months out of the twelve no dredger could possibly work; ... if you break up the bottom by a dredger and render it rough the quantity of sand deposited by a single tide would take a dredger many tides to remove."

Note.—Dredgers have been able to work, and they have removed the bar.

VIII.—EXTRACTS FROM AL ETTER BY MR. JAMES W. HARVEY TO THE *MANCHESTER GUARDIAN*, 27TH JUNE, 1884.

This writer showed that the engineering evidence quoted by Sir William Forwood in his letter was *ex parte*, and much of it was upset when the Liverpool witnesses were under cross-examination. This letter is mainly interesting because it gives a report of Admiral Fitzroy, the Acting Conservator in 1843, showing the wonderful erosions which were constantly taking place on the sides of the estuary.

IX.—EXTRACTS FROM A LETTER OF SIR WILLIAM B. FORWOOD TO THE *LIVERPOOL DAILY POST*, 9TH MAY, 1885.

He wrote: "The promoters in their present scheme have taken their canal along the margin of the estuary to a point above Eastham. From this entrance they propose to dredge a deep channel for $2\frac{1}{2}$ miles to Bromborough, and they also take wide powers under the 31st clause to dredge a channel 500 feet wide from Runcorn to Eastham, marching with and alongside the canal." Sir William went on to say that this differed from Mr. Lyster's plan, and would necessitate deep dredging, would do away with the fretting action, and stereotype a channel in the estuary, that the promoters ought to have taken their canal to deep water at Bromborough and adopted the higher sill proposed by Mr. Lyster. "But Manchester is more ambitious, and the deep sill proposed will permit vessels to enter at almost any state of the tide; therefore Liverpool has no option but to oppose the new scheme to the last, and to use every weapon in her armoury for this purpose." He feared "that the trade of this district may be saddled with £10,000,000 to £15,000,000 unproductive capital upon which, by its ultimate absorption by the railways, we shall be compelled to pay interest".

X.—EXTRACTS FROM A LETTER BY COUNCILLOR LEECH, TO THE *MANCHESTER EXAMINER AND TIMES* 10TH MAY, 1885.

The writer called attention to the attack made by the Liverpool Press on their eminent citizen, Sir James Picton, for deprecating further expenditure in opposing the Canal Bill, and noticed that cool-headed and reasoning Liverpool men were sick of Dock Board management, and would welcome a change. They saw that their quondam allies, the railway companies, were playing a game of their own, and would at any time foster their own ports of Garston, Holyhead and Fleetwood at the expense of Liverpool.

Referring to Sir William Forwood's statement that "the bar is in a delicate state," and that "our concern is to safeguard the navigation of the estuary," the writer recapitulated the evidence of Mr. Squarey, the solicitor of the Dock Board, who, in giving evidence before the Parliamentary Committee, said that though the Dock Board had received the dues for improving the river navigation they had never even tried to remove its greatest block, the bar. He said further: "Liverpool has at once become wonderfully mindful of the river and its bar. The Dock Board never cared to spend a shilling out of their ample income in improving the estuary and bar, nor did they trouble themselves to seek competent advice. At the same time it is well known the bar impedes commerce, and endangers valuable ships and lives; yet, with the risk of its being blocked up, and imperilling the whole trade of Lancashire, the Dock Board calmly folded their hands and looked on. They could go to the heavy expense of bringing Captain Eads from America, when they wanted to wreck the canal, yet it never crossed their minds previously to take counsel with the eminent man who successfully removed the bar from the mouth of the Mississippi." Again: "Liverpool wants a barge canal. Why? That she may continue to gather toll from all goods going outwards and inwards. . . . The veil is withdrawn, the hand is disclosed, and Liverpool has distinctly shown her last year's cry of 'the safety of the estuary was but a cloak'."

XI.—EXTRACTS FROM SIR WILLIAM B. FORWOOD'S SPEECHES TO THE LIVERPOOL CITY
    COUNCIL, MAY, 1885, AND TO A MEETING OF LIVERPOOL SHIPOWNERS AND
    MERCHANTS.

The Ship Canal Bill having passed the House of Lords, Sir William moved that the Council should oppose the Bill in the Commons. "We have witnessed the very strange anomaly, unprecedented in Parliamentary history, of a Bill passing through a Committee of Parliament with a majority of the Committee against the Bill." Sir William went on to say that the Lords Committee considered the Bill in sections. They passed the engineering portion by three votes to two, but when they came to the commercial section they were anxious to throw it out. "But inasmuch as they had already passed the Bill in regard to the engineering details, they felt they would be stultifying themselves in throwing out the Bill, and so they passed the Bill with the majority of the Committee against it." He justified further opposition by saying that, though in the previous session he had promised not to oppose a Ship Canal on the side of the estuary—carried out on Mr. Lyster's lines—now that the promoters had only carried out a portion of Mr. Lyster's suggestions, and were intending to dredge a channel 500 yards in width from Runcorn to Eastham, he should urge further opposition to the Bill. "If the charge for freightage on cotton be increased by only 5 per cent., the whole of the saving as regards Manchester will be gone. It is not likely that steamers will go 34 miles inland and through several locks, without any return cargo, on the same terms as they come to Liverpool. The advantages to be derived by Manchester are practically *nil*."

The shipowners and merchants of Liverpool, Sir William regretted to say, had hitherto been apathetic regarding the question of the Ship Canal. He, however, thought that the public had not been unmindful of the dangers to the port, but the general feeling had been that if the Bill was passed there was great improbability of the canal ever being made owing to the difficulty of raising the capital. He believed this was a great mistake. The capital would be obtained, not from capitalists, but from the working classes and small tradesmen of Lancashire who would be induced to invest their money in the greatest bubble ever thrust on the credulity of the British public since the South Sea Bubble. He had all along failed to see that Manchester had taken up the scheme in reality. They did not find the men of light and leading lending themselves to the scheme. The Manchester people had not adopted the suggestions made last year, that the canal should be constructed outside the estuary, because they were very ambitious and wished to make one which ships would be able to enter at all times and at all tides. To do this they were not only going to make a channel to the entrance of the Eastham Docks, but they were going to make one from Runcorn to Eastham. If the present scour in the upper estuary were diminished, Liverpool would be entirely closed as a port. If they did away with the vagaries of the estuary it must inevitably silt up. Liverpool did more than one-third of the total trade of the kingdom. What would become of trade if the port were closed.

XII.—Extracts from a Letter of Councillor Leech to the *Manchester Guardian*, 9th June, 1885.

"Sir William Forwood hopes to arouse and alarm the citizens of Liverpool. He can see no valid reason for making the canal, and is assured in his own mind that it means destruction to the shipping interests of Liverpool through damage done to the estuary and bar. He prophesies inadequate traffic, that only small ships will use the canal, and repudiates commercial jealousy. For years the Liverpool authorities have calmly watched the depth on the bar decline, and have done nothing. Now Sir William is ruffled because Manchester wishes to be progressive and insists the Mersey shall no longer be practically closed to please monopolists, but be made a great highway from the sea into the interior of England. When Liverpool was jealous of Birkenhead making docks and taking some of her trade, a similar cry was raised, 'damage to the river by contraction,' ending in destruction of her interests. Now even Liverpool admits the fears were groundless. Surely when the twin municipalities at the mouth of the river have enclosed about 1,200 acres for docks, thus narrowing the river, and blocking the influx of water, they should be the last people in the world to complain of a trained channel which does not narrow the estuarial basin, and is pronounced harmless by the only three practical engineers in Great Britain, who have carried out similar works; *viz.*, Messrs. Deas, Messent and Fowler of the Clyde, Tyne and Tees respectively."

XIII.—Extracts from Sir William B. Forwood's Letter to the *Liverpool Courier*, dated 14th August, 1885.

Herein Sir William described the Ship Canal as "an undertaking of such magnitude, and injurious to so many interests, that nothing short of a national necessity should have caused it to be authorised by Parliament. It is sinking an enormous capital to provide a most dangerous, difficult and costly waterway for the most costly and clumsy form of a canal barge or ocean steamer." He asserted that as neither sailing vessels nor liners could use the canal, Manchester could only compete for—

35 per cent. of the cotton trade;
26 per cent. of the grain trade;
18 per cent. of the timber trade;
6 per cent. of the provision trade.

Further that 2,500,000 tons was the sum total of the imports and exports of Manchester and the district 12 miles round that now came *via* Liverpool, and which was available for cartage. He was confident the railway companies would reduce their freights 5s. per ton rather than lose their business, and this would simply mean death to the canal. Sir William enumerated the onerous conditions forced on the Canal Company and said: "I must further tell them that if their estuary works prove at all injurious to the Mersey, the deep-water approaches will be interdicted by the Mersey Conservators, and the canal will only be available for barges and small coasting craft. These are not very cheery conditions for a hard-headed Lancashire man who wants a more solid security than mere sentiment. I cannot help thinking they will be glad to consider their task ended with the purchase of the Bridgewater Canal."

XIV.—Extracts from Mr. Alderman W. H. Bailey's Reply to Sir William B. Forwood in the *Examiner and Times*, 18th August, 1885.

"As Sir William Forwood has thought fit to publish a manifesto with the object of discrediting the Manchester Ship Canal as a commercial investment, it seems proper equal publicity should be given to the position which Sir William has always taken up with regard to the canal from the commercial point of view." He had said the people of Liverpool had no commercial jealousy, and that they had no objection to the canal if it would not do a serious injury to the Mersey. In 1884, before the Parliamentary Committee, he pledged Liverpool not to oppose in principle if the estuary were not interfered with, and said he had no objection to the scheme as far as Runcorn. He had said the scheme was a foolish one and fraught with disaster, but it was no affair of his or the people of Liverpool. Why should he now take such new-born interest, and seek to warn Manchester and Lancashire people from investing? "Within a week after the Bill has been passed, why should he do his best to kill the scheme by denouncing its commercial soundness?" He had professed to care nothing for the threatened opposition. If this were so, why take so much trouble to prevent the capital being raised? An insight into Sir William's fairness, reliability and fitness to guide the popular mind may be gleaned from his evidence before the Parliamentary Committees. In 1884-85 he ridiculed the idea of a cotton market in Manchester, and said it was an absurdity. In 1883 he owned to his own Counsel that it was quite possible to have a large cotton market in Manchester.

Before the Committee Sir William boldly stated that freights on cotton were higher by 2s. 6d. to 5s. per ton to Rouen than to Havre. On cross-examination he did not know what the freight would be, and after declaring charter parties excluded Rouen from Havre rates, he had to admit that when he applied to the brokers he had actually requested them to furnish him only with charter parties in which Rouen was excluded. Again, after declaring it was his invariable practice and experience to charge a higher rate in going to Rouen, one of his own charter parties was put in to show this was not the case. His excuse was that though issued by his own firm, it was by his London branch.

Sir William gave evidence before the Royal Commission in 1881, that 5s. per ton in the cost of freight would divert a trade. When before a Ship Canal Committee he declared that 5s. per ton saving in the carriage of cotton manufactures was infinitesimal.

In 1881 Sir William, before the Royal Commission, said the carriage of goods from Liverpool to Manchester could be done at 3d. per ton per mile, and he had offers from a carter to do it. Before the Ship Canal Committee a few years later he said the Ship Canal Company would have to pay 8d. per ton per mile for cartage from the docks, but he would not give the carter's name who had told him of the 8d. rate out of deference to his feelings. The trade of Liverpool in 1884 was estimated by Sir William as 25,000,000 tons, but when he wanted to show there was no trade for Manchester he accepted the railway estimates of 14,000,000 tons. So much for the fairness of Sir William's method of getting up evidence. "Most people will probably be of opinion that a gentleman who has had the hardihood to give on oath statements on the same subject which are absolutely either contradictory or altogether inconsistent with each other, is not likely to be a reliable witness when he ventures to prophesy the financial failure of the Ship Canal."

XV.—COUNCILLOR LEECH'S REPLY TO SIR WILLIAM B. FORWOOD IN THE *EXAMINER AND TIMES*, 20TH AUGUST, 1885.

"Our friends in Liverpool are doing their best to frighten the timid, and throw stones at the scheme. But it is a dangerous pastime for those who dwell in glass houses. Sir William Forwood has indulged in letters and speeches breathing nothing but hostility. In the past he affected to have no commercial fears, and now though comforting himself that the port of Liverpool is saved from destruction, he thinks it his duty to take a maternal interest in British investors, and judge what is or is not to be a paying investment. I shall take upon myself to question his qualification, and see if he is consistent, disinterested and, judging from his antecedents, fit to guide the public."

Sir William Forwood, after giving evidence before five separate Ship Canal Committees, and being roughly handled, nay, turned inside out, in the Lords Committee of 1885, shied the Commons Committee of the same year. Why? Because his Counsel thought it wise not to put him in the box, for they knew what awaited him. They simply asked that they might read his evidence given to the Lords Committee. Sir William Forwood would not confer with the promoters when asked to do so, but persisted in saying a 500 yards channel would be formed from Runcorn and damage Garston. After wasting time and money, he now says, " The measuer is as innocuous as any Ship Canal could be ". Where is his consistency?

To damage the prospects of the canal he had also said there were nine steam ferries and five locks, and that the railway bridges had to be lifted 75 feet. It turned out there were no steam ferries and only three locks, and the highest bridge was raised 47 feet. He was proved to be utterly wrong as to the saving on cotton, the cost of cartage, the percentage of the provision trade done by Manchester and the Irish cattle trade.

Sir William never missed an opportunity of charging Manchester with apathy about heavy railway rates, totally ignoring the fact that Mr. Peter Spence, of Manchester, and others gave most important evidence before the 1881 Railway Commission.

The very fact that a Parliamentary Committee presided over by such a shrewd man as Mr. W. E. Forster had passed the Bill, should inspire the public with confidence.

APPENDIX XI.

ARMS OF THE MANCHESTER SHIP CANAL COMPANY.

ARMS.—1st. Gules, three bendlets enhanced, or, on a chief, argent, thereon on waves of the sea, a steamship, proper (for Manchester).

2nd. Azure, semée of bees, volant, a shuttle between three garbs, or, on a chief of the last, a bale corded, proper, between two millrinds, sable (for Salford).

3rd. Argent, six lionettes, three, two, and one, gules (for Warrington).

CRESTS.—1st. A terrestial globe, semée of bees, volant, all proper.

2nd. A demi-lion, argent, therefrom flowing to the sinister, a flag.

3rd. A sword and mace in saltire, the sword enfiled with a wreath of laurel, both surmounted by a scroll bearing the inscription "Anno Regina Victoria".

SUPPORTERS.—On the dexter side an heraldic antelope, argent, attired, collared, and chain reflexed over the back, or; on the sinister side, a lion guardant, or, murally crowned, gules, each charged on the shoulder with a rose of the last.

MOTTO.—" Navigation and Commerce."

INDEX.

THE ABERDEEN UNIVERSITY PRESS LIMITED

Printed in the United States
By Bookmasters